Circular Economy and Sustainability

This book series aims at exploring the rising field of Circular Economy (CE) which is rapidly gaining interest and merit from scholars, decision makers and practitioners as the global economic model to decouple economic growth and development from the consumption of finite natural resources. This field suggests that global sustainability can be achieved by adopting a set of CE principles and strategies such as design out waste, systems thinking, adoption of nature-based approaches, shift to renewable energy and materials, reclaim, retain, and restore the health of ecosystems, return recovered biological resources to the biosphere, remanufacture products or components, among others.

However, the increasing complexity of sustainability challenges has made traditional engineering, business models, economics and existing social approaches unable to successfully adopt such principles and strategies. In fact, the CE field is often viewed as a simple evolution of the concept of sustainability or as a revisiting of an old discussion on recycling and reuse of waste materials. However, a modern perception of CE at different levels (micro, meso, and macro) indicates that CE is rather a systemic tool to achieve sustainability and a new eco-effective approach of returning and maintaining waste in the production processes by closing the loop of materials. In this frame, CE and sustainability can be seen as a multidimensional concept based on a variety of scientific disciplines (e.g., engineering, economics, environmental sciences, social sciences). Nevertheless, the interconnections and synergies among the scientific disciplines have been rarely and not in deep investigated. One significant goal of the book series is to study and highlight the growing theoretical links of CE and sustainability at different scales and levels, to investigate the synergies between the two concepts and to analyze and present its realization through strategies, policies, business models, entrepreneurship, financial instruments and technologies. Thus, the book series provides a new platform for CE and sustainability research and case studies and relevant scientific discussion towards new system-wide solutions. Specific topics that fall within the scope of the series include, but are not limited to, studies that investigate the systemic, integrated approach of CE and sustainability across different levels and its expression and realization in different disciplines and fields such as business models, economics, consumer services and behaviour, the Internet of Things, product design, sustainable consumption & production, bio-economy, environmental accounting, industrial ecology, industrial symbiosis, resource recovery, ecosystem services, circular water economy, circular cities, nature-based solutions, waste management, renewable energy, circular materials, life cycle assessment, strong sustainability, environmental education, among others.

Manfred Kircher • Thomas Schwarz

Editors

CO2 and CO as Feedstock

Sustainable Carbon Sources for the Circular Economy

 Springer

Editors
Manfred Kircher
KADIB
Frankfurt am Main, Hessen, Germany

Thomas Schwarz
Köln, Germany

ISSN 2731-5509　　　　　　　　ISSN 2731-5517　(electronic)
Circular Economy and Sustainability
ISBN 978-3-031-27813-6　　　　ISBN 978-3-031-27811-2　(eBook)
https://doi.org/10.1007/978-3-031-27811-2

Translation from the German language edition: "CO2 und CO – Nachhaltige Kohlenstoffquellen für die Kreislaufwirtschaft" by Manfred Kircher and Thomas Schwarz, © Springer-Verlag GmbH 2020. Published by Springer-Verlag Gmbh. All Rights Reserved.

This Springer imprint is published by the registered company Springer Nature Switzerland AG
The registered company address is: Gewerbestrasse 11, 6330 Cham, Switzerland

Paper in this product is recyclable.

Contents

CO2 and CO: Sustainable Carbon Sources for Circular Value Creation

1

Manfred Kircher, Cornelia Bähr, Dennis Herzberg, and Thomas Schwarz

Abstract

This book presents the volume of C1 gas streams available today in 35 chapters. Most conversion processes require hydrogen, so its supply is also discussed. Regarding the conversion processes, both chemo-catalytic and biotechnological methods and examples from industrial practice are presented. The integration into a future energy system and the possible contribution of C1 utilization to the further development of industrialized regions as well as ecological aspects and the framework conditions are also addressed. Finally, the state of the art and the industrial potential are summarized in a conclusion.

Keywords

C1 gas streams · Hydrogen · Conversion processes · Energy system

1.1 Introduction

Currently, there is an intensive search for ways to reduce carbon emissions—especially in sectors such as energy and steel and other energy-intensive industries. At the same time, the chemical industry is preparing for the "post-oil" era and

M. Kircher (✉)
KADIB, Frankfurt am Main, Hessen, Germany
e-mail: kircher@kadib.de

C. Bähr
b.value AG, Dortmund, Germany
e-mail: cornelia.baehr@b-value.de

D. Herzberg
CLIB - Cluster Industrial Biotechnology, Düsseldorf, Germany
e-mail: herzberg@clib-cluster.de

developing alternative carbon sources. The material use of (industrial) waste gas streams for the production of fuels and chemicals offers great solution potential for both challenges. This is because it reduces emissions and at the same time reduces the consumption of primary fossil raw materials. Since carbon-containing industrial waste gases are available in large volumes in Europe and do not compete with food, they can make a major contribution to reliable and sustainable raw material supplies and help close loops.

1.2   Problem

Energy and process industries need a reliably available and economically competitive raw material base. Today, electricity, heat, fuels and chemicals are produced worldwide and also in Germany predominantly on the basis of fossil raw materials. Gas and coal are the dominant sources for heat and electricity. Fuels (gasoline, diesel, kerosene) are predominantly products of petroleum refining, and in the chemical industry the most important raw material by volume is the petroleum fraction naphtha. At the same time, more than 95% of fossil carbon worldwide goes to the energy sector (11 billion tons of carbon annually from coal, gas, oil). Producing chemicals takes about 300 million tons (MT) of carbon (8% of oil and a small amount of gas and coal). For many decades, an infrastructure (logistics, industrial sites, etc.) has developed that is geared to these raw materials, which are reliably available in large quantities and can be easily transported.

Nevertheless, the development of alternatives for industry has long been an issue. The main driver was initially the finite nature of fossil resources (static range for oil 42, for gas 63 and for coal 340 years [1]). In recent years, however, climate change and the associated requirement to reduce greenhouse gas emissions have come to the fore. Both the energetic and the material use of fossil raw materials lead to the release of CO_2, because the energetic use (electricity, heat, fuel) causes CO_2 emission with the release of energy, and also the material use in the form of chemicals (plastics, paints, textiles, etc.) releases CO_2 in the production and after use of the products, e.g. in the course of energetic utilization in waste incineration plants. The increase in the CO_2 concentration of the atmosphere is considered one of the main causes of climate change.

Zero-emission energy and feedstock switching to bio-based carbon sources are therefore seen as important contributions to the Paris Climate Agreement (2015). This agreement aims to keep global warming below 2 °C and limit it to 1.5 °C. As soon as possible, greenhouse gas emissions are to peak and then be drastically reduced. For the second half of our century, "a balance is called for between anthropogenic emissions of greenhouse gases by sources and removals of such gases by sinks" [2]. The agreement entered into force on Nov. 4, 2016, and by May 2018, 90% of the countries at the World Climate Conference had ratified the agreement. It thus replaces the Kyoto Protocol of 1997, which would have expired in 2020. Based on this protocol, the EU had already set itself the goal of emitting at least 40% fewer greenhouse gases by 2030 than in 1990. In the Paris Agreement, it was agreed to reduce greenhouse gas emissions by at least 95% by 2050 compared with 1990. Accordingly, legislators are increasingly changing the framework conditions. Examples include preferential treatment of renewable energies, blending quotas for bio-based fuels (bioethanol, biodiesel) or the introduction of emissions certificates.

For energy generation, there is increasing investment in carbon-free alternatives (hydro and wind power, solar energy, geothermal power, in many countries also nuclear power), which is why there is generalized talk of a "decarbonization of the economy." However, this catchy term is highly oversimplified, because material recycling into organic chemistry products is and remains dependent on carbon. Also, no carbon-free alternatives are foreseeable in the medium term for heavy-duty and long-range fuels (trucks, ships, airplanes). Nevertheless, their carbon footprint can be mitigated by switching feedstocks to bio-based carbon sources. Plant biomass in particular, but also animal and marine biomass, is already being used in the energy sector (e.g. heat from wood, fuel from sugar and rapeseed oil, electricity from biogas), and biobased chemical products are also on the market. Their carbon footprint is reduced because the CO2 released into the atmosphere comes from biomass carbon and is photosynthetically recycled back into plant biomass via the natural carbon cycle. The carbon balance thus remains (theoretically) neutral. In theory, this is because the cultivation, harvesting, logistics and processing of biomass also release greenhouse gases. Such an economic model based on biomass is known as the bioeconomy and is now being pursued by more than 40 nations, including the EU. In Germany, the federal government has appointed the Bioeconomy Council for this purpose. Today, renewable biobased carbon sources are established in all sectors with a manageable but growing share. For the process industry, which will remain dependent on carbon-based feedstocks, these developments have extraordinary implications. From a purely technical perspective, bio-based carbon sources can indeed replace fossil feedstocks. In terms of supply, numerous studies [3, 4] also conclude that the global agricultural economy can provide food security and, together with forestry, supply industry with raw materials. Nevertheless, it must be emphasized that agricultural and forest land is limited, agriculture is a major emitter of greenhouse gases, biodiversity conservation requires land conservation, and the growing world population requires a doubling of food production by 2050. Overall, it is advisable to consider biobased carbon sources as a limited resource.

1.3 Concept of Circular Value Creation

To meet the required carbon demand sufficiently and sustainably, raw material efficiency must therefore be increased. In this context, it is obvious to consider the carbon flows of CO and CO2 emission, which have been neglected as a carbon source so far, i.e. to feed the gaseous carbon produced as residual material in technical processes into another process as raw material. Emission to the atmosphere is thereby avoided. In addition to emission gases, synthesis gas, which can be

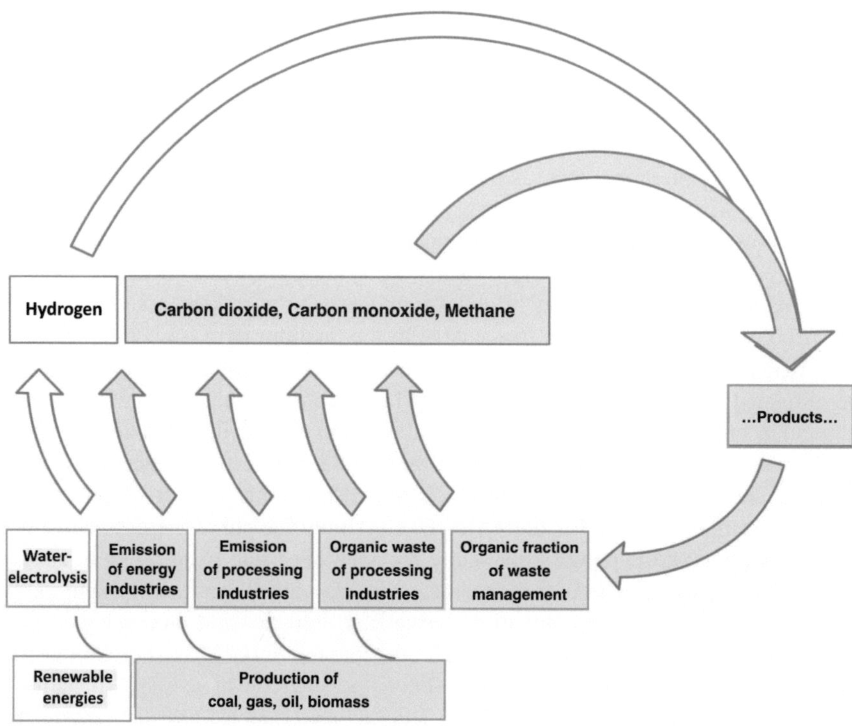

Fig. 1.1 The technical carbon cycle for carbon dioxide (CO_2) from emission gases from the energy and process industries and for carbon monoxide (CO) from organic wastes and side streams from the process and waste industries. The hydrogen required for the reduction of CO_2 and CO is provided by water electrolysis using renewable energy. Reprint rights: Not necessary

produced from any organic material such as municipal waste, can also be used. CO- and CO_2-using processes are therefore carbon sinks, as required by the Paris Climate Agreement. It is irrelevant whether the carbon source is bio-based or still of fossil origin until the raw material change is complete. For this very reason, the processes discussed here have particular potential for the current period of raw material change. This also applies to the provision of hydrogen, which is required as a reducing agent, because water electrolysis is a utilization option for the electricity peaks associated with the expansion of volatile energies (wind, photovoltaics). The technical carbon cycle made possible in this way can create a system of circular value creation that represents a further development of the currently prevailing principle of linear value creation (Fig. 1.1).

In Germany and Europe, CO- and CO_2-containing gas streams are available in large volumes. Especially for traditional industrial sites of high carbon emission, their use offers great opportunities. Carbon-containing gases become domestic raw materials that can be integrated into the existing infrastructure and help secure the future of local industrial regions. This will not least support the social acceptance of the required technologies.

This book focuses on such process gas streams for the production of chemicals and fuels. These gas streams occur, for example, in power generation and steel production, where the corresponding carbon comes from fossil sources, in biogas plants, which offer biogenic carbon, or in cement production and ore extraction, where the carbon is of mineral origin. When selecting suitable sources, the focus should be less on the origin of the carbon and more on the efficiency and sustainability of the corresponding conversion processes and the respective process chain. In addition to CO and CO2, methane (CH4) also represents an important C1 gas source, but it is only marginally considered in this book.

In order to realize the circular value creation outlined, companies in the manufacturing industrial sectors or the energy industry will in the future simultaneously become raw material suppliers and enter into completely new business relationships with companies in other sectors. It is easy to imagine that this will open up completely new bilateral business models, but also give rise to far more complex relationships. Innovative technical and economic solutions are needed to establish and shape these cross-industry relationships. These extend to linking entire sectors such as the energy industry and the process industry. This is why we also speak of cross-sector system solutions.

1.4 Aims of the Book

This book aims to draw attention to the opportunities and limitations associated with the use of carbon-containing (process) gases. So far, CO2 in particular is often perceived outside the technical community only as a climate killer, but not as a potential carbon source. In order to implement the circular value creation outlined, various technical and non-technical questions must be illuminated and answered. This book aims to contribute to this by providing a technically well-founded discussion of the potentials, but also the limitations, of the material use of carbon-containing gas streams by means of chemical and biotechnological processes and the options that can be opened up as a result. In this way, it is intended to provide not only experts but also interested laypersons with an introduction to the subject and to enable them to form a well-founded opinion on the opportunities and risks.

1.5 Structure of the Book

After an introduction to the possibilities of chemical and biotechnological conversion of C1 gases for material and energy use by Chap. 1, Chaps. 2–4 shed light on the current raw material situation. For example, Chap. 2 qualitatively and quantitatively describes the current industrial use of gas streams containing CO and CO2 and provides information on volumes used, processes from which the gas streams may originate, origin of the carbon, and current applications. CO and CO2 have high oxygen-to-carbon ratios and must be converted to hydrocarbons for their use as chemicals or fuels. For this reason, and because of its high energy content, hydrogen plays an elemental role in the conversion of C1 gases—particularly CO2. Chapters 3 and 4 therefore examine the sources from which the required hydrogen can be obtained and whether sufficient quantities can be made available for C1 gas utilization, especially for the production of large-volume products such as basic chemicals and energy carriers. In this context, Chap. 3 considers the state of the art of (electro)-chemical processes and their future technical potential, and Chap. 4 considers the state of the art of biotechnological processes. If a gas mixture of carbon monoxide and hydrogen as well as other gases such as CO2 or oxygen is present, it is referred to as synthesis gas. Due to its composition, it has a relatively high energy content and can therefore be used for the synthesis of basic chemicals. Chapter 5 describes the current state of the art for the production of such gas mixtures and, in addition to the processes, discusses the raw materials used, the quantities produced and the typical gas compositions depending on the processes and raw materials used.

Chapters 6–9 present current implementation technologies in chemical catalysis and biotechnology. To this end, Chap. 6 provides a basic overview of standard processes for the chemical catalytic conversion of CO and CO2, in particular Fischer-Tropsch synthesis, which is the benchmark for new (biotechnological) processes. Information on plant sizes and production capacities gives an impression of the importance of the processes used. Furthermore, advantages, but also general challenges and development needs are described. In addition to the raw materials used and potential raw materials, the focus is on the current value chains as well as the catalysts used on the process side, process conditions and respective challenges. Chapter 7, together with Chaps. 8 and 9, provides a basic overview of the possibilities of biotechnological C1 gas utilization as an alternative to chemical-catalytic processes. The two chapters describe the current state of the art and explain where particular application potentials exist for the different classes of biocatalysts and where biotechnological conversion offers advantages over chemical processes, especially Fischer-Tropsch synthesis. In this context, Chap. 7 focuses on the different biocatalysts (enzymes and whole-cell catalysis), presents the various microorganisms (classes) in question and compares their potential applications. While so-called photoautotrophic organisms derive their energy requirements for the utilization of CO and CO2 from light energy via photosynthesis, chemoautotrophs use energy-rich chemical compounds, which they metabolize for energy production. The potential for methane utilization is also discussed in brief. Using the groups of organisms Clostridia as chemoautotrophic producers and algae and cyanobacteria as photoautotrophs, the specific characteristics and fundamentals of the respective C1 gas-assimilating metabolic pathways are described. While chemoautotrophic organisms are more suitable for the production of large-volume chemicals and fuels, the production volume when using photoautotrophs is limited due to the light-dependent reaction. On the other hand, they allow access to more complex molecules that can be commercialized as higher-cost but lower-volume specialty chemicals. Chapters 8 and 9 explain the current state of the art in process engineering with regard to possible process concepts, the cultivation of chemo- and photoautotrophs and the necessary measurement and control technology. The challenges of the processes for the production of large-volume chemicals and fuels as well as the opportunities and limitations compared with chemical processes are also discussed. The chapters are rounded off with an overview of the current state of development and new reactor concepts.

Chapters 10–12 remain with biotechnological processes, but address specific aspects. For example, Chap. 10 focuses on the production of polyhydroxyalkanoates by gas fermentation; Chap. 11 deals with the cultivation of photosynthetic microalgae; and Chap. 12 addresses downstreaming issues.

Chapters 13–20 highlight concrete application possibilities for circular value creation based on CO- and CO2-containing gases from various existing industrial processes. Steel and cement production, power plants, chemical and bioprocesses (including biogas plants), (waste) incineration plants (MVA), gasification processes and factory farming are considered. The various gas-supplying processes are presented and integration options for chemical and biotechnological conversion processes are highlighted on the basis of realistic volume potentials and the respective framework conditions. Chapters 18 and 19 take a close look at municipal and industrial residues, the gasification of which can expand the range of raw materials. The chapters also presents corresponding gasification technologies.

In addition to the technical issues mentioned above, Chaps. 21–24 are devoted to various systemic and sustainability aspects. Chapter 21 addresses the fundamental issue of closing carbon cycles. Chapter 22 highlights ecological aspects such as the footprint of the chemical and biotechnological conversion of C1 gases to chemicals and energy sources/fuels, as well as possible contributions to emissions protection. The challenges and limitations of current accounting approaches are addressed as well. This is followed by Chap. 23, which examines the social dimension. The importance of C1 gases as a relevant domestic raw material source for regional development, in particular for the creation and safeguarding of jobs, is shown. On the example of the German state of North Rhine Westphalia as one of the leading national and European regions for energy, chemicals and heavy industry, such as the steel and cement industries, the political dimension of the issue will be presented— especially against the background of structural change in the lignite mining region of this state. A further focus will be on social acceptance. This will be analyzed in broad outline for the technology approaches, value chains and products considered and the system solutions described. Chapter 24 examines the influence of the regulatory framework on the establishment of new processes and value chains and their competitiveness. The aspects examined include: Funding opportunities, the importance of pilot and demonstration projects, incentives and hurdles posed by the regulatory framework and legislation.

Chapters 25–34 uses specific application, project and industry examples to illustrate the many possibilities for circular value creation through (electro)chemical conversion technologies as well as biotechnological approaches. For the various examples, the respective process concept, the current development status and the future potential are described.

The book ends in Chap. 35 with a final evaluation and summary. For this purpose, the various aspects and perspectives of the individual chapters are summarized, and an assessment is given of the contribution that chemical and biotechnological conversion technologies can make in terms of circular value creation in the three dimensions of economy, ecology and society. In addition, the potential of C1 gases as alternative raw materials is compared with other carbon sources such as biomass.

References

1. BGR (2010) Reserven, Ressourcen und Verfügbarkeit von Energierohstoffen. https://www.wiwi.uni-muenster.de/vwt/Veranstaltungen/Ausgewaehlte_Kapitel_der_Energiewirtschaft/WS1112/02a_globale-energiemrkte.pdf. Accessed 6 Nov 2022
2. UN (2015) Paris agreement. https://unfccc.int/sites/default/files/english_paris_agreement.pdf. Accessed 6 Nov 2022
3. Carus M, Piotrowski S (2009) Land use for bioplastics. Bioplast Mag 4:46–49
4. Souza GM, Victoria RL, Joly C, Verdade L (2015) Bioenergy and sustainability: bridging the gaps. SCOPE, Paris

CO2: Sources and Volumes

2

Katy Armstrong and Dennis Krämer

Abstract

The chapter presents industrial CO2 emission sites in Europe, so-called point sources, with volumes exceeding 0.1 Mt CO2. Among these, ethylene oxide, ammonia and hydrogen production plants provide very pure CO2 streams that can be captured at correspondingly low cost. In the spirit of industrial symbiosis, it makes sense to utilize such CO2 sources, located in chemical parks, for material purposes on site. However, the number and thus the CO2 supply of these plants in Europe is relatively small compared with sources from large coal-fired power plants, steel production and the manufacture of cement. While the latter have emission streams of lower CO2 concentration and thus higher capture costs, they are spread across Europe with more large plants and can thus enable cost advantages in terms of transport costs to future customers. Overall, a large number of CO2 sources with the technical potential for material utilization are available.

Keywords

CO2 emission sites · CO2 utilization · Chemical parks · Power plants · Steel production · Cement production

K. Armstrong
Department of Chemical and Biological Engineering, The University of Sheffield, Sheffield, UK

D. Krämer (✉)
DECHEMA - Gesellschaft für chemische Technik und Biotechnologie e.V., Frankfurt am Main, Germany
e-mail: kraemer@dechema.de

2.1   Introduction

CO_2 emissions in industry are mainly caused by the combustion of carbon-containing materials. Currently, these are mainly fossil raw materials such as crude oil, natural gas and coal. Due to anthropogenic CO_2 emissions, the CO_2 concentration in the atmosphere has increased from about 280 ppm to 400 ppm since the beginning of the industrial revolution. Climate scientists from the Intergovernmental Panel on Climate Change (IPCC) are certain, and have convinced the global community, that the increased CO_2 concentration in the atmosphere is closely linked to current climate changes. In order to prevent a further increase in the CO_2 concentration in the atmosphere, Germany has set itself the goal of becoming virtually climate-neutral by 2050. This can only be achieved in industry by refraining from burning fossil raw materials—in other words, by achieving "decarbonization" of industry. However, the chemical industry cannot be decarbonized in the literal sense of the word, since carbon—abbreviated to C in chemistry—is at the beginning of many value chains. In order to expand the raw material base of these value chains in the chemical industry, a great deal of research is currently being conducted into making CO_2 usable as a carbon source. On the one hand, CO_2 utilization pathways must be developed for this purpose, and on the other hand, CO_2 sources that could be considered as carbon sources for these processes must be identified and evaluated. CO_2-containing emissions from point sources that can be attributed to industrial processes and air from the atmosphere are the possible carbon sources.

From 2016 to 2018, the EU-funded Horizon2020 project "CarbonNext—The Next Generation of Carbon for the Process Industry" investigated which alternative carbon sources are available in principle for the process industry in Europe and how high the potential is to ultimately be able to tap these carbon sources for the process industry. The focus was on the material use of CO_2. The aim of the project was to show the entire value chain of processes that have a high volume potential for the material use of CO_2 in order to substitute fossil raw materials. The result of the project is a presentation of various CO_2 utilization paths that were evaluated as particularly convincing from an economic and ecological point of view. Furthermore, value chains of different processes between different industrial sectors (chemicals, cement, steel, etc.) were identified within the project. The project aims to inform decision-makers from industry and politics about the possible opportunities and the potential of alternative carbon sources and to raise awareness of options for action.

The first work package of the project identified and analyzed CO_2 point sources in Europe that could be considered as carbon sources for the process industry. The results of this report are presented in the following text.

2.2   Recording CO2 Sources (Data Collection)

In order to find out where and to what extent CO2 point sources exist in Europe, a data collection was essential as a first step. The data on alternative carbon sources were used to map the corresponding sources. For this purpose, data on CO2 emissions in 2014 from the European Pollutant Release Database of the European Environmental Agency were used and incorporated into a three-dimensional map. The European Pollutant and Release Transfer Register (E-PRTR) of the European Environmental Agency (EEA) records all point sources of CO2 that emit at least 0.1 Mt CO2 per year [1]. Data transfer to the EEA is mandatory for the respective emitters. The data are reported by each facility to authorities in their respective countries, which then check the data for quality before reporting it to the European Commission and the EEA. The register contains data from 30,000 industrial facilities from nine industrial sectors covering 65 economic activities. With the help of the interactive map, it is possible to break down the emissions in terms of the different industrial sectors. It also allows the sources of individual countries or regions to be viewed. The interactive map can thus be used to selectively generate quantitative statements about CO2 and CO emissions in Europe [1].

2.2.1   CO2 Sources in Europe

A total of 1779 Mt of CO2 was emitted to the atmosphere from 2000 industrial point sources in Europe in 2014. The sources can be divided into the following industrial sectors: Chemicals, Construction, Power Generation, Agriculture, Metals, Paper, Waste, and Others. In Table 2.1, the CO2 emissions are shown cumulatively by the individual sectors.

Figure 2.1 shows the individual point sources in Europe. From the distribution, it can be seen that CO2 sources are distributed throughout Europe. The largest sources are in Germany, the Benelux countries, Great Britain, and Poland. Coal-fired power plants are the largest emitters of CO2, although it should be noted that the data is from 2014 and some of the coal-fired power plants are now no longer in operation.

Table 2.1 CO2 emissions in the EU, categorized by sector, adapted from E-PRTP [1]

Sector	CO_2-emissions in 2014 [Mt]
Energies	1065.5
Chemistry	245.1
Metals (incl. Iron- and Steelindustry)	166.0
Construction (incl. Cement)	144.1
Paper	77.1
Waste	55.5
Mining	7.1
Agriculture and Food	5.9
Other	13.1

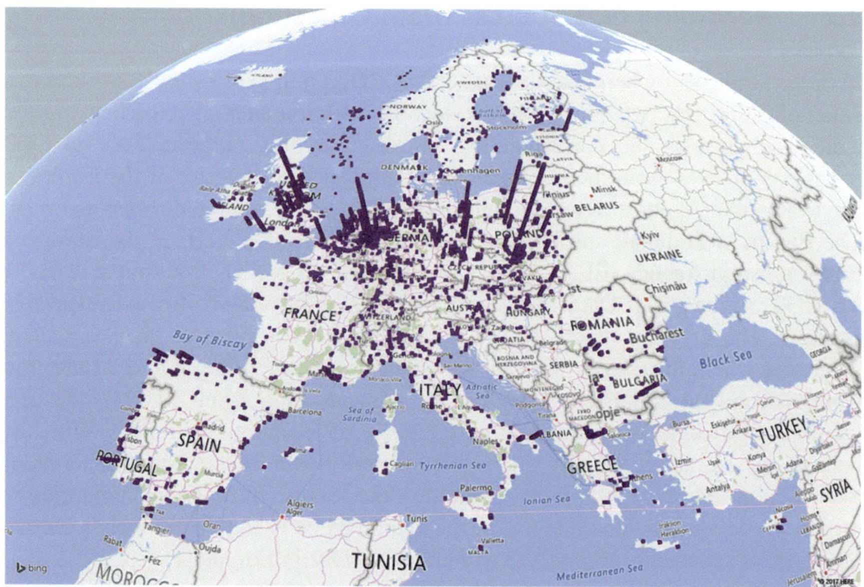

Fig. 2.1 CO2 point sources in the EU, adapted from E-PRTR [1]

Furthermore, large amounts of CO2 are produced in the chemical industry, the metal sector, the construction industry and paper production.

The amount of CO2 that could be used to manufacture products is estimated to be up to 7 Gt worldwide by 2030 in optimistic scenarios (Global CO2 Initiative, 2016), but other scenarios and estimates of around 1 Gt are more common. This estimate illustrates that the expectation for the material use of CO2, at least in the near future, does not necessarily have to target the major CO2 emitters as raw material suppliers, since at present there is no reason to fear a bottleneck in the supply of CO2 to the chemical industry. Even if all coal-fired power plants are shut down, there would still be enough CO2 for "large-scale" scenarios. For this reason, a hierarchy can be derived as to which CO2 point sources are most likely to be used for the process industry from an economic and ecological point of view.

2.2.2 Evaluation

When evaluating CO2 sources, a large part depends on how high the concentration of CO2 is in the corresponding flue gas. The higher the concentration, the lower the energy required for capture because smaller amounts of emitted gas must be processed to obtain the same amount of purified CO2 compared to more dilute sources. Furthermore, there are different gas composition requirements for different CO2 utilization technologies and thus for the treatment process. These factors affect cost, energy use, and the environment. Consequently, CO2 that is as pure as possible

Table 2.2 Potential CO2 sources in Europe (>0.1 Mt CO2/year), adapted from E-PRTR [1] and Naims (2016; [2])

CO_2-Source	CO_2-Concentration [%]	Emissions per Year [Mt CO_2/Year]	Cost [€/t CO_2]	Number of Pointsources over 0.1 Mt/Year Emissions
Hydrogen-production	70–100	5.3	30	15
Natural Gas Production	5–70	5.0	30	10
Ethylenoxid-production	100	17.7	30	6
Ammonia-produktion	100	22.6	33	27
Paper-production	7–20	31.4	58	35
Coal to Power (IGCC)	3–15	3.7	34	3
Iron and Steel	17–35	151.3	40	93
Cement	14–33	119.4	68	212
Total		356.4		

is the most attractive because, as a result, less energy is required to capture CO2 and this, in turn, is associated with a comparatively low carbon footprint.

Emissions from the energy sector—around 76% of CO2 emissions worldwide come from the combustion of coal and gas to generate electricity—are neglected in terms of the material use of CO2. This is due, on the one hand, to the fact that the CO2 concentration in flue gas is not particularly high (12–14 %) and, on the other hand, to the fact that, as a result of current climate policy, coal-fired power plants are to be shut down significantly by 2100.

Table 2.2 summarizes CO2 concentrations, emission volumes per year, and estimated capture costs per ton of CO2 from a range of CO2 emitters. The data in the table confirm that the lowest capture costs are for the most concentrated CO2 sources. The following sources have the highest CO2 concentrations: Hydrogen production, natural gas refining, ethylene oxide production, and ammonia production.

Two recent scientific papers discussed the evaluation of CO2 sources. The papers use different methods to identify the sources with the greatest economic potential. In "Economics of carbon dioxide capture and utilization—a supply and demand perspective" [2], economic viability is primarily used as the strongest criterion. In the publication "Selecting CO2 sources of CO2 utilization by environmental-merit-order curves" [3], CO2 sources are evaluated mainly by environmental aspects. Both studies are based on an intensive literature review. Naims [2] compares the cost of CO2 capture with the cost avoided. Von der Assen et al. [3] compare environmental impacts based on comparative life cycle assessment. There is overlap

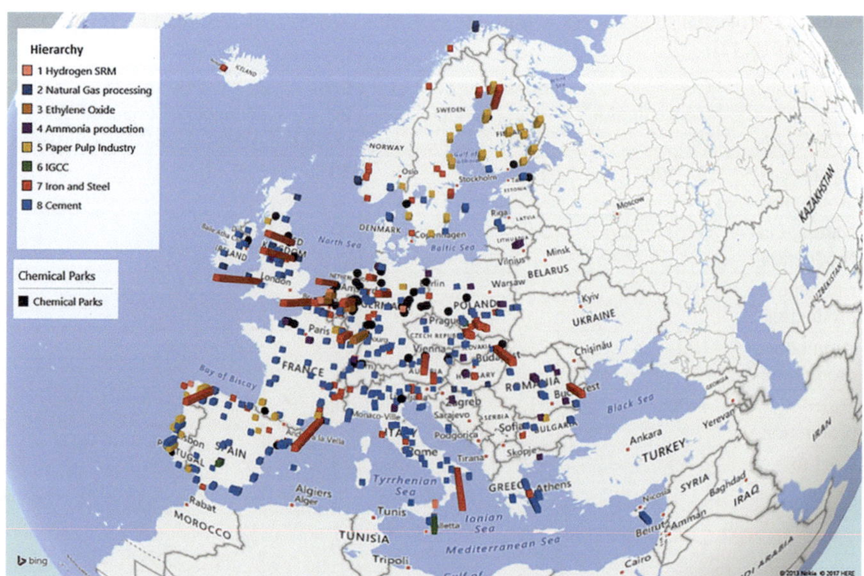

Fig. 2.2 Hierarchy of CO2-sources in the EU, adapted from E-PRTR [1]

between the studies; however, the same sources and capture technologies are not considered. Also, both studies conclude that the cleanest sources should be considered first.

Furthermore, large CO2 streams are generated in iron and steel production and in the cement industry, although the flue gases here contain lower CO2 concentrations. Provided that CO2 capture technologies develop further, such CO2 sources will also become attractive. In Fig. 2.2, CO2 sources for material use are listed, sorted according to their purity. The higher the CO2 purity of a source, the easier it is to capture and process.

2.3　Forecasts of the Developments of the Four "Most Important" CO2 Sources

2.3.1　Hydrogen Production

Currently, H2 consumption is split roughly equally between hydrotreating/hydrocracking by refineries and ammonia/nitrogen-based fertilizer production by the chemical industry [2]. The global H2 market is expected to grow at a compound annual growth rate (CAGR) of 5.99% from 2017 to 2021, mainly due to increasing demand for fertilizers [4]. The European H2 market is expected to grow at an annual CAGR of 3.5% through 2025, with a domestic consumption of 7 MT in 2015 [5]. About 96% of the global H2 production comes from steam reforming methane (SRM), is oil-based or originates from coal gasification. In Europe, there are 16 SRM plants, each emitting between 0.136 and 0.805 MT of CO2 per year [6]. In addition, many of these plants are located in close proximity to large European chemical parks. Emissions are expected to increase in proportion to the growth of the H2 market.

Without technological advances, it is likely that future production of H2 will be met by reforming natural gas and by electrolysis fueled by the electricity mix, some of which also comes from coal-fired power plants.

However, a switch to electrolysis of water using only renewables will result in a significant decrease in CO2 availability from this source.

2.3.2 Natural Gas Production Market

CO2 emissions from natural gas processing range from 0.1 to 1 MT per year per plant. Ten European plants are responsible for the emissions. Some of the largest natural gas processing plants are located in the UK and in close proximity to major chemical parks in Belgium, the Netherlands and Germany. The raw natural gas contains varying concentrations of CO2 depending on the origin of the gas, but it is often processed through a CO2 capture process to achieve pipeline quality [7]. Natural gas upgrading is considered one of the most important high-purity CO2 sources that do not require additional purification costs in carbon capture. Natural gas upgrading (and thus CO2 availability from this source) is expected to increase in the medium term as coal and natural gas-fired power generation continue to be used to balance intermittent renewable generation.

2.3.3 Ethylene Oxide Production

Ethylene oxide is used as an intermediate in the production of many industrial chemicals such as polymers and ethylene glycols. Ethylene oxide itself is used as a gas fumigant, disinfectant and sterilant for medical purposes. Six ethylene oxide plants producing more than 0.1 MT of CO2 per year per plant are listed in the E-PRTR and produce a combined output of 17.7 MT of CO2 per year [8]. Ethylene oxide production plants in Europe are located in Belgium, the Netherlands, and Germany; these countries also have large chemical parks that could use CO2 as feedstock. According to market reports, demand for ethylene oxide is expected to be high by 2022 [9].

2.3.4 Ammonia Production

The European fertilizer market is expected to grow at a constant CAGR of 2.5% during the forecast period from 2017 to 2022 [9]. CO2 emissions from ammonia production in Europe amounted to 22.6 MT in 2014, from 27 plants with emissions ranging from 0.1 to 3.2 MT per year per plant. Some of these plants are located near chemical parks. Ammonia is mainly used as a fertilizer and is produced by the Haber-Bosch process. CO_2 is produced during the manufacture of hydrogen, which is combined with nitrogen to produce ammonia. A report by the International Fertiliser Association shows that around 36% of the CO_2 removed from syngas during purification is used by industry. About 33% of the CO_2 is used for urea production, while the remaining CO_2 is sold for other purposes [9]. Therefore, the availability of CCU could be limited in the case of ammonia, as there would be competition with current CO_2 uses for urea production.

Although in the long term the use of natural gas (and thus the availability of CO_2 from purification) or steam reformed hydrogen production will decrease, the volumes from these four CO_2 sources are expected to remain the most attractive until 2030. They are also expected to be large enough in volume to meet CCU demand.

As described, CO_2 purity is an essential factor. Higher purity means higher CO_2 concentrations per equivalent volume, and thus lower costs. Purity also impacts compatibility with certain transportation and storage applications, as impurities can cause damage to pipelines.

Another criterion in the selection of CO_2 sources is the distance to the potential CO_2 off-taker. Transportation costs depend on the distance to be transported (which affects the type of transportation, i.e., trucked, local, or pipeline). For pipeline transportation, costs vary further depending on the length and diameter of the pipeline, the construction material used, and the route of the pipeline. A common CO_2 transportation network can also alleviate purity issues by allowing a mix of sources to deliver appropriate CO_2 quality to the end user. Thus, CO_2 sources located near potential off-takers are particularly attractive. Ideally, the CO_2 can be transported in a pipeline. As the distance increases, the costs—and also the CO_2 footprint—become higher.

Given infrastructures and short distances are consequently to be preferred. However, since many CO_2 utilization technologies are not yet available on a large scale, the only obvious scenario that can be assumed at present is that new plants will be integrated into existing industrial environments such as chemical parks. The large European chemical parks have thus been identified as top locations with existing infrastructure for CCU applications.

2.4   Conclusion

There is a wide range of CO_2 sources available that have the technical potential to cover all options for material use of CO_2 in the process industry in the near future. Primarily, sources should be used that have high CO_2 concentrations, since CO_2 capture is energy-intensive and costly. Large CO_2 streams are generated in iron, cement, and steel production. These sources may become very interesting at a later stage, especially as these industries look for ways to reduce their CO_2 emissions and therefore intensively research new, more efficient capture technologies.

The interactive maps can be used to identify suitable locations where industrial plants for the material use of CO2 could preferentially develop. In the spirit of industrial symbiosis, CO2 sources from chemical parks should be used for material purposes directly on site.

References

1. European Pollutant Release and Transfer Register (2019). https://ec.europa.eu/environment/industry/stationary/e-prtr/legislation.htm. Accessed 12 Aug 2022
2. Naims H (2016) Economics of carbon dioxide capture and utilization - a supply and demand perspective. Environ Sci Pollut Res 23(22):22226–22241
3. von der Assen N, Müller LJ, Steingrube A, Voll P, Bardow A (2016) Selecting CO_2 sources for CO_2 utilization by environmental-merit-order curves. Environ Sci Technol 50(3):1093–1101
4. TechNavio (Infinity Research Ltd) (2017) Global Hydrogen Generation Market 2017–2021
5. CertifHy (2015) Overview of the market segmentation for hydrogen across potential customer groups, based on key application areas. http://www.certifhy.eu/images/D1_2_Overview_of_the_market_segmentation_Final_22_June_low-res.pdf. Accessed 11 Aug 2022
6. International Energy Agency (IEA) (2012) Energy technology perspectives 2012
7. Baker RW, Lohhandwala K (2008) Natural gas processing with membranes: an overview. Ind Eng Chem Res 47:2109–2121
8. ICIS (2013). https://www.icis.com/resources/news/2013/04/13/9658385/chemical-profile-europe-ethylene-oxide/. Accessed 12 Aug. 2022
9. Mordor Intelligence (2017). https://www.mordorintelligence.com/industry-reports/europe-fertilizers-market?gclid=Cj0KCQjwqM3VBRCwARIsAKcekb2UeWMFz5758xh80N_17i20nRrKe0OASR0KH0Y6kVqGqOelJMrA4wQaAju-EALw_wcB. Accessed 12 Aug 2022

Conventional Processes for Hydrogen Production

Fausto Gallucci, Jose Antonio Medrano, and Emma Palo

Abstract

Hydrogen is used extensively in the chemical industry for producing a wide variety of higher-value products, while also finding applications in heat and power generation and as a carbon-free energy carrier. Since hydrogen does not occur as pure hydrogen gas on Earth, the gas must be produced from fossil or renewable sources. Hydrogen is nowadays produced primarily from fossil fuels, usually by steam methane reforming (SMR) or coal gasification, in centralized units and is used and consumed mainly for ammonia and methanol synthesis processes. In petroleum refining, hydrogen is produced in very small quantities as a byproduct of naphtha reforming and oil fractionation, and is consumed mostly in hydrocracking to enhance product quality or to remove sulfur compounds from petroleum. Hydrogen can also be produced from renewable sources. In particular, hydrogen can be produced by biomass or waste gasification or by water electrolysis using electricity from renewable sources. This chapter provides a comprehensive overview of the major existing technologies for hydrogen production, including typical operating conditions and manufacturers, compares existing technologies, and presents future trends in hydrogen production.

Keywords

Hydrogen · Steam methane reforming (SMR) · Coal gasification · Petroleum refining · Water electrolysis · Hydrogen from biomass · Hydrogen from waste

F. Gallucci (✉) · J. A. Medrano
Department of Chemical Engineering, Eindhoven University of Technology, Eindhoven, Netherlands
e-mail: f.gallucci@tue.nl; J.A.Medrano.Jimenez@tue.nl

E. Palo
KT – Kinetics Technology S.p.A, Rome, Italy
e-mail: e.palo@kt-met.it

© The Author(s), under exclusive license to Springer Nature Switzerland AG 2023
M. Kircher, T. Schwarz (eds.), *CO2 and CO as Feedstock*, Circular Economy and Sustainability, https://doi.org/10.1007/978-3-031-27811-2_3

3.1 Introduction

According to the US Department of Energy [1], more than 50 Mt. of hydrogen (H_2) is produced annually worldwide, about 95% of which is from fossil fuels. H_2 is mostly produced in centralized plants, where it is usually directly used in downstream processes [2]. Worldwide, about 90% of all hydrogen produced is directly consumed again in process chains: 50% by the ammonia industry for the production of fertilizers, 25% by petroleum refineries for processes such as hydrocracking, hydrodesulfurization or -isomerization, and 10% for the production of methanol, one of the most important feedstocks worldwide. These three processes consume about 3000 MNm^3H_2/h, and hydrogen production plants with capacities ranging from 1 to 100 t/h are installed worldwide.

Other sectors consume hydrogen on a much smaller scale, these are for example the food industry (30–200 Nm^3H_2/h), the pharmaceutical industry (50–400 Nm^3H_2/h) or the metal industry (100–600 Nm^3H_2/h). These sectors generally do not have their own on-site hydrogen production facilities and therefore import hydrogen from intermediaries, resulting in significant cost increases.

Hydrogen production costs are largely determined by feedstock's cost and quality, as well as the hydrogen production process chosen [3]. In general, as also stated by the U.S. Department of Energy, processes for hydrogen production can be classified into four different technology pathways: thermochemical, electrolytic, photocatalytic water splitting, and biological processes [1]. However, only the first two processes are relevant in worldwide hydrogen production. Electrolytic processes are well developed, already commercialized, and account for about 5% of the total hydrogen produced worldwide. This technology typically uses electricity from renewable sources and produces CO_x-free H_2. Thermochemical processes, on the other hand, involve technologies that require heat to produce hydrogen. Such technologies can use a variety of feedstocks, such as natural gas, coal, or biofuels, and even wastes, and have been well developed and commercially applicable for many decades.

For several years, hydrogen has increasingly been viewed not only as an intermediate for the production of most basic chemicals. Instead, there is growing interest in the potential use of hydrogen in power generation, as a replacement for currently used fossil fuels (i.e., the use of hydrogen as an energy carrier). The relevant technology for this has already been developed and is known as a fuel cell. Fuel cells are at the heart of the so-called hydrogen economy, which is becoming increasingly attractive because hydrogen has a much higher energy density than conventional fuels, produces only water and heat in its end use (combustion), and thus has no carbon footprint at the point of use.

One of the main problems with H_2 production is still the large environmental impact, as production involves significant CO_2 emissions as by-product. For example, the ammonia industry alone is responsible for over 300 MT of CO_2 emissions annually [4]. Most of the newly installed hydrogen production plants have currently achieved efficiencies that can reduce the CO_2 footprint to just over 10% of the theoretical value. To further reduce these emissions when using fossil fuels, carbon capture technologies should be integrated into (or downstream) the process, which in turn leads to higher hydrogen production costs [5, 6]. Therefore, it is not surprising that many environmental agencies are already forecasting the production of hydrogen from renewable energy sources in the near term.

Chapters 3 and 4 describe the main hydrogen production processes from different feedstocks and draw an overall comparison between existing technologies. It also outlines an overview of the major environmental agencies' vision for the hydrogen economy in the near future and the future trends observed at the research level.

3.2 Hydrogen Production from Various Raw Materials

3.2.1 Natural Gas as a Fuel Source for Hydrogen Production

Methane is the feedstock with the highest H/C ratio, consequently natural gas has been the preferred feedstock for hydrogen production so far. Based on this feedstock, various technologies have been developed and commercialized for many decades, such as methane steam reforming (SMR), autothermal reforming (ATR), and partial oxidation of methane (POX). Among these, SMR is the oldest and most developed technology for hydrogen production. In addition, new technologies for hydrogen production using methane as feedstock have been proposed in recent years. Among these, dry reforming of methane or methane decomposition appear to be particularly interesting ways to convert methane into hydrogen and also, as in the case of methane decomposition, into solid carbon, an important value-added product. However, the technologies have not yet found commercial application.

In methane steam reforming, natural gas from the grid is first purified of impurities, usually sulfur compounds, in a hydrodesulfurization phase. These sulfur compounds are converted to H_2S, which is typically adsorbed in a ZnO bed. The natural gas is then mixed with steam at a specific ratio of steam to carbon (S/C of 2.5–3) to prevent carbon deposition on the catalyst surface. Depending on the application and feedstock composition, the gas mixture may first be pre-reformed at intermediate temperatures over a Ni-based catalyst to remove heavier hydrocarbons. The pre-reformed gas mixture is then sent to the fired steam reformer reactor where the natural gas is converted to syngas (a gas mixture of CO and H_2) over a Ni catalyst at high temperatures (850–920 °C) and pressure of 25–35 bar (Eq. 3.1).

Various reformer configurations are commercially available. For example, the company KT—Kinetics Technology (KT) offers top-fired tubular reformers. In these reformers, the heat to be supplied for the endothermic reaction comes from burners installed on the ceiling of the reformer box, which fire downward in co-current with the process gas flowing downward through catalytically active tubes made of high quality alloys [7]. A similar configuration is used by other suppliers such as CB&I [8]. Other technology suppliers, such as Haldor Topsoe [9], sell reformers with burners on the sides to achieve a more homogeneous temperature along the reforming section. Another existing reformer configuration is the so-called terrace wall furnaces, as supplied by companies such as Amec Foster Wheeler [10]. In this configuration, the furnaces are located at different levels within the reactor to obtain a more homogeneous distribution of the heat flux.

The syngas produced in the reforming zone is cooled in the process gas boiler by producing high/medium pressure steam, which in turn are partially used as process steam to achieve the desired S/C ratio and as export steam (advantage in the compound process). To maximize hydrogen efficiency and overall carbon conversion, a water gas shift (WGS; Eq. 3.2) is installed in the methane steam reforming process. Similar to the reformer section, the configuration and operating conditions of the WGS section are influenced by the particular technology provider and the final application of the hydrogen produced. For example, Haldor Topsoe installs a single adiabatic WGS reactor at intermediate temperatures, which is sufficient for good reforming efficiency and full heat integration of the process. Other suppliers instead use two WGS reactors in series. The first stage is a high-temperature WGS reactor that typically uses an Fe-Cr catalyst. The inlet temperature of this reactor is typically 320–350 °C, leading to an increase in reaction kinetics [11]. Then, the H_2-rich syngas is cooled with heat recovery and fed into a low-temperature WGS reactor operating at temperatures around 200 °C. This reactor is usually filled with Cu-Zn catalysts and shifts the thermodynamic equilibrium towards carbon conversion [12].

$$\text{Methane steam reforming}: CH_4 + H_2O \rightleftharpoons CO + 3H_2 \quad \Delta H^0_{298K} = 206 \ kJ/mol$$

$$(3.1)$$

$$\text{Water gas shift reaction}: CO + H_2O \rightleftharpoons CO_2 + H_2 \Delta H^0_{298K} = -41 \ kJ/mol$$

$$(3.2)$$

Then, the H_2-rich gas mixture is cooled to ambient temperature to condense the vapor and then sent to a pressure swing adsorption (PSA) unit, where about 85–90% of the hydrogen can be recovered with purity above 99.999%. This hydrogen is finally compressed to delivery conditions, and only a small portion of it is recycled to the process for the hydrodesulfurization stage.

The exhaust gas from the PSA unit is returned to the incinerator where it is mixed with fuel and burned. Flue gases generated in the radiant box flow through the convective heat recovery section and are further emitted through a common stack. Depending on the reformer configuration, this section is installed at the bottom of the reformer (for top-fired reformers) or in the upper part. The heat content of the flue gas is used partly to preheat process streams and partly to generate and heat high/ medium pressure steam. For a schematic diagram of the SMR process, see Fig. 3.1.

This technology has been mature for several decades and therefore arguably offers little scope for significant technological innovation or cost reduction of the process. It is likely to remain the most widely installed method of hydrogen production in the world for decades to come.

Other technologies with a large market share include ATR and POX. In autothermal reforming (ATR), oxygen in sub-stoichiometric quantities (pure oxygen or atmospheric oxygen) is fed to a burner from the top of the reformer together with methane and steam. In this thermal section, the oxygen is consumed by burning

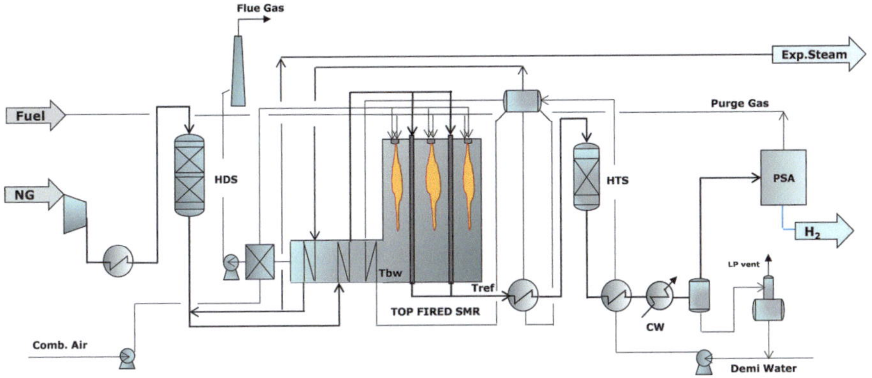

Fig. 3.1 Schematic drawing of the methane steam reforming process

some of the methane, thereby generating the heat needed in the catalytic reforming section. The hot gas from the thermal section flows through a fixed bed filled with reforming catalysts. Compared to SMR, the entire catalyst is housed in a single unit rather than being distributed among multiple tubular reformers. ATR technology is relatively simple and significantly reduces the OPEX cost of the process because the required steam-to-carbon ratios are much lower (S/C $\approx$ 0.6–1), although oxygen costs should also be considered when calculating OPEX costs. Since the oxygen present reduces the selectivity for hydrogen, the H_2/CO ratio of the syngas is lower compared to SMR, resulting in lower reforming efficiency [13, 14]. On the other hand, the compact heat integration of ATR leads to a much smaller footprint, so this technology is well referenced nowadays. The main drawback is the need for (almost) pure oxygen. This oxygen is produced in a costly process in a cryogenic air separation unit (ASU).

Many licensors around the world supply ATR systems. Haldor Topsoe offers its SynCORTM ATR reactor [15], which is used for many applications as a syngas production unit for further GtL or methanol synthesis. Johnson Matthey, in collaboration with BP (British Petroleum), is another company that can also supply ATR units, although the main purpose of the technology is syngas production [16].

A similar technology to ATR is partial methane oxidation (POX), in which oxygen is introduced along with methane in the stoichiometry leading to partial oxidation (as per reaction 3.3), reducing full combustion. This process is exothermic and is carried out at high temperatures, which reduces the effort and energy required to maintain reaction temperatures and heat integration. The process is normally run at temperatures up to 1100 °C and does not require a catalyst [17]. However, to increase hydrogen efficiency and lower the operating temperature, the use of a catalyst is often proposed in the literature, generally nickel-based [18]. Due to the higher temperatures, this process, unlike the SMR and ATR processes, allows the use of different reactants, which increases its versatility (also due to the possible variant without catalyst). Nevertheless, the need for pure oxygen remains the main drawback of this technology, since an ASU adjacent to the plant is required, similar to the ATR technology. Despite the good heat integration achieved by this technology, the low hydrogen yields limit its hydrogen production compared to SMR.

$$\text{Partial oxidation of methane}: \quad CH_4 + \frac{1}{2}O_2$$

$$\rightleftharpoons CO + 2H_2 \, \Delta H^0_{298K} = -35.6 \, kJ/mol \qquad (3.3)$$

A comparison of the different strategies shows that the methane steam reforming process does not require the addition of pure oxygen, operates at lower temperatures, and produces a syngas with higher H_2/CO ratios, implying higher hydrogen reforming efficiencies [14]. However, since the reforming reaction is endothermic, the need for external heat supply leads to expensive reformers and high carbon emissions, which is the main drawback of SMR.

3.2.2   Coal

Coal is the main fuel source in China, where as much coal was consumed in 2017 as in the rest of the world combined, as BP recently reported in its statistical overview of the world energy market [19]. Other countries with large coal consumption are the United States, Japan, or India. The main use of coal worldwide is combined heat and power. However, in many countries there is already a smooth transition to natural gas-fired power plants for electricity generation, favored by the decreasing cost of natural gas and lower CO_2 emissions. This, in turn, has led to an increased use of coal for chemical production, where it is used as a feedstock to produce syngas during coal gasification or partial oxidation.

The composition of coal varies depending on the region of origin and consists of carbon (70–90% by weight), hydrogen (4–6% by weight) and oxygen. Its quality is defined by the proportion of volatile components. Coal gasification technology includes processes for direct reaction of coal with oxygen and steam. It was developed in the nineteenth century as a technology for the production of town gas. It gained maturity in the 1930s and, with various further developments, was introduced into the 1950s for the production of a wide variety of products. The driver of this development was the Fischer-Tropsch reaction. Since then, various companies have used coal gasification as a process for hydrogen production [20].

The advantage of gasification compared to other technologies is that any carbon-based feedstock can be easily processed as in this way. This means that not only coal can be used as feedstock, but also petroleum residues, wastes or biomass. The main product obtained by gasification is synthesis gas, and to a small extent, depending on the selected process conditions, by-products such as CO_2 and CH_4 are produced. This syngas is commonly used for power generation in gas turbines and is nowadays the most efficient technology for low-cost power generation, which can also be coupled with carbon sequestration units [21]. The second main application of gasification is the production of liquid fuels by the Fischer-Tropsch reaction using the syngas produced during gasification. When the product is hydrogen rather than syngas, a WGS reaction and a PSA unit are installed.

In coal gasification, dry coal particles are introduced into the gasifier and contacted with steam and oxygen (pure or from the air). Typically, the reaction is carried out at temperatures in the range of 950–1150 °C and at moderate pressure (typically 5–10 bar). During the gasification process, a complex chemistry takes place. In the first part, while the coal is heated in the gasifier, progressive vaporization of the coal occurs, producing light hydrocarbons, phenols, oils, and tars [22]. The conversion of these gas products, in contact with oxygen and steam at elevated temperatures, is the result of combined methanation, water gas conversion reaction, partial oxidation, and methane steam reforming reactions, and the final composition of the gas products is highly dependent on the process conditions. High temperatures lead to partial oxidation reactions, which result in increased conversion to CO and H_2 and limit the yields of methane, CO_2, and water. On the other hand, lower temperatures result in higher amounts of CO_2 and water in the product spectrum. The pressure level is typically adjusted to the required downstream processing. Generally, the pressure is kept low when syngas formation is desired. To stop product formation, the gasifier outlet gas is typically quenched with recycled syngas or water to increase the yield toward the desired product distribution. For ATR and POX technologies, oxygen should be added to the system, which requires an ASU.

In the literature, the use of a catalyst is often suggested to improve the process efficiency of gasification by reducing tar content and increasing hydrogen production. In particular, Fe- and Ca-based catalysts affect the gasification rate, while Ni-based catalysts favor the reforming reactions, maximizing syngas production [21]. However, they are quickly deactivated in the process and washed out of the system or settle along with the ash (slag) formed during gasification. This slag should be removed continuously to avoid damage to the gasifier [23].

The syngas produced during gasification contains a large amount of impurities originating from the coal used. Fine inorganic particles, volatiles, H_2S, NH_3 or HCN are common impurities that require a subsequent cleaning step of the syngas products. The syngas purification is a very important part of the process and consists of several stages where all impurities are removed. The quenched syngas first passes through cyclones and filters to remove the particles. This is followed by gas scrubbing, where particulate matter, ammonia and chlorides are removed [24]. Then, depending on the coal composition, absorbers are required to remove traces of mercury and other heavy inorganic metals. Depending on the operating conditions of the gasifier, some of the sulfur content is in the form of carbonyl sulfide (COS). Removal of COS is accomplished by hydrolysis, converting all the COS to H_2S. After this step, only acids remain as impurities of the syngas, which are removed in a sour gas removal section. The syngas should contain a maximum of 1 ppm sulfur, and physical solvent processes are preferred over chemical solvents for chemical synthesis applications. Typical physical solvents are Selexol (DME) or Rectisol (methanol), which operate at relatively high pressures and in a cryogenic process (around 0 °C) [25].

Efficient gasification plants for syngas production are available on the market worldwide. The Shell Coal Gasification Process (SCGP) [26] can convert a wide range of coals into syngas. In particular, Shell has two SCPG technologies in its portfolio, depending on the cooling technology chosen: one technology uses syngas for quenching and is a preferred option for IGCC power generation due to its high efficiency. On the other hand, if syngas is preferred as the reactant for Fischer-Tropsch synthesis, water is used for quenching, resulting in lower investment costs [27]. Shell is one of the largest suppliers of gasifiers. The technology includes membranous reactor walls from which water vapor and oxygen are supplied, resulting in good mixing with the coal feed. This gasifier operates at about 1500 to 1600 °C and at 20 to 35 bar, resulting in a syngas that contains only very small amounts of CO_2 and no liquid or gaseous hydrocarbons [27].

GE Energy produces a downdraft gasifier that operates at very high temperatures (1200–1500 °C) and 20 bar to produce a rich syngas [28]. CB&I supplies an upward pressure gasifier that uses slurry coal as feedstock. Compared to other gasifiers, the CB&I gasifier has a two-stage feed system. At the bottom, about 70% of the slurry coal is fed together with oxygen (purity 95%) and the synthesis gas produced from it is mixed with the second slurry coal in the absence of oxygen, triggering endothermic reactions that form hydrocarbons [29]. These hydrocarbons are returned to the cycle in the first stage at the end of the process. Siemens uses a similar technology to GE Energy [30], while KBR uses a circulating fluidized bed reactor [31].

Air Liquide applies fixed bed gasification technology (Lurgi FBDBTM), which is suitable for low-grade coals with high water or ash content that cannot be economically used in the entrained flow process. This technology is well referenced and operated economically. It allows capacities between 40,000 and 120,000 Nm^3/h of dry syngas [32].

3.2.3 Oil

Crude oil is the most important fossil energy source in many countries. It is easy to transport and has a high energy density. In 2015, 93 million barrels per day were consumed, with the USA and China being the main consuming countries [33]. A wide variety of fractionation cuts can be obtained from petroleum. In particular, most of the crude oil is used for the production of liquefied petroleum gas (LPG), gasoline, diesel, jet fuel, and fuel oil.

In refineries, crude oil is distilled into various fractions. Light hydrocarbons (LPG) and a mixture of kerosenes, naphthenes and short-chain aromatics (C4-C11) are recovered from the upper fraction. This complex mixture, also called crude gasoline (naphtha), is the main source for gasoline production and contains a large amount of impurities such as sulfur, salt or metals that must be removed before further use.

Naphtha is mainly reformed to produce high octane gasoline, which is now preferred over low octane gasoline and diesel fuels in terms of reducing carbon emissions. During the naphtha reforming process, hydrogen (2 vol% fraction) is produced along with a mixture of aromatics, kerosenes, isoparaffins, and cycloalkanes [34]. This high-purity hydrogen is formed during the various dehydrogenation and aromatization reactions of aromatics production and is recovered in flash drums in the so-called recovery area. The hydrogen thus produced as a by-product is subsequently consumed again directly in situ in other processes in the refinery. In this context, hydrocracking and hydrotreating are two processes that require large amounts of hydrogen to break down long-chain hydrocarbons into shorter and hydrogenated products and to desulfurize the products. Hydrocracking is one of the most important operations in the production of diesel fuel, while hydrotreating is an important basic operation to avoid catalyst poisoning and to provide fuels with extremely low pollutant levels.

In addition, naphtha is the most important raw material for the production of chemicals. Steam cracking of naphtha is a standard process that converts naphtha into a mixture of short-chain hydrocarbons and especially olefins, such as ethylene and propylene. This chemical conversion produces a certain amount of hydrogen as a byproduct, which is subsequently used in the refinery to remove impurities or upgrade low-quality petroleum fractions [35].

Unlike coal or natural gas as feedstocks from which the hydrogen is intentionally produced, hydrogen is not the main product of the process in the petroleum refinery. Instead, it is produced as a by-product in large quantities, making it an important contributor to global hydrogen production.

In recent decades, increasing consumption of diesel fuel, on the one hand, and stricter pollution regulations, on the other, have led to a massive increase in the demand for hydrogen in the oil refinery. As a result, the hydrogen production of oil refineries is not sufficient to meet all the captive hydrogen demand required in the hydrocracking and hydrotreating stages. Therefore, additional steam methane reformers or gasification plants with capacities of 100,000 to 200,000 Nm^3/h are currently being built alongside the refineries for dedicated hydrogen production. This is shown in Fig. 3.2, the overall view of an oil refinery [35], and also in the recently published report of the US energy information reproduced in Fig. 3.3 [36].

In the petroleum refinery, hydrogen is produced during naphtha reforming, and UOP offers UOP CCR Platforming™ [37], a process for producing high-octane products suitable for gasoline and BTX production. In addition, the CCR platform produces high purity hydrogen as a byproduct that can be used in any other part of the process platform. The CCR unit achieves high efficiencies and enables broad application flexibility.

Refineries are in operation around the world. The largest and located in India is the Jamnagar refinery owned by Reliance (India) with a crude capacity of 1.24 million barrels per day [38]. In the same region, Reliance also owns the sixth largest refinery in the world with a capacity of 580,000 barrels of crude oil per day [38]. On Jurong Island (Singapore), many companies have installed their refineries, including BASF [39], ExxonMobil [40], Shell [41], or Petrochemical Complex of Singapore [42]. Not all refinery sites have very large capacities. For example, Repsol (Spain) owns the La Pampilla refinery in Peru, which supplies low-sulfur diesel at a capacity of 117,000 barrels per day [43]. Shell Martinez in California (USA) processes up to

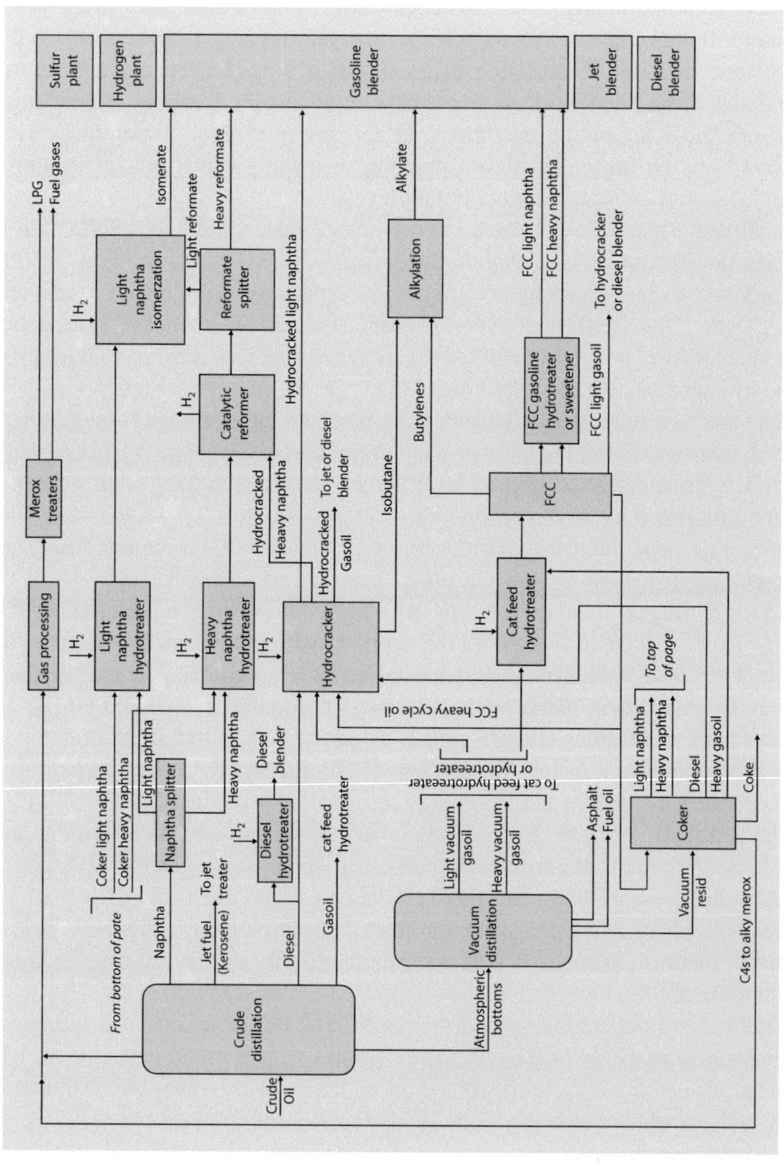

Fig. 3.2 Overview diagram of a petroleum refinery. © Elsevier

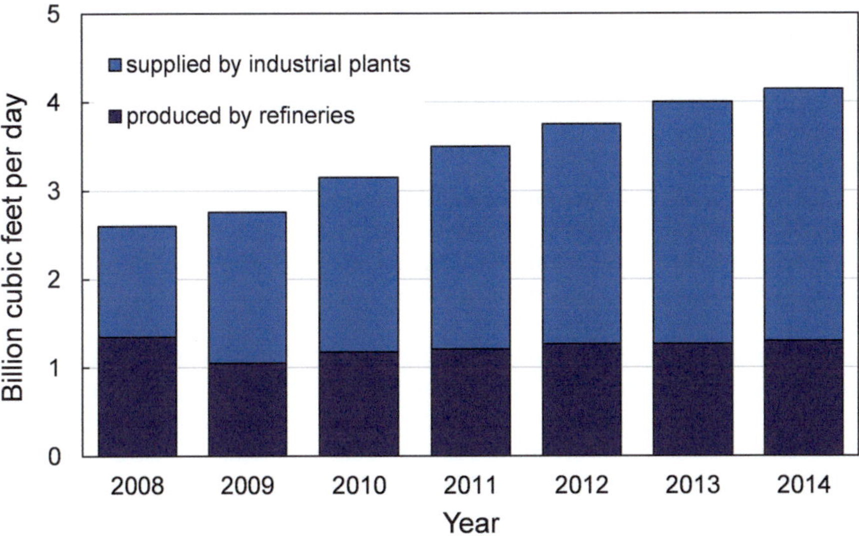

Fig. 3.3 Hydrogen demand of oil refinery (reprinted from [36])

165,000 barrels per day and produces a variety of products such as gasoline, jet fuel, and LPG [44]. This list could be extended by many more refineries installed worldwide by different suppliers, such as BP (UK), ENI (Italy) or Marathon Petroleum (USA).

3.2.4 Biomass

Fossil fuel depletion, together with concerns about the large amount of anthropogenic CO_2 emissions, has led to increased use of biomass as a feedstock for combined heat and power (CHP), with the possibility of adaptation toward hydrogen production. However, the cost of the electricity produced cannot yet compete with conventional coal gasification [45]. However, because biomass can be produced locally, it also has the advantage of reducing the need to import oil, coal, or gas. Biomass consists of long carbon compounds, mainly cellulose, lignin, and hemicellulose. If it is to be used as a feedstock for hydrogen production, its physicochemical properties must be known in advance in order to select the optimal operating conditions for the process. Biomass consists mainly of carbon and, to a lesser extent, oxygen, hydrogen and other components, such as nitrogen, calcium, sulfur or chloride. In this case, the moisture content determines the quality of the biomass and strongly influences the overall energy balance. For example, undried wood pellets have an energy density about five times lower compared to domestic coal, and even up to ten times lower compared to natural gas or coal as used in gasifiers [46].

The conversion of biomass to hydrogen can be done mainly in two ways. The thermochemical route involves pyrolysis and gasification of biomass, requires biomass pretreatment, and can provide high hydrogen capacities. Biological hydrogen production is less energy intensive and even more environmentally friendly compared to thermal processes. However, the difficulty of achieving constant hydrogen production in this process is a major challenge. Therefore, only thermochemical conversion is discussed in this chapter.

Biomass gasification is currently the technology of choice for producing hydrogen from renewable sources without an important technological breakthrough, although the cost is still higher than natural gas reforming or coal gasification [45]. Any biomass used for hydrogen production must be pretreated. This conditioning consists in drying the biomass to a moisture content below 15% by volume and obtaining a solid material with a particle size and composition as homogeneous as possible. Subsequently, this biomass can be fed directly into a gasifier reactor as for coal, where similar chemical reactions take place. First, the celluloses, hemicelluloses and lignin decompose to form coke and volatiles. The coke is then gasified, and depending on the operating conditions, the gas product composition can vary greatly. Biomass gasification is also a proven technology in commercial gasifiers. The low hydrogen yields achieved by this process currently limit its industrialization [47–49].

Nowadays, all commercial biomass gasification technologies are intended for combined heat and power (CHP) generation, although there are also an increasing number of projects for fuel and syngas production. Among the various suppliers of biomass gasification plants, Holz Energie (Germany) supplies CHP gasifiers of different sizes that use wood as feedstock and provide a minimum operation time of 7500 h per year [50]. Syncraft (Austria) also supplies wood to CHP gasifiers, resulting in electrical efficiencies of 30% and fuel utilization efficiencies of up to 92% with any carbon footprint [51]. Other companies, such as Prodesa (Spain), have their main economic target in the production of biomass pellets for use in gasifiers [52]. All of these companies are expected to bring to market biomass gasification technologies that are also intended for syngas production. Up to 20 different collaborative projects are currently being pursued in IEA Bioenergy Task 33, with the main goal of producing syngas from biomass using gasification technology [53]. Some biomass gasifier projects are already in operation, while others are still in the planning phase, as can be seen from the 2016 status report of IEA-Bioenergy Task 33 [54].

3.2.5 Water Electrolysis

The use of electricity to decompose water into oxygen and hydrogen is a technology that is becoming increasingly important given the decreasing availability of fossil fuels and the potential to use electricity from renewable sources. Water electrolysis is a mature technology that has already been commercialized and holds a share of about

4–5% of the total market [13]. This technology is preferred when ultra-pure hydrogen is needed and can be considered as an interesting strategy for energy storage.

In water splitting, a direct current is applied between the anode and the cathode, both coated with noble metals and separated by an electrolyte. Overall, this process leads to the formation of ions and cations that recombine at both electrodes, forming pure H_2 at the cathode side and pure oxygen at the anode side according to the overall reaction 3.4. On an industrial scale, the two most developed technologies for water electrolysis are, first, the alkaline electrolyzer using a potassium hydroxide solution (KOH) as the electrolyte [55] and, second, the solid polymer electrolyte electrolyzer (SPE), commonly known as proton exchange membrane (PEM) electrolyzer, in which a membrane is used as the electrolyte [56]. Water electrolysis can typically achieve efficiencies in the range of 60 to 80%, based on heating value. Even higher efficiencies can be achieved using steam (or high temperature water). From a thermodynamic point of view, the electrolysis of 1 ton of water results in 888.5 kg of oxygen and 111.5 kg of hydrogen [13]. At atmospheric pressure, this amount of hydrogen is equivalent to 1237 m^3. One of the main limitations of this technology is still the production of oxygen as a by-product, for which nowadays there is a limited demand on an industrial scale compared to hydrogen [13].

$$\text{Total reaction } H_2O \rightarrow \frac{1}{2}O_2 + H_2 \tag{3.4}$$

In the alkaline electrolyzer, the ions are transported from the cathode to the anode by means of a membrane with ion transport capacity. The two main reactions (3.5 and 3.6) that take place in this system correspond to hydrogen production at the cathode and oxygen prod uction at the anode. In the PEM electrolyzer, a solid membrane acts as the electrolyte, transporting protons from the anode to the cathode side, where H_2 is deposited. The reactions that take place in a PEM electrolyzer differ from those of the alkaline electrolyzer and are given in reactions 3.7 and 3.8.

Alkaline electrolyzer:

$$\text{Hydrogen production at the cathode } 2H_2O + 2e^- \rightarrow H_2 + 2OH^- \tag{3.5}$$

$$\text{Oxygen production at the anode } 2OH^- \rightarrow \frac{1}{2}O_2 + H_2O + 2e^- \tag{3.6}$$

PEM electrolyzer:

$$\text{Anode reaction } 2H_2O \rightarrow O_2 + 4H^+ + 4e^- \tag{3.7}$$

$$\text{Cathode reaction } 2H^+ + 4e^- \rightarrow H_2 \tag{3.8}$$

On an industrial scale, multiple electrolysis cells are stacked to meet the H_2 production requirements demanded by the end user. Only low temperature (60–80 °C) electrolysis is commercially available, including alkaline, PEM or AEM (anion exchange membrane) electrolysis. However, the latter technology is hardly used because it is relatively new, and both its capacity and the purity of the hydrogen obtained are far from those of alkaline and PEM electrolysers. The main characteristics of these electrolysers were recently described by Shell in the 2017 Hydrogen Study [57] and are summarized in Table 3.1.

Alkaline electrolyzers are the most commercialized in the world due to their maturity. McPhy supplies alkaline electrolyzers in two sizes. The small capacity units (up to 12 Nm^3/h H_2) are designed for small light operations and can deliver H_2 pressures up to 8 bar without the need for a compressor. Large capacity units (10–400 Nm^3/h) deliver H_2 at pressures up to 30 bar and are suitable for the steel industry or power-to-gas processes [58]. Hydrogenics is another supplier for alkaline electrolysers (HySTAT®). The patented IMET® ion exchange membrane is made of inorganic materials [59] and delivers capacities up to 500 Nm^3/h. Hydrogenics also offers PEM electrolyzers (HyLYZERTM) using an ionically conductive solid polymer.

Siemens is another supplier of PEM electrolyzers for ultra-pure hydrogen production. The SILYZER module has a rated capacity of 225 Nm^3H_2/h and a lifetime of over 80,000 h. It starts up in less than 10 s and delivers hydrogen at pressures up to 35 bar [60]. ArevaH2Gen also supplies PEM electrolysers, with units ranging from one to four stacks with two different capacities. In these units, hydrogen reaches 99.999% purity and the system includes water purification systems that allow tap water to be used [61]. ITM Power has developed its HGAS-PEM electrolyzer, which is intended for gas-to-electricity applications and delivers pressurized (80 bar) H_2 in just one second [62]. Recently, ITM Power has started a project with Shell to install the world's largest hydrogen electrolyzer. The electrolyzer will have a capacity of 10 MW and will produce pure H_2 at the Rheinland Refinery (Wesseling site) in Germany [63].

The production of hydrogen from electricity is still largely influenced by the cost of electricity, as the required electricity is in the range of 4.2 to 4.5 kWh/Nm^3H_2 [56]. From an economic point of view, conventional fossil fuel-based technologies are still preferred. However, water electrolysis has found its market for small scale applications, with high hydrogen purity requirements.

3.2.6   Comparison of the Different Technologies

The technology chosen for hydrogen production depends heavily on various factors such as the cost and availability of the energy source, hydrogen production capacity, or hydrogen purity. Despite these factors, all the technologies described in the previous section each have specific advantages and disadvantages that should be taken into account when selecting the preferred technology. These characteristics are reflected in Table 3.2.

Table 3.1 Characteristics of commercially available electrolysers, recorded in the Shell Hydrogen Study [57]. *n.a.* not available

Electrolzer	Temperature [°C]	Electrolyte	Plant size [kW]	[Nm^{3}H$_2$/h]	Purity H$_2$ [%]	System costs [€/kW]	Life expectancy [h]	Maturity
Alkaline electrolier	60–80	KOH	1.8–530)	0.25–760	>99.5	1000–1200	60,000–90,000	Commercially used in industry for 100 years
PEM	60–80	Solid membrane	0.2–1150	0.01–240	>99.9	1900–2300	20,000–60,000	Commercially available for medium and small applications (<300 kW)
AEM	60–80	Polymer membrane	0.7–4.5	0.1–1	≈ 99.4	n. a.	n. a.	Commercially available for limited applications

Table 3.2 Overview of possible advantages and disadvantages of the main hydrogen production technologies available on the market

Technology	Strengths	Disadvantages
Natural gas as energy sources		
• Steam reforming of Methan	• Well known technology • High energy efficiency (LHV) • High H_2/CO-relation after reforming • Inexpensive process for hydrogen production	• High CO_2-emissions • Energy-intensive technology due to endothermic reactions • External heat supply necessary
• Autothermal reforming	• No heat supply • Low methane slip • Compact reactor design	• Low hydrogen reforming efficiency • ASU required
• Partial oxidation	• Low methane slip • No catalyst required • Low reactor volumes • High feedstock flexibility	• ASU required • Low H_2/CO ratio • Reaction is more difficult to control
Coal gasification	• High efficiency • Well established technology	• Purification of the synthesis gas required
Oil as energy source	• Hydrogen is a by-product • Simple purification from the gas mixture	• High CO_2 emissions • Low production capacities
Gasification of biomass	• Renewable energy source • Significantly lower greenhouse gas emissions • Allows modernization of existing coal gasification plants	• Lower efficiency compared to coal gasification • Raw material must be prepared • Operation under high pressure is limited [57].
Water splitting	• Clean technology if electricity used comes from renewable resources • Possibility to store electricity as hydrogen • High purity of hydrogen	• Electricity consumption • Higher costs compared to reforming natural gas

3.3   Perspective

CO_2 emissions are one of the greatest challenges facing humanity. The Intergovernmental Panel on Climate Change (IPCC), given the impact of persistent greenhouse gases in the atmosphere, has developed various scenarios based on atmospheric CO_2 concentrations and recommended measures to mitigate emissions of greenhouse gases in order to limit global warming to a maximum of 2 °C by 2100 [64]. These scenarios are primarily based on a shift from fossil fuel use to power generation from renewable sources, including short- and medium-term carbon capture and storage (CCS) measures [65, 66]. This has increased research efforts to develop novel technologies for hydrogen production.

Chemical looping reforming is a promising technology for syngas production and carbon capture. Chemical looping (CL) is a process in which a chemical reaction is divided into two steps that occur in two different reactors. In CL reforming, a solid metal oxide circulates between two reactors operated under different reaction conditions. In the so-called air reactor, the solid metal is oxidized with air and then transferred to a second reactor, the fuel reactor, where it is reduced by partial oxidation with methane [67–69]. This solid transfers oxygen from one reactor to the other, replaces the need for a furnace as in steam reforming, and avoids the need for ASU as in POX and ATR technologies. The process can also run on coal, making it more versatile compared to other conventional units. Recent assessments of this technology can be found in the literature [70, 71].

Membrane reactors offer an interesting technology for direct separation of hydrogen from the reaction. In a membrane reactor, a reaction product is selectively removed from the catalytic bed, allowing the process to be run beyond thermodynamic constraints in equilibrium reactions such as fuel reforming [72–74]. H_2-selective membranes have been developed in the last decades using palladium or Pd alloys. This technology has already been tested at laboratory scale and demonstrated by companies such as Tokyo gas [75]. Direct recovery of H_2 from the reactor allows operation at lower temperatures, which has a direct impact on energy requirements. In addition, the downstream units (WGS and PSA) can be eliminated, which also reduces the CAPEX of the process. The techno-economic analysis of this concept has shown its potential, and recent developments and future trends can also be found in the literature [6]. A similar strategy is the selective adsorption of CO_2 on solid adsorbent materials, which modify the thermodynamic equilibrium of the process and lead to higher hydrogen production efficiencies [76].

These technologies can be powered by renewable energy sources such as biogas [77]. Biogas is a gas mixture consisting mainly of CH4 and CO_2, which is produced during the anaerobic digestion of waste. When biogas is injected into chemical-looping systems, capture of the injected CO_2 results in an overall negative carbon dioxide emission because this CO_2 comes from biomass.

Chemical reforming (or CO_2 reforming) of methane is also an interesting technology where methane reacts directly with CO_2 to produce H_2 and CO with a 1 to 1 molar ratio. It is considered a technology for the future, although it requires high temperatures, and the hydrogen yield is rather low [78, 79]. Therefore, this approach has not yet found application on an industrial scale.

Another alternative increasingly considered is the direct decomposition of methane into hydrogen and solid carbon. This process requires high temperatures and has only been performed on a laboratory scale. The high added value of the solid carbon produced as a byproduct can make this process interesting from an economic point of view [80, 81].

Another strategy to avoid the energy-intensive methane steam reforming process is the further development of catalytic partial oxidation (CPO). Although this process has been known for some time, the rapid deactivation of the catalyst has limited commercialization to date. Recently, more research has focused on developing new catalyst formulations that deliver high hydrogen production rates without catalyst deactivation [21, 82, 83].

In almost all of these technologies, methane is the main energy source for hydrogen production. Increasingly, therefore, biomethane can be expected to be produced either from biogas or from a methanation reaction with syngas produced from biomass or syngasification. Alternatively, bioethanol can be used as an energy source, as already used for power generation in several countries. Bioethanol can be produced by fermentation of sugars and, when reformed, results in similar syngas compositions to SMR. The advantage of using biomethane or bioethanol is that existing technologies (SMR, ATR, POX) as well as those currently under development (chemical looping or membrane reactors) do not require significant technological adjustments and therefore existing production facilities can be easily upgraded. Increased use of biogenic resources is also expected as a result of declining oil reserves. In this case, Fischer-Tropsch synthesis could become the main route for the production of gasolines, diesel or olefins. This process uses synthesis gas as a reactant, resulting in a large increase in hydrogen production.

An alternative to biomass gasification is the use of waste to produce hydrogen. Compared to biomass gasification, this could be a more favorable strategy because biomass, such as side streams from forestry, must first be collected, transported, and pretreated before they can be used in gasifiers. A household waste-based process is already available on the market [84] and converts household waste into syngas after pretreatment and gasification [85].

As for water electrolysis, the high cost of electricity generation is still the main drawback. A further increase in the use of renewable energy sources, such as solar or wind energy, will lead to a decrease in the cost of electricity and thus to a decrease in the production cost of hydrogen. In many cases, the electricity used is not from renewable sources, as it is a discontinuous energy source. In this scenario, water electrolysis is expected to serve not only as a source of hydrogen production, but also as an energy storage technology. The use of excess electricity, such as wind energy overnight, could be compensated by the production of hydrogen through water electrolysis. This hydrogen could later be used for energy and power generation by fuel cells when needed [86, 87].

3.4   Conclusions

This chapter has provided a comprehensive overview of the main commercial routes to hydrogen production. Today, about 95% of the total hydrogen production still comes from fossil fuels, mainly from methane steam reforming. Only the remaining share of total hydrogen production comes from alternative sources, mainly from water splitting. Hydrogen is consumed in the chemical industry primarily in the processes used to produce ammonia and methanol. This hydrogen is typically produced via reforming processes (SMR or ATR) or partial oxidation technologies (POX) and, to a lesser extent, coal gasification. These technologies are exclusively for dedicated production of hydrogen. In the petrochemical industry, where about 25% of the hydrogen produced worldwide is consumed, hydrogen is produced as a by-product of naphtha reforming. This hydrogen is directly consumed in further process steps of the oil refinery. Due to more restrictive regulations regarding contaminants and pollutants in gasoline and diesel fuels, which require additional purification steps and thus hydrogen input, there is a shortage of hydrogen in the refining sector today. Some of the hydrogen needed is supplied externally by installing hydrogen production plants near refineries.

With the depletion of fossil fuels and the increase of greenhouse gases in the atmosphere, alternative ways and technologies should be found to produce hydrogen through more sustainable processes. Membrane reactors, chemical looping, dry reforming, or methane decomposition are the technologies that will become market-ready in the next few decades and can replace the established technologies. It is expected that water splitting, which is already available or almost commercialized, as well as biomass gasification will contribute more and more to the worldwide hydrogen production. Therefore, a completely different scenario regarding hydrogen production processes can be expected in the coming decades.

References

1. US Department of Energy (2018). https://www.energy.gov/. Accessed 10 Mar 2018
2. Rostrup-Nielsen JR (2000) New aspects of syngas production and use. Catal Today 63:159–164
3. da Silva VT, Mozer TS, da Costa Rubim Messeder dos Santos D, da Silva César A (2017) Hydrogen: trends, production and characterization of the main process worldwide. Int J Hydrog Energy 42:2018–2033
4. Rafiqul I, Weber C, Lehmann B, Voss A (2005) Energy efficiency improvements in ammonia production - perspectives and uncertainties. Energy 30:2487–2504
5. Ball M, Wietschel M (2009) The future of hydrogen – opportunities and challenges. Int J Hydrog Energy 34:615–627
6. Spallina V, Pandolfo D, Battistella A, Romano MC, van Sint AM, Gallucci F (2016) Techno-economic assessment of membrane assisted fluidized bed reactors for pure H_2 production with CO_2 capture. Energy Convers Manag 120:257–273
7. Kinetics Technologies (2018) KT Reformers. http://www.kt-met.com/en/business/hydrogen-syngas-technology. Accessed 10 Mar 2018
8. McDermott (2018) Steam methane reformers. https://www.cbi.com/What-We-Do/Technology/Gas-Processing/Hydrogen-Synthesis-Gas-Production/Steam-Methane-Reformers. Accessed 10 Mar 2018
9. Haldor Topsoe (2018) Tubular radiant wall steam reformer. https://www.topsoe.com/products/tubular-radiant-wall-steam-reformer. Accessed 10 Mar 2018
10. Amec Foster Wheeler (2018) Terrace Wall™ steam reformers. https://www.amecfw.com/products/fired-heaters/steam-reformers/terrace-wall-steam-reformers. Accessed 10 Mar 2018
11. Ratnasamy C, Wagner JP (2009) Water gas shift catalysis. Catal Rev 51:325–440
12. Holladay JD, Hu J, King DL, Wang Y (2009) An overview of hydrogen production technologies. Catal Today 139:244–260
13. Armaroli N, Balzani V (2011) The hydrogen issue. ChemSusChem 4:21–36
14. Angeli SD, Monteleone G, Giaconia A, Lemonidou A (2014) State-of-the-art catalysts for CH4 steam reforming at low temperature. Int J Hydrog Energy 39:1979–1997
15. Haldor Topsoe (2018) Autothermal reformer. https://www.topsoe.com/products/syncortm-autothermal-reformer-atr. Accessed 10 Mar 2018
16. Johnson Matthey (2018) Reforming technologies (ATR, GHR, SMR). http://www.jmprotech.com/core-technologies-reforming-ATR-GHR-SMR. Accessed 10 Mar 2018
17. Basini L, Aasberg-Petersen K, Guarinoni A, Østberg M (2001) Catalytic partial oxidation of natural gas at elevated pressure and low residence time. Catal Today 64:9–20

18. Basile F, Basini L, Amore MD, Fornasari G, Guarinoni A, Matteuzzi D, Piero GD, Trifirò F, Vaccari A (1998) Ni/Mg/Al anionic clay derived catalysts for the catalytic partial oxidation of methane: residence time dependence of the reactivity features. J Catal 173:247–256
19. BP (2017) BP statistical review of world energy 2017. http://www.bp.com/content/dam/bp/en/corporate/pdf/energy-economics/statistical-review-2017/bp-statistical-review-of-world-energy-2017-full-report.pdf. Accessed 10 Mar 2018
20. DOE/NETL (2018) Gasification systems. https://www.netl.doe.gov/research/coal/energy-systems/gasification/gasifipedia/history-gasification. Accessed 10 Mar 2018
21. Navarro RM, Peña MA, Fierro JLG (2007) Hydrogen production reactions from carbon feedstocks: fossil fuels and biomass. Chem Rev 107:3952–3991
22. Barnes I (2011) Next generation coal gasification technology, 1–49
23. Kopyscinski J, Schildhauer TJ, Biollaz SMA (2010) Production of synthetic natural gas (SNG) from coal and dry biomass – a technology review from 1950 to 2009. Fuel 89:1763–1783
24. DOE/NETL (2018) Syngas contaminant removal and conditioning. https://www.netl.doe.gov/research/coal/energy-systems/gasification/gasifipedia/cleanup. Accessed 10 Mar 2018
25. DOE/NETL (2018) Acid gas removal. https://www.netl.doe.gov/research/coal/energy-systems/gasification/gasifipedia/syngas. Accessed 10 Mar 2018
26. DOE/NETL (2018) Shell gasifiers. https://www.netl.doe.gov/research/coal/energy-systems/gasification/gasifipedia/shell. Accessed 10 Mar 2018
27. Shell (2018) Shell gasification technology. https://www.shell.com/business-customers/global-solutions/gasification-licensing/coal-gasification.html. Accessed 10 Mar 2018
28. DOE/NETL (2018) GE energy gasifier. https://www.netl.doe.gov/research/coal/energy-systems/gasification/gasifipedia/ge. Accessed 10 Mar 2018
29. DOE/NETL (2018) CB&I E-Gas™ gasifiers. https://www.netl.doe.gov/research/coal/energy-systems/gasification/gasifipedia/egas. Accessed 10 Mar 2018
30. DOE/NETL (2018) Siemens gasifiers. https://www.netl.doe.gov/research/coal/energy-systems/gasification/gasifipedia/siemens. Accessed 10 Mar 2018
31. DOE/NETL (2018) KBR transport gasifiers. https://www.netl.doe.gov/research/coal/energy-systems/gasification/gasifipedia/kbr. Accessed 10 Mar 2018
32. Air Liquide (2018) Lurgi FBDB™ gasifier. https://www.engineering-airliquide.com/lurgi-fbdb-fixed-bed-dry-bottom-gasification. Accessed 10 Mar 2018
33. International Energy Agency Oil (2018) Öl. https://www.iea.org/about/faqs/oil/. Accessed 10 Mar 2018
34. Rahimpour MR, Jafari M, Iranshahi D (2013) Progress in catalytic naphtha reforming process: a review. Appl Energy 109:79–93
35. Fraser S (2014) Distillation in refining. In: Górak A, Schoenmakers H (eds) Distillation: operation and application. Elsevier, Amsterdam, pp 155–190
36. EIA (2018) USA refinery hydrogen consumption. https://www.eia.gov/todayinenergy/detail.php?id=24612. Accessed 10 Mar 2018
37. Honeywell UOP (2018) UOP CCR Platforming™ Process. https://www.uop.com/processing-solutions/refining/gasoline/naphtha-reforming/. Accessed 10 Mar 2018
38. Reliance (2018) Petroleum refining and marketing. http://www.ril.com/OurBusinesses/PetroleumRefiningAndMarketing.aspx. Accessed 10 Mar 2018
39. BASF Singapore (2018). https://www.basf.com/sg/en.html. Accessed 10 Mar 2018
40. ExxonMobil Jurong (2018) Singapore refinery. http://www.exxonmobil.com.sg/en-sg/company/business-and-operations/operations/singapore-refinery-overview. Accessed 10 Mar 2018
41. Shell Jurong (2018) Shell Jurong Island. https://www.shell.com.sg/about-us/projects-and-sites/shell-jurong-island.html. Accessed 10 Mar 2018
42. Petrochemical Complex of Singapore (2018) Jurong Island. http://www.pcs.com.sg/singapore-petrochemical-complex/jurong-island/. Accessed 10 Mar 2018
43. Repsol (2018) Peru's most important refinery: La Pampilla. https://www.repsol.energy/en/about-us/where-we-work/la-pampilla-refinery/index.cshtml. Accessed 10 Mar 2018

44. Shell (2018) Shell Martinez refinery. https://www.shell.us/about-us/projects-and-locations/martinez-refinery/about-shell-martinez-refinery.html. Accessed 10 Mar 2018

45. IEA (2007) IEA energy technology essentials - biomass for power generation and CHP. High Temp 1–4

46. Balat H, Kirtay E (2010) Hydrogen from biomass - present scenario and future prospects. Int J Hydrog Energy 35:7416–7426

47. Zhang L, Xu C, Champagne P (2010) Overview of recent advances in thermo-chemical conversion of biomass. Energy Convers Manag 51:969–982

48. Molino A, Chianese S, Musmarra D (2016) Biomass gasification technology: the state of the art overview. J Energy Chem 25:10–25

49. Sikarwar VS, Zhao M, Clough P, Yao J, Zhong X, Memon MZ, Shah N, Anthony EJ, Fennell PS (2016) An overview of advances in biomass gasification. Energy Environ Sci 9:2939–2977

50. Holz Energy (2018) Gasifier. http://www.holzenergie-wegscheid.de/home.html. Accessed 10 Mar 2018

51. Syncraft (2018) Gasifiers. http://syncraft.at/index.php/en/menu-products-en/menu-holzgaskraftwerk-en. Accessed 10 Mar 2018

52. Prodesa (2018). https://prodesa.net/. Accessed 10 Mar 2018

53. IEA Bioenergy/Task 33 (2018) Work scope, approach and industrial involvement. http://task33.ieabioenergy.com/content/taks_description. Accessed 10 Mar 2018

54. Hrbek J (2016) Status report on thermal biomass gasification in countries participating in IEA Bioenergy Task 33. Statusbericht

55. Zeng K, Zhang D (2017) Recent progress in alkaline water electrolysis for hydrogen production and applications. Prog Energy Combust Sci 36:307–326

56. Carmo M, Fritz DL, Mergel J, Stolten D (2013) A comprehensive review on PEM water electrolysis. Int J Hydrog Energy 38:4901–4934

57. Shell (2017) Shell hydrogen study 2017. https://www.shell.com/energy-and-innovation/the-energy-future/future-transport/hydrogen/_jcr_content/par/textimage_1062121309.stream/1496312627865/46fec8302a3871b190fed35fa8c09e449f57bf73bdc35e0c8a34c8c5c53c5986/shell-h2-study-new.pdf. Accessed 10 Mar 2018

58. McPhy (2018) Electrolyzers. http://mcphy.com/en/our-products-and-solutions/electrolyzers/. Accessed 10 Mar 2018

59. Hydrogenics (2018) Electrolysis. http://www.hydrogenics.com/technology-resources/hydrogen-technology/electrolysis/. Accessed 10 Mar 2018

60. Siemens (2018) Electrolyzer. http://www.industry.siemens.com/topics/global/en/pem-electrolyzer/silyzer/Documents/2017-04-Silyzer200-boschure-en.pdf. Accessed 10 Mar 2018

61. Areva H2Gen (2018) AREVA H2Gen hydrogen generators. http://www.arevah2gen.com/en/products-services/hydrogen-generators. Accessed 10 Mar 2018

62. ITM-Power (2018) HGas. http://www.itm-power.com/product/hgas. Accessed 10 Mar 2018

63. ITM-Power (2018) Project with shell. http://www.itm-power.com/news-item/10mw-refinery-hydrogen-project-with-shell. Accessed 10 Mar 2018

64. IPCC (2014) Climate change 2014: synthesis report. Contribution of working groups I, II and III to the fifth assessment report of the intergovernmental panel on climate change

65. IPCC (2005) IPCC special report on carbon dioxide capture and storage. Cambridge University Press, Cambridge

66. IEA (2010) Energy technology perspectives: scenarios and strategies to 2050. OECD/IEA, Paris

67. Medrano JA, Potdar I, Melendez J, Spallina V, Pacheco-Tanaka DA, van Sint AM, Gallucci F (2018) The membrane-assisted chemical looping reforming concept for efficient H_2 production with inherent CO_2 capture: experimental demonstration and model validation. Appl Energy 215:75–86

68. Medrano JA, Spallina V, van Sint AM, Gallucci F (2014) Thermodynamic analysis of a membrane-assisted chemical looping reforming reactor concept for combined H_2 production and CO_2 capture. Int J Hydrog Energy 39:4725–4738

69. Rydén M, Lyngfelt A, Mattisson T (2006) Synthesis gas generation by chemical-looping reforming in a continuously operating laboratory reactor. Fuel 85:1631–1641
70. Tang M, Xu L, Fan M (2015) Progress in oxygen carrier development of methane-based chemical-looping reforming: a review. Appl Energy 151:143–156
71. Adanez J, Abad A, Garcia-Labiano F, Gayan P, de Diego LF (2012) Progress in chemical-looping combustion and reforming technologies. Prog Energy Combust Sci 38:215–282
72. Gallucci F, Fernandez E, Corengia P, van Sint AM (2013) Recent advances on membranes and membrane reactors for hydrogen production. Chem Eng Sci 92:40–66
73. Gallucci F, Medrano JA, Fernandez E, Melendez J, van Sint AM (2017) Advances on high temperature Pd-based membranes and membrane reactors for hydrogen purification and production. J Membr Sci Res 3:142–156
74. Fernandez E, Medrano JA, Melendez J, Parco M, van Sint AM, Gallucci F, Pacheco Tanaka DA (2016) Preparation and characterization of metallic supported thin Pd-Ag membranes for high temperature hydrogen separation. Chem Eng J 305:182–190
75. Shirasaki Y, Yasuda I (2013) Membrane reactor for hydrogen production from natural gas at the Tokyo Gas Company: a case study. In: Basile A (ed) Handbook of membrane reactors. Woodhead Publishing Limited, Sawston, Cambridge, pp 487–507
76. Aloisi I, Jand N, Stendardo S, Foscolo PU (2016) Hydrogen by sorption enhanced methane reforming: a grain model to study the behavior of bi-functional sorbent-catalyst particles. Chem Eng Sci 149:22–34
77. Ugarte P, Durán P, Lasobras J, Soler J, Menéndez M, Herguido J (2017) Dry reforming of biogas in fluidized bed: process intensification. Int J Hydrog Energy 42:13589–13597
78. Usman M, Wan Daud WMA, Abbas HF (2015) Dry reforming of methane: influence of process parameters – a review. Renew Sust Energ Rev 45:710–744
79. Abdullah B, Abd Ghani NA, Vo DVN (2017) Recent advances in dry reforming of methane over Ni-based catalysts. J Clean Prod 162:170–185
80. Upham DC, Agarwal V, Khechfe A, Snodgrass ZR, Gordon MJ, Metiu H, McFarland EW (2017) Catalytic molten metals for the direct conversion of methane to hydrogen and separable carbon. Science 358:917–921
81. Abbas HF, Wan Daud WMA (2010) Hydrogen production by methane decomposition: a review. Int J Hydrog Energy 35:1160–1190
82. Ashcroft AT, Cheetham AK, Green MLH, Vernon PDF (1991) Partial oxidation of methane to synthesis gas using carbon dioxide. Nature 352:225–226
83. Christian Enger B, Lødeng R, Holmen A (2008) A review of catalytic partial oxidation of methane to synthesis gas with emphasis on reaction mechanisms over transition metal catalysts. Appl Catal A Gen 346:1–27
84. Enerkem (2018). http://enerkem.com/. Accessed 10 Mar 2018
85. Iaquaniello G, Centi G, Salladini A, Palo E (2017) Waste as a source of carbon for methanol production. In: Basile A, Dalena F (eds) Methanol, science and engineering. Elsevier B.V., Amsterdam, pp 95–111
86. Kendall K (2017) Hydrogen fuel cells. In: Abraham MA (ed) Encyclopedia of sustainable technology. Elsevier, Oxford, pp 305–316
87. Eftekhari A, Fang B (2017) Electrochemical hydrogen storage: opportunities for fuel storage, batteries, fuel cells, and supercapacitors. Int J Hydrog Energy 42:25143–25165

Alternative Biological and Biotechnological Processes for Hydrogen Production

4

Thomas Happe and Christina Marx

Abstract

This chapter reviews the current state of research on biological and bioinspired hydrogen production. Nitrogenases and hydrogenases play a special role as key enzymes of hydrogen evolution in prokaryotic and eukaryotic organisms. The diversity of cell-specific metabolic pathways is illuminated by describing fermentative, photofermentative as well as photobiological hydrogen production. Limitations within the complex networks can serve as possible starting points to further develop *in vivo* systems for hydrogen production. As an example, current research on the optimization of bottlenecks in bacteria, cyanobacteria, and microalgae will be highlighted. Furthermore, novel concepts for artificial or semi-enzymatic catalysis are described.

Keywords

biological hydrogen production · enzyme · microorganism · nitrogenase · hydrogenase

Solar energy is the largest source of energy currently available to the world. 120,000 TW (1 TW = 1012 W) of solar energy hits the earth per year, of which, minus natural reflection, about 70% is absorbed by the earth's surface and the atmosphere [1]. Depending on the geographical location, on average about 170 W/m^2 per year of solar radiation reaches the Earth's surface [2]. About a quarter of this energy is available to plant systems as photosynthetically active radiation at the Earth's

T. Happe (✉)
Ruhr-Universität Bochum, Bochum, Germany

C. Marx
SolarBioproducts Ruhr, WFG Herne mbH, Herne, Germany

© The Author(s), under exclusive license to Springer Nature Switzerland AG 2023
M. Kircher, T. Schwarz (eds.), *CO2 and CO as Feedstock*, Circular Economy and Sustainability, https://doi.org/10.1007/978-3-031-27811-2_4

Table 4.1 Overview of hydrogen-producing metabolic pathways and their organisms

Type of H_2 production	Energy source	Electron source	Organism
H_2 production during dark fermentation	Organic molecules (organic acids, sugar)	Organic molecules (organic acids, sugar)	Enterobacteria, Clostridia
H_2 production of organisms with anoxygenic photosynthesis	Light	Organic acids, H_2S, H_2	(Sulfur-)Purple bacteria
H_2 production of organisms with oxygenic photosynthesis	Light	Water	Cyanobacteria, unicellular green algae

surface [3, 4]. Thus, the annual radiation exceeds the total global energy demand of about 15 TW per year and provides the ideal energy source to produce food, fuel, and oxygen [5].

Biological and technical processes can produce hydrogen by achieving zero emission of greenhouse gases. This includes photocatalytic production, which enables direct hydrogen production from sunlight and water, biomass gasification or water electrolysis using renewable energy production. Since the combustion of hydrogen as a fuel in turn produces water, solar hydrogen production is particularly relevant considering rising carbon emission levels and the usage of solar energy as an inexhaustible resource. In addition, hydrogen has an energy density up to three times higher compared to other hydrocarbon fuels [6]. Steam reforming of natural gas is currently the cheapest method for hydrogen production, with a market price of about 1 € per kg of hydrogen, while the market prices for ecological hydrogen production from biomass or by alkaline electrolysis are 7 € and 3 € per kg of hydrogen, respectively, and therefore have a similar competitive potential as solar hydrogen production [7].

Besides the possibility of using photochemical fuel cells or photovoltaic systems for photoconversion of water, photosynthetic or (photo)fermentative microorganisms can be used. The Leopoldina, which as the German National Academy of Sciences deals with important future social issues from a scientific point of view, independently of economic or political interests, also made the following recommendation in its statement [7], to which the author of this chapter also contributed:

> Given the nearly unlimited availability of water and sunlight, the production of hydrogen by direct photolytic splitting of water using phototrophic microorganisms could provide an ideal energy carrier that is renewable as well as environmentally friendly and sustainable. However, whether and when the formation of hydrogen by natural photosynthetic systems will become technically operational is an open question and the subject of ongoing fundamental research.

Biological models of photocatalysts for hydrogen production are often found in the world of microorganisms and include both prokaryotic representatives and a few eukaryotic unicellular organisms. Table 4.1 shows an overview of hydrogen-producing metabolic pathways found in various organisms. Since these microbial systems use sunlight as an energy source, they require less total energy for catalytic H_2 generation [4]. Biological hydrogen production probably serves to balance energy within the cell. In the metabolic pathways of some organisms, enzymes are used to reduce protons to hydrogen using excess electrons. Gaseous hydrogen, in turn, can be used as a source of electrons when needed to provide electrons for anabolic metabolic pathways [8].

Unlike catalysts used in the chemical industry, biocatalysts are active at more moderate temperatures and pressures and therefore have the potential as climate-neutral hydrogen producers. The key enzyme in biological hydrogen production is hydrogenase, which catalyzes the simple reaction of protons and electrons to form molecular hydrogen (Eq. 4.1). Depending on the metal content of the active site, a distinction is made between ferrous [Fe], ferrous-ferrous [FeFe], and nickel-ferrous [NiFe] hydrogenases [9, 10]. Since [FeFe] hydrogenases are among the most efficient hydrogen producers within the hydrogenase groups, they are of great biotechnological interest. Hydrogen is also produced in a side reaction during the reduction of atmospheric dinitrogen to ammonium under the action of the key enzyme of biological nitrogen (N_2) fixation, nitrogenase.

$$2\,H^+ + 2\,e^- \rightarrow H_2 \tag{4.1}$$

4.1 Biological Metabolic Pathways of Bacteria for Hydrogen Generation

4.1.1 Hydrogen Production in Fermentative Bacteria

In nature, a variety of metabolic processes are found under anoxic conditions, in which either hydrogen is oxidized or protons are reduced to molecular hydrogen by the enzyme hydrogenase. A prominent example is the well-known intestinal bacterium *Escherichia coli*, which evolves hydrogen under anaerobic conditions as part of mixed acid fermentation. In the fermentative metabolic pathway, pyruvate is converted to acetyl-CoA and formate, the latter of which is subsequently metabolized to CO_2 and H_2 by the enzyme formate hydrogenlyase (FHL), a multiprotein complex consisting of the formate dehydrogenase (Fdh-H) and the [NiFe] hydrogenase [11, 12]. Genomically encoded in *E. coli* are several hydrogenases, of which Hyd-3 and Hyd-4 presumably act together with Fdh-H in two distinct FHL-based metabolic pathways whose activity is influenced by both formate concentration and extracellular pH. The other two hydrogenases, Hyd-1 and Hyd-2, oxidize hydrogen during glucose fermentation at neutral or low pH on the outside of the cytoplasmic membrane [13].

Examples of H_2 generation using [FeFe] hydrogenases can be found in strictly anaerobic bacteria of the genus Clostridium. During butyric acid fermentation, these organisms oxidize pyruvate first to acetyl-CoA and further to acetate. The released electrons are transferred to ferredoxins by the pyruvate:ferredoxin oxidoreductase. Subsequently, the reduced ferredoxin transfers its electrons to a hydrogenase, where protons are converted to elemental hydrogen [14]. In another metabolic pathway, the electrons originate from NADH, which is formed during glycolysis. NADH is subsequently used by a cytosolic hydrogenase coupled to a NADH:ferredoxin oxidoreductase to reduce protons to hydrogen [15]. Moreover, another metabolic pathway has already been described for some organisms such as *C. kluyveri*, *Moorella thermoacetica* or *Termotoga maritima*, in which a so-called confurcating hydrogenase oxidizes two electron donors simultaneously, being reduced ferredoxin and NADH [16–18]. These electrons allow the reduction of protons to hydrogen.

The use of prokaryotic microorganisms for biohydrogen production has great potential (a good review can be found in Stephen et al. 2017 [19]), as the bacteria can use a variety of solid waste or wastewater as a source of nutrients [20–22]. In the biological treatment of solid waste and biomass under anaerobic conditions, Clostridia are the dominant bacteria within the mixed microbial population. Many studies for a possible application of the Clostridium system showed that mixed cultures have comparable efficiency to pure cultures [23]. A maximum production of 437 ml H_2/g hexose was achieved in a semi-continuous reactor with a waste mixture [24].

4.1.2 Hydrogen Production in Photofermentative Bacteria

While acetogenic thermophilic or mesophilic bacteria produce hydrogen through dark fermentative production pathways, (non-sulfur) purple bacteria can use a photofermentative pathway for hydrogen production. Due to anoxygenic photosynthesis using only one photosynthetic reaction center, purple bacteria such as *Rhodobacter (R.) sphaeroides* and *R. capsulatus* have been the subject of intense scientific investigation [25, 26].

Although ATP molecules are generated under anoxic conditions in this light-driven reaction, an additional organic source is necessary to produce sufficient reaction equivalents for nitrogen fixation [27]. In addition to glucose, alcohols, and glycerol, short-chain and volatile fatty acids can also be used. This range of usable carbon sources enables the utilization of food waste or wastewater from food production for sustainable hydrogen production via photofermentation [28]. Given the availability of a carbon source as well as light, dinitrogen can be reduced to ammonium by nitrogenase under anaerobic conditions, consuming 16 ATP molecules (Eq. 4.2). Hydrogen is produced as a byproduct of this reaction with an average production rate of up to 10.5 mmol H_2 L-1 h-1 [29]. If little or no dinitrogen is available, the process of nitrogen fixation by nitrogenase stops. Nevertheless, an alternative reaction (Eq. 4.3) occurs in which the ATP molecules are used entirely for hydrogen production. This reaction of nitrogenase is thus significantly more efficient for hydrogen production.

$$N_2 + 8\,H^+ + 8\,e^- + 16\,ATP \rightarrow 2\,NH_3 + H_2 + 16\,ADP + 16\,P_i \qquad (4.2)$$

$$8\,H^+ + 8\,e^- + 16\,ATP \rightarrow 4\,H_2 + 16\,ADP + 16\,P_i \qquad (4.3)$$

For photofermentative hydrogen production, a high carbon/nitrogen ratio is particularly important. Studies showed that a ratio of 25 carbon molecules to one nitrogen molecule results in the highest possible hydrogen production [30].

However, the influence of abiotic factors on the overall metabolism of non-sulfur purple bacteria should not be underestimated. Fluctuations in pH or utilization of different carbon sources can already lead to a change in hydrogen metabolism [31]. Another relevant factor is the membrane-bound hydrogen uptake system "Hup" (uptake hydrogenase), which, in addition to nitrogenase, significantly influences hydrogen metabolism. HupSL hydrogenase, which is one of the [NiFe] hydrogenases, catalyzes the oxidation of hydrogen, and the resulting electrons are subsequently transferred to ferredoxins and cytochromes [32, 33]. Since HupSL hydrogenase is activated by the availability of molecular hydrogen, its expression negatively affects the overall balance of hydrogen production. Genetic manipulations allow inactivation of the hup genes and thus the entire Hup system. These altered strains may have a significantly higher hydrogen production rate compared to the wild type. The target of such manipulation can be the hupSL gene, but also all other auxiliary proteins essential for hydrogenase synthesis and maturation [34, 35]. In addition to these factors, however, the entire regulatory network of hydrogen and nitrogen metabolism should also be considered, manipulation of which can produce up to more than 20% more hydrogen in *R. sphaeroides*, for example [18]. This can additionally be more than doubled, reaching hydrogen production rates of up to 142 ml $L^{-1}\,h^{-1}$ when the modified *R. sphaeroides* strain is grown on glutamate as the sole nitrogen source.

This complex network illustrates how many different factors influence hydrogen metabolism and thus reveals an attractive starting point for further genetic studies to make photofermentative hydrogen production as efficient as possible. This approach is made particularly interesting by the possibility of using released hydrogen gas directly for fuel cells due to its purity [36]. Great potential is also shown by organisms that, like *R. palustris*, can use lignin monomers as a source of electrons for hydrogen production [37, 38].

4.2   Light-Dependent Metabolic Pathways in Organisms with Oxygenic Photosynthesis

In addition to the photofermentative bacteria mentioned above, many genera of Gram-negative cyanobacteria can also produce hydrogen photobiologically [39]. Which enzyme is responsible for hydrogen production depends on whether the species in question is able to fix nitrogen or not [40]. As Fig. 4.1 illustrates, nitrogenase activity under nitrogen deficiency conditions leads to hydrogen production as part of nitrogen fixation. Nitrogen fixation occurs within filamentous

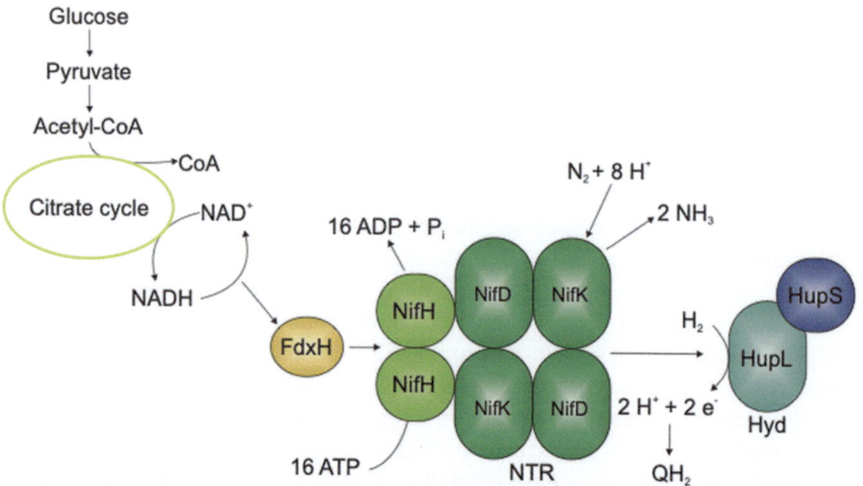

Fig. 4.1 Hydrogen metabolism in nitrogen-fixing cyanobacteria. Hydrogen production depends on the activity of nitrogenase (Nif), which is composed of dinitrogenase reductase (NifH) and dinitrogenase (NifDK). During the conversion of atmospheric nitrogen, ammonium is formed with the consumption of 16 ATP. Hydrogen is formed as a by-product, which can subsequently be recycled by the uptake hydrogenase HupSL

cyanobacteria in so-called heterocysts. These specialized cells have thick-walled cell inclusions that prevent the diffusion of oxygen into the cell and form an anaerobic microcompartment within an aerobic environment [41]. The hydrogen formed by nitrogenase is only a by-product and can subsequently be oxidized by the organism through a so-called uptake hydrogenase (encoded by the hupSL gene). If this hydrogenase is switched off, the cells produce significantly more hydrogen [42]. Cyanobacteria, which cannot fix nitrogen, possess a bidirectional hydrogenase that is used *in vivo* primarily for H_2 production in the light. Here, the enzyme is directly linked to the photosynthetic electron transport chain. Recent studies in the model cyanobacterium *Synechocystis sp.* PCC 6803 indicate that the bidirectional hydrogenase is coupled to ferredoxin and flavodoxin and reduces protons to hydrogen under anoxic conditions [43].

As mentioned above, hydrogen production rates can be optimized by inactivating the uptake hydrogenase HupSL. This results in up to a sevenfold increase in hydrogen production in nitrogen-fixing species [42, 44–46]. Since the processes of photosynthesis and cellular respiration are linked, the absence of quinone oxidase can cause up to a twelvefold increase in the activity of Hox hydrogenase under light conditions [47]. Presumably, both enzyme reactions act as electron valves when, under low oxygen concentrations, metabolic processes are not balanced or PSII activity is not in equilibrium with that of PSI. An up to 140-fold increase in hydrogen production rate could also be induced by inactivating nitrate assimilation [48]. In addition to direct manipulation of factors relevant to metabolic pathways (metabolic engineering), alteration of the heterocyst rate reveals another alternative for increasing hydrogen production [49]. The optimized heterocyst-to-vegetative cell rate allows photobiological hydrogen production under high light conditions and high cell densities. An alternative is the isolation of previously unknown strains such as Cyanothece, which have a significantly higher hydrogen production rate than known strains and may serve as future model organisms. The wild-type strain *Cyanothece sp.* ATCC51142 shows high hydrogen production even under aerobic conditions, which can be further enhanced by high CO_2 concentrations and addition of glycerol. A particularly active nitrogenase enzyme is responsible for the high hydrogen production rate of up to 465 µmol H_2/mg Chl/h [50]. Cyanobacteria such as *Lyngbya aestuarii* and *Microcoleus chthonoplastes* are also known for their high hydrogen production rates [51].

Isolation of novel wild-type strains, as well as genetic manipulation and alteration of metabolic pathways of already characterized species, allow optimized and sustainable biohydrogen production. By intensifying these approaches, the potential of feasible and efficient hydrogen production by cyanobacteria will be strengthened. In combination with technical advances in cultivation, cyanobacteria can be a promising alternative for a future environmentally compatible, but also competitive energy source.

4.2.1 Hydrogen Metabolism in Eukaryotic Microalgae

Organisms that use energy from photosynthesis to produce hydrogen use sunlight as their sole energy source and water as their sole electron source. In 2000, Anastasios Melis of the University of California, Berkeley, made a discovery about light-dependent hydrogen evolution in the green alga *Chlamydomonas reinhardtii*. He explored that this alga accumulates larger amounts of hydrogen in the absence of sulfur in a photobioreactor [52]. Thus, green hydrogen production forms an emission-free, economically attractive system, which is therefore of biotechnological interest and has been the subject of intensive research in recent years. Based on this, knowledge of physiological, genetic, and biochemical processes in the cell can be used to optimize natural hydrogen production. Studies of the structural and catalytic mechanisms of enzymes strengthen the potential of artificial or semisynthetic processes for industrial applications. This section is devoted to the hydrogen metabolism of photosynthetic eukaryotic organisms, as this is a particularly attractive route for obtaining a clean energy carrier.

While photosynthesis in cyanobacteria occurs within the thylakoids in the cytoplasm of the cell, this process is anchored in the thylakoid membranes in the chloroplast of green microalgae. The photosynthetic electron transport pathway, illustrated in Fig. 4.2, consists of two processes, a light reaction and a dark reaction. Whereas in the light reaction, water splitting at the PSII leads to the acquisition of electrons. Subsequently, they are transferred via the plastoquinone (PQ) pool, the cytochrome b_6f complex (Cyt b_6f, also plastoquinol plastocyanin oxidoreductase), and the PSI to ferredoxin for ATP and NADPH generation. Energy from the light reaction is used for the dark reaction, where CO_2 fixation and storage in the form of

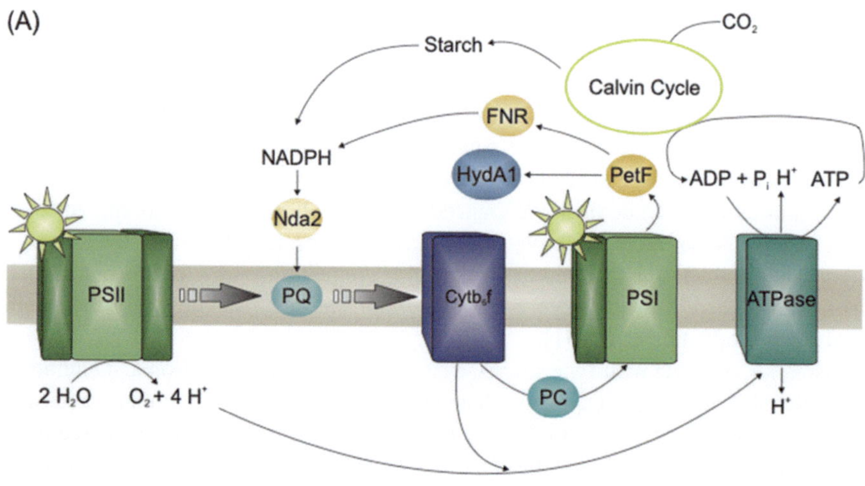

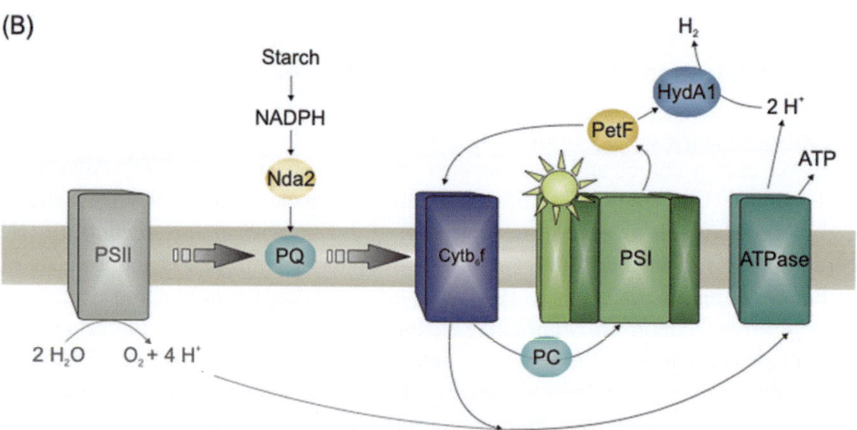

Fig. 4.2 Pathways of photosynthetic hydrogen production in green microalgae. (**a**) The reduction equivalents generated by water splitting at photosystem II (PSII) are transferred within the photosynthetic electron transport chain. Photosystem I (PSI) reduces ferredoxin (PetF), which subsequently reduces $NADP^+$ to NADPH by ferredoxin-$NADP^+$ reductase (FNR). Electrons from starch degradation are thereby transferred from NADPH to the plastoquinone (PQ) pool by a type II NADPH dehydrogenase. (**b**) Under anaerobic conditions in the light, PSII is only slightly active. Via PSI, electrons are transferred to PetF and consequently to HydA1. Via starch degradation, additional electrons can be injected via the plastoquinone pool

starch (or glycogen in cyanobacteria) takes place. In addition to the water-splitting PSII, the electron-supplying PSI is also essential for photobiological hydrogen production. The protons and electrons generated during water oxidation are subsequently converted to hydrogen by the hydrogenase HydA1 [52, 53]. In addition to this light-dependent, direct pathway of hydrogen production through the photosynthetic electron transport chain (PET), the generated energy can also be initially stored in the form of carbohydrates and later serve as a substrate for hydrogen production (Fig. 4.2). The coupling of oxidative degradation of organic substrates with PET allows PSII-independent electron transfer to PetF. In this process, electrons from starch degradation are transferred from NADPH to the PQ pool via a type II NADPH dehydrogenase [54–57]. However, this indirect pathway leads to a tenfold lower hydrogen production rate [58]. Apart from these two light-dependent pathways, a dark fermentative pathway of hydrogen production can also occur. Here, electrons are provided from starch degradation, and pyruvate is oxidized by pyruvate:ferredoxin oxidoreductase. Subsequently, PetF is reduced similarly to the previously described light-dependent pathway [59–62].

Hydrogen production occurring under oxygenic photosynthesis is only transient as hydrogenase is inactivated by oxygen [63, 64]. However, under light conditions, hydrogenase activity can be maintained by scavenging with inert gas or by sulfur starvation, since due to light damage and insufficient recycling of the D1 major subunit, PSII increasingly reduces $NADP^+$ to NADPH [65–67]. The lack of PSII activity in turn results in oxygen production being lower than oxygen consumption from respiration [52]. Within 24 hours, gas-tight sulfur-deficient cultures can thus reach an anaerobic or hypoxic state. Adaptation of metabolism in green microalgae such as *Chlamydomonas (C.) reinhardtii* results in the expression of hydrogenase [68, 69], among others. Since oxidative phosphorylation cannot occur in mitochondria under anaerobic conditions in the light without oxygen, over-reduction in the chloroplast leads to a decrease in ATP concentration in addition to the light damage already mentioned above. In this situation, hydrogenase acts as an electron and proton sink, the excess of which is caused by PSII activity in the direct pathway or by starch degradation in the indirect metabolic pathway and can subsequently be converted to hydrogen [70]. If hydrogen production also takes place in addition to starch degradation, the organic products formate, ethanol, and acetate are also produced during the photofermentation process [71–73]. In continuous cultivation systems, hydrogen production rates of 2 ml of H_2 per liter of wild-type algal culture per hour can be achieved under sulfur deficiency [52].

In addition to the previously described sulfur deficiency, nitrogen limitation induces hydrogen production under light conditions [74]. Nitrogen deficiency further causes starch and lipid accumulation in cells [75]. Under these deficiency conditions, PSII activity is maintained slightly longer, so that hydrogen production starts with a 2-day delay compared to sulfur deficiency. However, since starch degradation occurs at a lower rate, the hydrogen production rate remains significantly lower. Among different nutrient limitations tested, magnesium deficiency in particular shows an interesting alternative [76]. Under these conditions, the PSII activity level remains higher, so that the metabolism of the cell is not damaged too much. After about 35 hours, gas-tight cultures have reached the anaerobic state and produce hydrogen for more than a week. Compared to sulfur starvation, up to 60% higher hydrogen yield can be achieved by magnesium limitation [77].

4.2.2 Current Limitations and Opportunities of Biological Hydrogen Production Using Microalgae

In terms of hydrogen metabolism, *C. reinhardtii* is the best characterized microalga. Furthermore, a diverse array of genetic tools is available to manipulate this alga, as well as other structural information about the photosynthetic machinery. As previously described, *C. reinhardtii* is not only capable of direct or indirect oxygenic photolysis but features a fermentation process that occurs in the dark. Of all these metabolic pathways, light-dependent hydrogen production is the process with the most efficient proton-to-fuel conversion.

Although the potential for sunlight-to-hydrogen energy conversion efficiency in microalgae is theoretically very high (about 10%), expression of hydrogenase is induced only under anaerobic or microoxic conditions. By the process of oxygenic photosynthesis, during which oxygen is enriched, the hydrogenase is inactivated after a short time (60-90 seconds) [63]. In addition to oxygen sensitivity, several metabolic pathways have a significant influence on the hydrogen production rate. For example, electrons can be used by reduced ferredoxin (Fdx) for NADPH production by ferredoxin-NADP$^+$ oxidoreductase. Thus, this enzyme competes with HydA1, which uses the electrons for hydrogen production. Moreover, since reduced ferredoxin is the main electron acceptor from PSI, it acts as a versatile electron donor, transferring electrons from the photosynthetic electron transport chain to other various reactants, such as the enzymes glutamate synthase and fatty acid desaturase, which significantly affect lipid as well as amino acid biosynthesis. Since ferredoxin also accepts electrons from pyruvate:ferredoxin oxidoreductase (PFR1) and is likewise the electron donor for the hydrogenase HydA1, ferredoxin PetF forms a link from starch metabolism to hydrogen metabolism [61, 78, 79]. In addition to PetF (Fdx1), five other isoforms of ferredoxin exist in *C. reinhardtii*, all of which are thought to be localized in the chloroplast [80]. Although Fdx2 can perform similar functions to Fdx1 to a minor extent under anaerobic conditions, PetF represents the major set screw for optimizing hydrogen production due to its role as the primary electron donor for NADPH and hydrogen production. Electrons recycled for the reduction of NADP$^+$ contribute significantly to CO_2 fixation and thus ultimately to biomass production. Replacing the natural hydrogenase with a ferredoxin-hydrogenase fusion represents an effective optimization approach. In this modified strain, the fusion protein increases the transfer of electrons from ferredoxin:NADP$^+$ oxidoreductase to hydrogenase, and the hydrogen production rate increases [81].

Another important candidate to ensure efficient electron utilization for hydrogen production is ferredoxin:NADP$^+$ oxidoreductase itself, as the enzyme shares electrons between linear (LEF) and cyclic electron flow (CEF) due to its stromal or thylakoid membrane-associated dynamic localization [82]. Association of the enzyme with PSI or stromal localization presumably increases LEF. In association with Cyt b$_6$f, mainly CEF takes place, with PGRL1 and PGR5 probably also playing an important role [83, 84]. In the corresponding mutants, LEF is significantly responsible for hydrogen production. In addition, they exhibit higher oxygen consumption leading to increased PSII stability and faster attainment of the anaerobic state [85]. The *pgr5* mutant exhibits the highest hydrogen production rate known to date, reaching up to 850 ml of H_2 per liter of algal culture within the first nine days under sulfur deficiency conditions.

Previous hydrogen production rates with a *C. reinhardtii* wild-type strain over six days in a sulfur-free medium result in only a lower efficiency for the conversion of light to hydrogen - despite optimized growth conditions - of up to 1.6% [86]. Microalgae can achieve solar energy conversion efficiencies of about 3% by culturing in bioreactors, achieving the highest values for photosynthetic organisms [87]. However, in short-term exponential growth phases, this efficiency of wild-type strains can increase up to threefold [88] but remains well below the efficiency of solar cells. An attractive alternative is the use of engineered algal strains, as the *pgr5* mutant shows up to tenfold higher hydrogen production rate compared to the wild-type strain [85]. As concluded 2012 by the Leopoldina study, direct photolytic splitting of water using phototrophic microorganisms may provide an alternative renewable energy source in the long term, but its technical applicability needs to be further developed.

In contrast to silicon cells, which absorb a large portion of light, photosynthetic organisms use only a limited range of the light spectrum. Taking these characteristics into account, the efficiency of photosynthesis can theoretically only reach a maximum efficiency of 12% without respiration [2]. Therefore, a possible optimization approach of photosynthesis is especially in the field of light-absorbing pigments to absorb also non-visible light [89]. Some cyanobacteria possess chlorophyll variants that allow absorption of shortwave infrared radiation. These could be used to create designer cells with two photosystems that usefully complement each other in their absorption properties [2]. In addition, all photosynthetically active organisms have light-harvesting complexes in which solar energy is captured by pigments and transferred to the photosystems. Under strong light conditions, up to 80% of the absorbed energy is wasted; this energy can also cause damage within the cell that must subsequently be repaired by repair mechanisms. Reducing light antennas could thus lead to more even light use within a bioreactor and avoid shadowing effects at high cell densities. Reducing phycobilisomes in cyanobacteria could also raise the PSII-to-PSI ratio from 0.1 to 0.4, improving it toward the LEF, which is important for hydrogen production [90]. Furthermore, increasing CO_2 concentrations within the atmosphere could be used to improve the turnover rate of the key carbon fixation enzyme, ribulose-1,5-bisphosphate carboxylase/oxygenase (RuBisCO), thus optimizing the efficiency of photosynthesis [91]. However, this would be detrimental to efficient hydrogen production. *C. reinhardtii* strains deficient in one of the subunits of RuBisCO can exhibit up to 15-fold higher hydrogen production rates under sulfur starvation, as PSII protein complexes decrease in quantity and efficiency [70, 92].

Although [FeFe]-hydrogenases are up to 100-fold more efficient than [NiFe]-hydrogenases, inactivation by oxygen has a negative impact on the balance, as anaerobic systems are often more complex and therefore expensive. The anaerobic state of the cells is not optimal for a continuous cultivation system. In addition to heterologous expression of oxygen-tolerant hydrogenases, optimization of [FeFe] hydrogenase to a more oxygen-stable variant is an ongoing research goal [93]. *In vivo*, a possible approach has already been demonstrated by expressing leghemoglobin, which is present in the symbiotic nitrogen-fixing root nodules of legumes to protect the oxygen-sensitive nitrogenase, in *C. reinhardtii* [94]. Synthesis of leghemoglobin in chloroplasts demonstrated more stable transcript levels of hydrogenase-encoding genes and higher hydrogen evolution rates in the strains. Furthermore, the expression of the hydrogenase genes can be enhanced to synthesize more active enzyme in the cell. However, under high hydrogen partial pressure, hydrogenase can also consume hydrogen in the reversible reaction, thus reducing the net balance.

An important factor in hydrogen production is the starch metabolism that provides the electron source for the indirect, PSII-independent pathway. In addition, it can probably contribute to maintaining anoxic conditions in the direct pathway through respiratory respiration. However, analysis of various starch mutants also showed that some of these strains were able to produce similar amounts of hydrogen as the parent strain despite the lack of starch accumulation [95]. By inhibition of PSII the hydrogen production rate is significantly reduced since electrons cannot be injected via the plastoquinone pool. Under light conditions, PSII inhibition has no effect on hydrogen production. Here, the process is likely driven by the cyclic electron flow to PSI via the proton gradient [95].

Although the indirect pathway is more attractive for hydrogen production because no oxygen is produced by this process that inactivates the hydrogenase, it is very low in effectiveness. To optimize indirect photoproduction, type II NADPH dehydrogenase (Nda2) was produced in the chloroplast. Under nutrient limitation, the activity of the non-photochemical reduction of plastoquinone could be increased, allowing more hydrogen to be produced via the indirect pathway. In this scenario, the indirect pathway has similar efficiency to the direct pathway [57]. Alternative approaches aim to achieve partial or complete inactivation of PSII. This can be carried out by mutagenesis altering essential amino acid residues or by reduction of individual PSII subunits [96, 97]. A genetic switch can also be used to regulate the abundance of the D2 protein in the cell. This allows a 2-phase system for accumulation of starch reserves and hydrogen production. In the anaerobic phase, up to 20 μmol H_2 per liter of algal culture (max. 1 mmol H_2 mol-1 Chl s-1) are produced [98].

4.3   *In vitro* Processes with Biocatalysts: Semi- and Biosynthetic Hydrogen Catalysts

Fundamental knowledge about biological principles is important when a synthetic catalyst should be replaced by an enzyme that could convert a reaction much more efficiently. Biocatalysts have the advantage that metal atoms, which are essential for enzyme activity, are abundant, sustainable, and cheap. Enzymes such as hydrogenases are also excellent electrocatalysts that can be used to carry out reversible reactions. Iron atoms, as well as sometimes nickel atoms, are required for the active site of hydrogenases; thus, hydrogenases exhibit properties like platinum. Also, the water-oxidizing manganese complex of PSII acts at lower overvoltage and with higher conversion rates than any other known synthetic catalyst. The intrinsic metal cofactor is integrated into the structural network of amino acids to form the protein binding pocket and tunnel for substrate, product, protons, and electrons. The specific structure together with the essential cofactors are probably also largely responsible for the oxygen tolerance of [NiFe]-hydrogenases [99]. In this case, the iron-sulfur cluster (4Fe-3S) is coordinated by six instead of four cysteine residues near the [NiFe] active site [100].

4.3.1 (Semi-)enzymatic Systems for H_2 Production and Their Efficiency

Responsible for the catalytic activity of hydrogenase is the so-called H-cluster (FeS-cluster). This cofactor consists of a cubic [4Fe4S] center ($4Fe_H$) covalently linked to a [2Fe] center ($2Fe_H$) via a cysteine bridge [101–103]. Here, based on the position relative to the subcluster, a proximal (Fe_p) and a distal iron atom (Fe_d) exist. The two iron atoms are coordinated by three CO and two CN^- ligands, which adjust the redox potential of the H-cluster. Fe_p and Fe_d are linked by an azadithiolate bridge (adt, $(SCH_2)_2NH$), which is involved in the proton transfer mechanism between the enzyme surface and the active site. Like all known [FeFe]-hydrogenases from microalgae, exemplified by the hydrogenase from *Clostridium pasteurianum* shown in Fig. 4.3, HydA1 possesses the catalytically active H domain within its protein structure [103]. HydA1 is often used for the studies because of its relatively simple structure compared to other noneukaryotic [FeFe]-hydrogenases [104]. For example, CpI, [FeFe] hydrogenase 1 from *C. pasteurianum*, has two additional N-terminal binding motifs for a [4Fe4S] cluster and a [2Fe2S] cluster (Fig. 4.3). Specific maturation factors and the FeS cluster assembly machinery are required for 2Fe subcluster biosynthesis and catalytic activity [105, 106]. Endogenous maturation machinery may prove beneficial when hydrogenases are isolated from the native organism. However, standard expression and maturation systems are also known to require heterologous coproduction of the hydrogenase apoprotein and the three H-cluster maturases in *E. coli* [107, 108]. Recently, it was shown that chemical mimicry of the [2Fe] subcluster with an azadithiolate bridge can mature apo-hydrogenases such as HydA1 and CpI into a hydrogen-producing enzyme independent of other auxiliary proteins. They show comparable activities to the native or recombinant proteins and are also functional *in vivo* [109, 110].

In vitro systems are particularly suited to enable semi-artificial hydrogen production, but at the same time offer the possibility to transfer successful approaches to *in vivo* applications. For example, recombinant fusion between the small stromal subunit PsaE of PSI from *Synechocystis* sp. and the subunit MBH of the oxygen-tolerant [NiFe] hydrogenase from *R. eutropha* enables efficient electron transfer between PSI and the directly attached hydrogenase [111]. This hybrid complex MBH-PsaE can produce up to 3000 μmol mg Chl^{-1} h^{-1} hydrogen by immobilization

Fig. 4.3 Enzymatic structure of the catalytic center of [FeFe] hydrogenase Cp1 from *Clostridium pasteurianum*. (**a**) The protein shell of CpI (green) determined by crystal structure has specific side residues (blue) involved in proton transfer. (**b**) Model of the reaction center of CpI. The H cluster consists of a [4Fe4S] cluster connected to the [2Fe2S] cluster via a cysteine. This cluster is coordinated by three CO⁻, two CN⁻ ligands, and an azadithiolate bridge

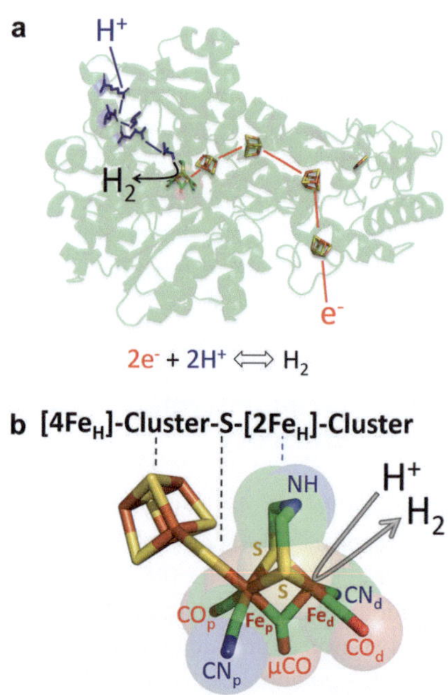

at a gold electrode [112]. In addition to the fusion of the electron transfer agent to the hydrogenase, another starting point is the linkage of physiologically relevant cofactors to support the electron transfer. In this regard, a photosensitizer (dye) was introduced through the Fe-S coordination links between the F_B[4Fe-4S] cluster of PSI and the distal [4Fe-4S] cluster of hydrogenase; this resulted in successful electron transfer and consequent hydrogen production under light conditions [113]. The development of biological-organic hybrid nanoconstructs can produce up to 2200 μmol mg Chl^{-1} h^{-1} [114]. Both homogeneous, molecular-based, and heterogeneous, semiconductor-based approaches are currently under investigation for the development of an efficient water-splitting catalyst. In the homogeneous system, electrons are transferred to the catalyst directly by a photosensitizer or via a mediator such as methyl viologen or PetF. Studies with different photosensitizers revealed effective light-dependent hydrogen production by an *in vitro* system containing the synthetic 5-carboxy-eosin in combination with HydA1 [115]. A high conversion rate of 577 ml H_2 s-1 per mole of [FeFe]-hydrogenase was achieved. Diketopyrrolopyrroles (DPP) also showed good properties as organic photosensitizers on metal oxide surfaces and were successfully used for hydrogen production with hydrogenase [116]. Since light-driven enzymatic catalysis can often be costly, toxic as well as fragile due to the attachment of a protein to a photosensitizer, carbon materials have been developed that can serve as low-cost light-collecting photosensitizers [117]. For hydrogen production with a hydrogenase, the surface was positively charged by treating ammonium to transfer excited electrons to the negatively charged enzyme. After 24 h, up to 43,000 mol H_2 (mol

H_2ase)$^{-1}$ are produced. The combination of enzymes with semiconductor systems can lead to highly active photochemical electron transfer at a light-collecting electrode surface. Thus, hydrogenase was successfully established on TiO_2-coated p-Si photocathodes for the photoreduction of protons to hydrogen [118]. Further development of this principle by using a stabilizing mesoporous separating layer could further increase the conversion rates many times [119]. These and other recently developed artificial systems offer the opportunity to obtain active and economically affordable catalysts. A study by the Leopoldina [7] also gave a positive recommendation for research fields based on artificial systems due to their high innovation potential.

4.4   Conclusion

Learning from nature about hydrogen production means understanding and copying the biological principles of hydrogen metabolism and the enzymes involved in it, thus making these principles usable for technical systems. In particular, the biological catalyst hydrogenase reduces protons under standard conditions at significantly higher rates than any synthetic chemical substance. Based on this natural model, synthetic or engineered units could be developed into biomimetic or bioinspired systems in the future, especially in combination with artificial photosynthesis. Another approach uses living cell for H_2 production. Systematic screening could identify novel microorganisms whose hydrogen production rates may exceed anything previously known. Hydrogen producers could by further enhanced by targeted genetic or metabolic engineering. The development of cultivation systems will support that hydrogen production systems become economically competitive. Although consideration of conversion efficiencies is useful for comparing different hydrogen production systems, all associated costs should be considered in the LCA.

References

1. Molina A, Falvey M, Rondanelli R (2017) A solar radiation database for Chile. Sci Rep 7(1): 14823
2. Blankenship RE, Tiede DM, Barber J, Brudvig GW, Fleming G, Ghirardi M, Gunner MR, Junge W, Kramer DM, Melis A, Moore TA, Moser CC, Nocera DG, Nozik AJ, Ort DR, Parson WW, Prince RC, Sayre RT (2011) Comparing photosynthetic and photovoltaic efficiencies and recognizing the potential for improvement. Science 332(6031):805–809
3. Smil V (2008) Energy in Nature and Society: General Energetics of Complex Systems. MIT Press, Cambridge
4. Oey M, Sawyer AL, Ross IL, Hankamer B (2016) Challenges and opportunities for hydrogen production from microalgae. Plant Biotechnol J 14(7):1487–1499
5. Hoffert MI, Caldeira K, Benford G, Criswell DR, Green C, Herzog H, Jain AK, Kheshgi HS, Lackner KS, Lewis JS, Lightfoot HD, Manheimer W, Mankins JC, Mauel ME, Perkins LJ, Schlesinger ME, Volk T, Wigley TM (2002) Advanced technology paths to global climate stability: energy for a greenhouse planet. Science 298(5595):981–987

6. Gupta SK, Kumari S, Reddy K, Bux F (2013) Trends in biohydrogen production: major challenges and state-of-the-art developments. Environ Technol 34(13–16):1653–1670
7. Leopoldina – Nationale Akademie der Wissenschaften (2012) Bioenergie: Möglichkeiten und Grenzen. Empfehlungen. Kurzzusammenfassung und Empfehlungen
8. Rey FE, Heiniger EK, Harwood CS (2007) Redirection of metabolism for biological hydrogen production. Appl Environ Microbiol 73(5):1665–1671
9. Vignais PM, Billoud B, Meyer J (2001) Classification and phylogeny of hydrogenases. FEMS Microbiol Rev 25(4):455–501
10. Vignais PM, Billoud B (2007) Occurrence, classification, and biological function of hydrogenases: an overview. Chem Rev 107(10):4206–4272
11. Trchounian K, Poladyan A, Vassilian A, Trchounian A (2012) Multiple and reversible hydrogenases for hydrogen production by Escherichia coli: dependence on fermentation substrate, pH and the F_0F_1-ATPase. Crit Rev Biochem Mol Biol 47(3):236–249
12. Sargent F (2016) The model [NiFe]-hydrogenases of Escherichia coli. Adv Microb Physiol 68: 433–507
13. Trchounian K, Trchounian A (2013) Escherichia coli multiple [Ni-Fe]-hydrogenases are sensitive to osmotic stress during glycerol fermentation but at different pHs. FEBS Lett 587(21):3562–3566
14. Fang HH, Liu H (2002) Effect of pH on hydrogen production from glucose by a mixed culture. Bioresour Technol 82(1):87–93
15. Vardar-Schara G, Krab IM, Yi G, Su WW (2007) A homogeneous fluorometric assay platform based on novel synthetic proteins. Biochem Biophys Res Commun 361(1):103–108
16. Schut GJ, Adams MW (2009) The Iron-hydrogenase of Thermotoga maritima utilizes ferredoxin and NADH synergistically: a new perspective on anaerobic hydrogen production. J Bacteriol 191(13):4451–4457
17. Wang S, Huang H, Kahnt J, Thauer RK (2013) A reversible electron-bifurcating ferredoxin- and NAD-dependent [FeFe]-hydrogenase (HydABC) in Moorella thermoacetica. J Bacteriol 195(6):1267–1275
18. Wang X, Yang H, Zhang Y, Guo L (2014) Remarkable enhancement on hydrogen production performance of Rhodobacter sphaeroides by disrupting spbA and hupSL genes. Int J Hydrog Energy 39(27):14633–14641
19. Stephen AJ, Archer SA, Orozco RL, Macaskie LE (2017) Advances and bottlenecks in microbial hydrogen production. Microb Biotechnol 10(5):1120–1127
20. Gray CT, Gest H (1965) Biological formation of molecular hydrogen. Science 148(3667): 186–192
21. Chitapornpan S, Chiemchaisri C, Chiemchaisri W, Honda R, Yamamoto K (2013) Organic carbon recovery and photosynthetic bacteria population in an anaerobic membrane photo-bioreactor treating food processing wastewater. Bioresour Technol 141:65–74
22. Boboescu IZ, Ilie M, Gherman VD, Mirel I, Pap B, Negrea A, Kondorosi É, Bíró T, Maróti G (2014) Revealing the factors influencing a fermentative biohydrogen production process using industrial wastewater as fermentation substrate. Biotechnol Biofuels 7(1):139
23. Li C, Fang HHP (2007) Fermentative hydrogen production from wastewater and solid wastes by mixed cultures. Crit Rev Environ Sci Technol 37(1):1–39
24. Valdez-Vazquez I, Ríos-Leal E, Esparza-García F, Cecchi F, Poggi-Varaldo HM (2005) Semi-continuous solid substrate anaerobic reactors for H_2 production from organic waste: mesophilic versus thermophilic regime. Int J Hydrog Energy 30(14):1383–1391
25. Yakunin AF, Fedorov AS, Laurinavichene TV, Glaser VM, Egorov NS, Tsygankov AA, Zinchenko VV, Hallenbeck PC (2001) Regulation of nitrogenase in the photosynthetic bacterium Rhodobacter sphaeroides containing draTG and nifHDK genes from Rhodobacter capsulatus. Can J Microbiol 47(3):206–212
26. Kim DH, Lee JH, Kang S, Hallenbeck PC, Kim EJ, Lee JK, Kim MS (2014) Enhanced photo-fermentative H_2 production using Rhodobacter sphaeroides by ethanol addition and analysis of soluble microbial products. Biotechnol Biofuels 7:79

27. Rupprecht J, Hankamer B, Mussgnug JH, Ananyev G, Dismukes C, Kruse O (2006) Perspectives and advances of biological H_2 production in microorganisms. Appl Microbiol Biotechnol 72(3):442–449

28. Ghosh S, Dairkee UK, Chowdhury R, Bhattacharya P (2017) Hydrogen from food processing wastes via photofermentation using purple non-sulfur bacteria (PNSB) – a review. Energy Convers Manag 141:299–314

29. Miyake J, Kawamura S (1987) Efficiency of light energy conversion to hydrogen by the photosynthetic bacterium Rhodobacter sphaeroides. Int J Hydrog Energy 12(3):147–149

30. Androga DD, Özgür E, Eroglu I, Gündüz U, Yücel M (2011) Significance of carbon to nitrogen ratio on the long-term stability of continuous photofermentative hydrogen production. Int J Hydrog Energy 36(24):15583–15594

31. Hädicke O, Grammel H, Klamt S (2011) Metabolic network modeling of redox balancing and biohydrogen production in purple nonsulfur bacteria. BMC Syst Biol 5(150)

32. Vignais PM, Toussaint B (1994) Molecular biology of membrane bound H_2 uptake hydrogenases. Arch Microbiol 161(1):1–10

33. Bernhard M, Buhrke T, Bleijlevens B, De Lacey AL, Fernandez VM, Albracht SP, Friedrich B (2001) The H_2 sensor of Ralstonia eutropha. biochemical characteristics, spectroscopic properties, and its interaction with a histidine protein kinase. J Biol Chem 276(19): 15592–15597

34. Franchi E, Tosi C, Scolla G, Penna GD, Rodriguez F, Pedroni PM (2004) Metabolically engineered Rhodobacter sphaeroides RV strains for improved biohydrogen photoproduction combined with disposal of food wastes. Mar Biotechnol 6(6):552–565

35. Öztürk Y, Yücel M, Daldal F, Mandacı S, Gündüz U, Türker L, Eroğlu İ (2006) Hydrogen production by using Rhodobacter capsulatus mutants with genetically modified electron transfer chains. Int J Hydrog Energy 31(11):1545–1552

36. Nakada E, Nishikata S, Asada Y, Miyake J (1999) Photosynthetic bacterial hydrogen production combined with a fuel cell. Int J Hydrog Energy 24(11):1053–1057

37. Lenz O, Bernhard M, Buhrke T, Schwartz E, Friedrich B (2002) The hydrogen-sensing apparatus in Ralstonia eutropha. J Mol Microbiol Biotechnol 4(3):255–262

38. Larimer FW, Chain P, Hauser L, Lamerdin J, Malfatti S, Do L, Land ML, Pelletier DA, Beatty JT, Lang AS, Tabita FR, Gibson JL, Hanson TE, Bobst C, Torres JL, Peres C, Harrison FH, Gibson J, Harwood CS (2004) Complete genome sequence of the metabolically versatile photosynthetic bacterium Rhodopseudomonas palustris. Nat Biotechnol 22(1):55–61

39. Lopes Pinto FA, Troshina O, Lindblad P (2002) A brief look at three decades of research on cyanobacterial hydrogen evolution. Int J Hydrog Energy 27(11):1209–1215

40. Schütz K, Happe T, Troshina O, Lindblad P, Leitão E, Oliveira P, Tamagnini P (2004) Cyanobacterial H_2 production - a comparative analysis. Planta 218(3):350–359

41. Kumar K, Mella-Herrera RA, Golden JW (2010) Cyanobacterial heterocysts. Cold Spring Harb Perspect Biol 2(4):a000315. https://doi.org/10.1101/cshperspect.a000315

42. Happe T, Schütz K, Böhme H (2000) Transcriptional and mutational analysis of the uptake hydrogenase of the filamentous cyanobacterium Anabaena variabilis ATCC 29413. J Bacteriol 182(6):1624–1631

43. Gutekunst K, Chen X, Schreiber K, Kaspar U, Makam S, Appel J (2014) The bidirectional NiFe-hydrogenase in Synechocystis sp. PCC 6803 is reduced by flavodoxin and ferredoxin and is essential under mixotrophic, nitrate-limiting conditions. J Biol Chem 289(4):1930–1937

44. Masukawa H, Mochimaru M, Sakurai H (2002) Disruption of the uptake hydrogenase gene, but not of the bidirectional hydrogenase gene, leads to enhanced photobiological hydrogen production by the nitrogen-fixing cyanobacterium Anabaena sp. PCC 7120. Appl Microbiol Biotechnol 58(5):618–624

45. Yoshino F, Ikeda H, Masukawa H, Sakurai H (2007) High photobiological hydrogen production activity of a Nostoc sp. PCC 7422 uptake hydrogenase-deficient mutant with high nitrogenase activity. Mar Biotechnol 9(1):101–112

46. Khetkorn W, Lindblad P, Incharoensakdi A (2012) Inactivation of uptake hydrogenase leads to enhanced and sustained hydrogen production with high nitrogenase activity under high light exposure in the cyanobacterium Anabaena siamensis TISTR 8012. J Biol Eng 6(1):19

47. Gutthann F, Egert M, Marques A, Appel J (2007) Inhibition of respiration and nitrate assimilation enhances photohydrogen evolution under low oxygen concentrations in Synechocystis sp. PCC 6803. Biochim Biophys Acta 1767(2):161–169

48. Baebprasert W, Jantaro S, Khetkorn W, Lindblad P, Incharoensakdi A (2011) Increased H_2 production in the cyanobacterium Synechocystis sp. strain PCC 6803 by redirecting the electron supply via genetic engineering of the nitrate assimilation pathway. Metab Eng 13(5):610–616

49. Masukawa H, Sakurai H, Hausinger RP, Inoue K (2017) Increased heterocyst frequency by patN disruption in Anabaena leads to enhanced photobiological hydrogen production at high light intensity and high cell density. Appl Microbiol Biotechnol 101(5):2177–2188

50. Bandyopadhyay A, Stöckel J, Min H, Sherman LA, Pakrasi HB (2010) High rates of photobiological H_2 production by a cyanobacterium under aerobic conditions. Nat Commun 1:139

51. Kothari A, Potrafka R, Garcia-Pichel F (2012) Diversity in hydrogen evolution from bidirectional hydrogenases in cyanobacteria from terrestrial, freshwater and marine intertidal environments. J Biotechnol 162(1):105–114

52. Melis A, Zhang L, Forestier M, Ghirardi ML, Seibert M (2000) Sustained photobiological hydrogen gas production upon reversible inactivation of oxygen evolution in the Green Alga Chlamydomonas reinhardtii. Plant Physiol 122(1):127–136

53. Melis A, Happe T (2001) Hydrogen production. Green algae as a source of energy. Plant Physiol 127(3):740–748

54. Godde D, Trebst A (1980) NADH as electron donor for the photosynthetic membrane of Chlamydomonas reinhardtii. Arch Microbiol 127(3):245–252

55. Bamberger ES, King D, Erbes DL, Gibbs M (1982) H_2 and CO_2 evolution by anaerobically adapted Chlamydomonas reinhardtii F-60. Plant Physiol 69(6):1268–1273

56. Mus F, Cournac L, Cardettini V, Caruana A, Peltier G (2005) Inhibitor studies on non-photochemical plastoquinone reduction and H_2 photoproduction in Chlamydomonas reinhardtii. Biochim Biophys Acta 1708(3):322–332

57. Baltz A, Dang KV, Beyly A, Auroy P, Richaud P, Cournac L, Peltier G (2014) Plastidial expression of type II NAD(P)H dehydrogenase increases the reducing state of plastoquinones and hydrogen photoproduction rate by the indirect pathway in Chlamydomonas reinhardtii. Plant Physiol 165(3):1344–1352

58. Chochois V, Dauvillée D, Beyly A, Tolleter D, Cuiné S, Timpano H, Ball S, Cournac L, Peltier G (2009) Hydrogen production in Chlamydomonas: photosystem II-dependent and -independent pathways differ in their requirement for starch metabolism. Plant Physiol 151(2):631–640

59. Grossman AR, Catalanotti C, Yang W, Dubini A, Magneschi L, Subramanian V, Posewitz MC, Seibert M (2011) Multiple facets of anoxic metabolism and hydrogen production in the unicellular green alga Chlamydomonas reinhardtii. New Phytol 190(2):279–288

60. Atteia A, van Lis R, Tielens AG (1827) Martin WF (2013) Anaerobic energy metabolism in unicellular photosynthetic eukaryotes. Biochim Biophys Acta 2:210–223

61. Noth J, Krawietz D, Hemschemeier A, Happe T (2013) Pyruvate:ferredoxin oxidoreductase is coupled to light-independent hydrogen production in Chlamydomonas reinhardtii. J Biol Chem 288(6):4368–4377

62. van Lis R, Baffert C, Couté Y, Nitschke W, Atteia A (2013) Chlamydomonas reinhardtii chloroplasts contain a homodimeric pyruvate:ferredoxin oxidoreductase that functions with FDX1. Plant Physiol 161(1):57–71

63. Ghirardi ML, Togasaki RK, Seibert M (1997) Oxygen sensitivity of algal H_2-production. Appl Biochem Biotechnol 63–65:141–151

64. Melis A, Seibert M, Happe T (2004) Genomics of green algal hydrogen research. Photosynth Res 82(3):277–288

65. Davies JP, Yildiz FH, Grossman A (1996) Sac1, a putative regulator that is critical for survival of Chlamydomonas reinhardtii during sulfur deprivation. EMBO J 15(9):2150–2159
66. Wykoff DD, Davies JP, Melis A, Grossman AR (1998) The regulation of photosynthetic electron transport during nutrient deprivation in Chlamydomonas reinhardtii. Plant Physiol 117(1):129–139
67. Chen HC, Newton AJ, Melis A (2005) Role of SulP, a nuclear-encoded chloroplast sulfate permease, in sulfate transport and H_2 evolution in Chlamydomonas reinhardtii. Photosynth Res 84(1–3):289–296
68. Happe T, Kaminski A (2002) Differential regulation of the Fe-hydrogenase during anaerobic adaptation in the green alga Chlamydomonas reinhardtii. Eur J Biochem 269(3):1022–1032
69. Mus F, Dubini A, Seibert M, Posewitz MC, Grossman AR (2007) Anaerobic acclimation in Chlamydomonas reinhardtii: anoxic gene expression, hydrogenase induction, and metabolic pathways. J Biol Chem 282(35):25475–25486
70. Hemschemeier A, Fouchard S, Cournac L, Peltier G, Happe T (2008) Hydrogen production by Chlamydomonas reinhardtii: an elaborate interplay of electron sources and sinks. Planta 227(2):397–407
71. Winkler M, Heil B, Heil B, Happe T (2002) Isolation and molecular characterization of the [Fe]-hydrogenase from the unicellular green alga Chlorella fusca. Biochim Biophys Acta 1576(3):330–334
72. Kosourov S, Seibert M, Ghirardi ML (2003) Effects of extracellular pH on the metabolic pathways in sulfur-deprived, H_2-producing Chlamydomonas reinhardtii cultures. Plant Cell Physiol 44(2):146–155
73. Hemschemeier A, Happe T (2005) The exceptional photofermentative hydrogen metabolism of the green alga Chlamydomonas reinhardtii. Biochem Soc Trans 33(Pt 1):39–41
74. Philipps G, Happe T, Hemschemeier A (2012) Nitrogen deprivation results in photosynthetic hydrogen production in Chlamydomonas reinhardtii. Planta 235(4):729–745
75. Juergens MT, Disbrow B, Shachar-Hill Y (2016) The relationship of triacylglycerol and starch accumulation to carbon and energy flows during nutrient deprivation in Chlamydomonas reinhardtii. Plant Physiol 171(4):2445–2457
76. Volgusheva A, Kukarskikh G, Krendeleva T, Rubin A, Mamedov F (2015) Hydrogen photoproduction in green algae Chlamydomonas reinhardtii under magnesium deprivation. RSC Adv 5(8):5633–5637
77. Volgusheva AA, Jokel M, Allahverdiyeva Y, Kukarskikh GP, Lukashev EP, Lambreva MD, Krendeleva TE, Antal TK (2017) Comparative analyses of H_2 photoproduction in magnesium- and sulfur-starved Chlamydomonas reinhardtii cultures. Physiol Plant 161(1):124–137
78. Winkler M, Kuhlgert S, Hippler M, Happe T (2009) Characterization of the key step for light-driven hydrogen evolution in green algae. J Biol Chem 284(52):36620–36627
79. Hemschemeier A, Happe T (2011) Alternative photosynthetic electron transport pathways during anaerobiosis in the green alga Chlamydomonas reinhardtii. Biochim Biophys Acta 1807(8):919–926
80. Winkler M, Hemschemeier A, Jacobs J, Stripp S, Happe T (2010) Multiple ferredoxin isoforms in Chlamydomonas reinhardtii - their role under stress conditions and biotechnological implications. Eur J Cell Biol 89(12):998–1004
81. Yacoby I, Tegler LT, Pochekailov S, Zhang S, King PW (2012) Optimized expression and purification for high-activity preparations of algal [FeFe]-hydrogenase. PLoS One 7(4): e35886. https://doi.org/10.1371/journal.pone.0035886
82. Joliot P, Johnson GN (2011) Regulation of cyclic and linear electron flow in higher plants. Proc Natl Acad Sci U S A 108(32):13317–13322
83. Hertle AP, Blunder T, Wunder T, Pesaresi P, Pribil M, Armbruster U, Leister D (2013) PGRL1 Is the elusive ferredoxin-plastoquinone reductase in photosynthetic cyclic electron flow. Mol Cell 49(3):511–523
84. Mosebach L, Heilmann C, Mutoh R, Gäbelein P, Steinbeck J, Happe T, Ikegami T, Hanke G, Kurisu G, Hippler M (2017) Association of ferredoxin:$NADP^+$ oxidoreductase with the

photosynthetic apparatus modulates electron transfer in Chlamydomonas reinhardtii. Photosynth Res 134(3):291–306

85. Steinbeck J, Nikolova D, Weingarten R, Johnson X, Richaud P, Peltier G, Hermann M, Magneschi L, Hippler M (2015) Deletion of proton gradient regulation 5 (PGR5) and PGR5-like 1 (PGRL1) proteins promote sustainable light-driven hydrogen production in Chlamydomonas reinhardtii due to increased PSII activity under sulfur deprivation. Front Plant Sci 6:892

86. Giannelli L, Scoma A, Torzillo G (2009) Interplay between light intensity, chlorophyll concentration and culture mixing on the hydrogen production in sulfur-deprived Chlamydomonas reinhardtii cultures grown in laboratory photobioreactors. Biotechnol Bioeng 104(1):76–90

87. Wijffels RH, Barbosa MJ (2010) An outlook on microalgal biofuels. Science 329(5993): 796–799

88. Janssen M, Tramper J, Mur LR, Wijffels RH (2003) Enclosed outdoor photobioreactors: light regime, photosynthetic efficiency, scale-up, and future prospects. Biotechnol Bioeng 81(2): 193–210

89. Chen M, Blankenship RE (2011) Expanding the solar spectrum used by photosynthesis. Trends Plant Sci 16(8):427–431

90. Bernát G, Waschewski N, Rögner M (2009) Towards efficient hydrogen production: the impact of antenna size and external factors on electron transport dynamics in Synechocystis PCC 6803. Photosynth Res 99(3):205–216

91. Zhu XG, de Sturler E, Long SP (2007) Optimizing the distribution of resources between enzymes of carbon metabolism can dramatically increase photosynthetic rate: a numerical simulation using an evolutionary algorithm. Plant Physiol 145(2):513–526

92. Pinto TS, Malcata FX, Arrabaça JD, Silva JM, Spreitzer RJ, Esquível MG (2013) Rubisco mutants of Chlamydomonas reinhardtii enhance photosynthetic hydrogen production. Appl Microbiol Biotechnol 97(12):5635–5643

93. Chang CH, King PW, Ghirardi ML, Kim K (2007) Atomic resolution modeling of the ferredoxin:[FeFe] hydrogenase complex from Chlamydomonas reinhardtii. Biophys J 93(9): 3034–3045

94. Wu S, Xu L, Huang R, Wang Q (2011) Improved biohydrogen production with an expression of codon-optimized hemH and lba genes in the chloroplast of Chlamydomonas reinhardtii. Bioresour Technol 102(3):2610–2616

95. Chochois V, Constans L, Dauvillée D, Beyly A, Solivérès M, Ball S, Peltier G, Cournac L (2010) Relationships between PSII-independent hydrogen bioproduction and starch metabolism as evidenced from isolation of starch catabolism mutants in the green alga Chlamydomonas reinhardtii. Int J Hydrog Energy 35(19):10731–10740

96. Scoma A, Krawietz D, Faraloni C, Giannelli L, Happe T, Torzillo G (2012) Sustained H_2 production in a Chlamydomonas reinhardtii D1 protein mutant. J Biotechnol 157(4):613–619

97. Lin HD, Liu BH, Kuo TT, Tsai HC, Feng TY, Huang CC, Chien LF (2013) Knockdown of PsbO leads to induction of HydA and production of photobiological H_2 in the green alga Chlorella sp. DT. Bioresour Technol 143:154–162

98. Surzycki R, Cournac L, Peltier G, Rochaix JD (2007) Potential for hydrogen production with inducible chloroplast gene expression in Chlamydomonas. Proc Natl Acad Sci U S A 104(44): 17548–17553

99. Fritsch J, Scheerer P, Frielingsdorf S, Kroschinsky S, Friedrich B, Lenz O, Spahn CM (2011) The crystal structure of an oxygen-tolerant hydrogenase uncovers a novel iron-sulphur centre. Nature 479(7372):249–252

100. Goris T, Wait AF, Saggu M, Fritsch J, Heidary N, Stein M, Zebger I, Lendzian F, Armstrong FA, Friedrich B, Lenz O (2011) A unique iron-sulfur cluster is crucial for oxygen tolerance of a [NiFe]-hydrogenase. Nat Chem Biol 7(5):310–318

101. Peters JW (2009) Carbon monoxide and cyanide ligands in the active site of [FeFe]-hydrogenases. Met Ions Life Sci 6:179–218

102. Nicolet Y, Piras C, Legrand P, Hatchikian CE, Fontecilla-Camps JC (1999) Desulfovibrio desulfuricans iron hydrogenase: the structure shows unusual coordination to an active site Fe binuclear center. Structure 7(1):13–23. https://doi.org/10.1016/S0969-2126(99)80005-7
103. Peters JW, Schut GJ, Boyd ES, Mulder DW, Shepard EM, Broderick JB, King PW (1853) Adams MW (2015) [FeFe]- and [NiFe]-hydrogenase diversity, mechanism, and maturation. Biochim Biophys Acta 6:1350–1369
104. Lubitz W, Ogata H, Rüdiger O, Reijerse E (2014) Hydrogenases. Chem Rev 114(8): 4081–4148
105. Posewitz MC, Smolinski SL, Kanakagiri S, Melis A, Seibert M, Ghirardi ML (2004) Hydrogen photoproduction is attenuated by disruption of an isoamylase gene in Chlamydomonas reinhardtii. Plant Cell 16(8):2151–2163
106. Sawyer A, Bai Y, Lu Y, Hemschemeier A, Happe T (2017) Compartmentalisation of [FeFe]-hydrogenase maturation in Chlamydomonas reinhardtii. Plant J 90(6):1134–1143
107. Kuchenreuther JM, Grady-Smith CS, Bingham AS, George SJ, Cramer SP, Swartz JR (2010) High-yield expression of heterologous [FeFe] hydrogenases in Escherichia coli. PLoS One 5(11):e15491
108. Yacoby I, Pochekailov S, Toporik H, Ghirardi ML, King PW, Zhang S (2011) Photosynthetic electron partitioning between [FeFe]-hydrogenase and ferredoxin:NADP$^+$-oxidoreductase (FNR) enzymes in vitro. Proc Natl Acad Sci U S A 108(23):9396–9401
109. Berggren G, Adamska A, Lambertz C, Simmons TR, Esselborn J, Atta M, Gambarelli S, Mouesca JM, Reijerse E, Lubitz W, Happe T, Artero V, Fontecave M (2013) Biomimetic assembly and activation of [FeFe]-hydrogenases. Nature 499(7456):66–69
110. Esselborn J, Lambertz C, Adamska-Venkates A, Simmons T, Berggren G, Noth J, Siebel J, Hemschemeier A, Artero V, Reijerse E, Fontecave M, Lubitz W, Happe T (2013) Spontaneous activation of [FeFe]-hydrogenases by an inorganic [2Fe] active site mimic. Nat Chem Biol 9(10):607–609
111. Schwarze A, Kopczak MJ, Rögner M, Lenz O (2010) Requirements for construction of a functional hybrid complex of photosystem I and [NiFe]-hydrogenase. Appl Environ Microbiol 76(8):2641–2651
112. Krassen H, Schwarze A, Friedrich B, Ataka K, Lenz O, Heberle J (2009) Photosynthetic hydrogen production by a hybrid complex of photosystem I and [NiFe]-hydrogenase. ACS Nano 3(12):4055–4061
113. Lubner CE, Knörzer P, Silva PJ, Vincent KA, Happe T, Bryant DA, Golbeck JH (2010) Wiring an [FeFe]-hydrogenase with photosystem I for light-induced hydrogen production. Biochemistry 49(48):10264–10266
114. Lubner CE, Applegate AM, Knörzer P, Ganago A, Bryant DA, Happe T, Golbeck JH (2011) Solar hydrogen-producing bionanodevice outperforms natural photosynthesis. Proc Natl Acad Sci U S A 108(52):20988–20991
115. Adam D, Bösche L, Castañeda-Losada L, Winkler M, Apfel UP, Happe T (2017) Sunlight-dependent hydrogen production by photosensitizer/hydrogenase systems. ChemSusChem 10(5):894–902
116. Warnan J, Willkomm J, Ng JN, Godin R, Prantl S, Durrant JR, Reisner E (2017) Solar H$_2$ evolution in water with modified diketopyrrolopyrrole dyes immobilised on molecular Co and Ni catalyst-TiO$_2$ hybrids. Chem Sci 8(4):3070–3079
117. Hutton GA, Reuillard B, Martindale BC, Caputo CA, Lockwood CW, Butt JN, Reisner E (2016) Carbon dots as versatile photosensitizers for solar-driven catalysis with redox enzymes. J Am Chem Soc 138(51):16722–16730
118. Lee CY, Park HS, Fontecilla-Camps JC, Reisner E (2016) Photoelectrochemical H$_2$ evolution with a hydrogenase immobilized on a TiO$_2$-protected silicon electrode. Angew Chem Int Ed Engl 55(20):5971–5974
119. Leung JJ, Warnan J, Nam DH, Zhang JZ, Willkomm J, Reisner E (2017) Photoelectrocatalytic H$_2$ evolution in water with molecular catalysts immobilised on p-Si via a stabilising mesoporous TiO$_2$ interlayer. Chem Sci 8(7):5172–5180

Production of Synthesis Gas

5

Johannes Booz, Dominik Höhner, and Stefan Burmester

Abstract

Today, synthesis gas is mainly produced from natural gas by steam reforming. Important large-scale products derived from syngas are ammonia for the production of fertilizers and methanol for organic chemical products. The composition of syngas depends on the feedstock, the gasification medium, the gasifier type and the reaction conditions. In principle, the fluidized bed gasifier developed for the gasification of coal is also suitable for the gasification of biomass. However, biomass requires a high degree of purification of the synthesis gas produced.

Keywords

Synthesis gas · Ammonia · Methanol · Gasifier · Gasification of biomass

5.1 Introduction

In principle, the generic term "synthesis gas"covers all gases that are used for subsequent chemical synthesis. In process technology, due to the importance of large-scale syntheses, this term essentially refers to industrially produced gases containing CO and H2. These large-scale syntheses include ammonia synthesis, Fischer-Tropsch synthesis for the production of hydrocarbons, methanol synthesis, oxo synthesis, and other processes (Table 5.1). In the meantime, the term synthesis gas has become so widely used that product gases from the gasification of coal, biomass and other feedstocks, which are subsequently not used for synthesis but, for example, in a power-heat process, are sometimes also referred to as synthesis gas.

J. Booz · D. Höhner · S. Burmester (✉)
Concord Blue Engineering GmbH, Herten, Germany
e-mail: hoehner@concordblue.de; sb@concordblue.de

Table 5.1 Overview of applications in which synthesis gas is produced

Composition synthesis gas [H_2: CO]	Application	Product
H_2 raw	Ammonia synthesis	NH_3
1–2: 2	Fischer-Tropsch-processes	LPG, Naphtha, Diesel
2: 1	Dimethyl ether and gasoline synthesis	DME, Gasoline
2: 1	Alcohol synthesis	Methanol, Ethanol
3: 1	Methanisation	SNG, Methan
1: 1	Oxosynthesis	Aldehydes

LPG liquified petroleum gas, *DME* dimethyl ether, *SNG* synthetic natural gas

Suitable feedstocks for the production of synthesis gases include solid energy sources, such as coal and biomass, as well as liquids and gases, such as crude oil and natural gas. The material conversion of solid energy sources into a gas is generally referred to as "gasification," while the conversion of gases and liquids is called "reforming," but in some cases also gasification. In addition to these terms, special designations such as "cracked gas" (synthesis gas from the conversion of natural gas and other hydrocarbons with steam), water gas (synthesis gas from the conversion of coke with steam) or generator gas (synthesis gas from the conversion of coal with air) are also used.

Until the middle of the last century, synthesis gas was mainly produced from coal. Subsequently, gaseous and liquid fuels became increasingly popular as feedstocks. The advantage of these materials is that the hydrogen content in the synthesis gas produced is higher, and for many syntheses it is precisely a high hydrogen content that is required. In addition, the plants for liquid and gaseous feedstocks are often less complex, and the handling of these materials is simpler.

5.2 Feedstocks/Quantities Produced

Synthesis gas is produced by the chemical conversion of fossil fuels, renewable organic raw materials and waste. Gaseous or vaporizable feedstocks, such as natural gas, light gasoline, refinery gas, coke oven gas, and solid or liquid feedstocks, such as heavy oils, petroleum distillation residues, petroleum coke, biomass, and coal, can serve as feedstocks [1].

The choice of feedstock is highly dependent on the prevailing price, availability of the feedstock, and the downstream application of the syngas. In countries such as South Africa, India, or China, syngas production by coal gasification is widely used due to the large availability of coal. In Germany, on the other hand, most of the syngas is produced by steam reforming of natural gas, although coal gasification has historically been highly developed in Europe [2]. Natural gas is preferred as feedstock because the production of syngas by steam reforming has been tested on a large scale and makes more economic and ecological sense in Germany than production

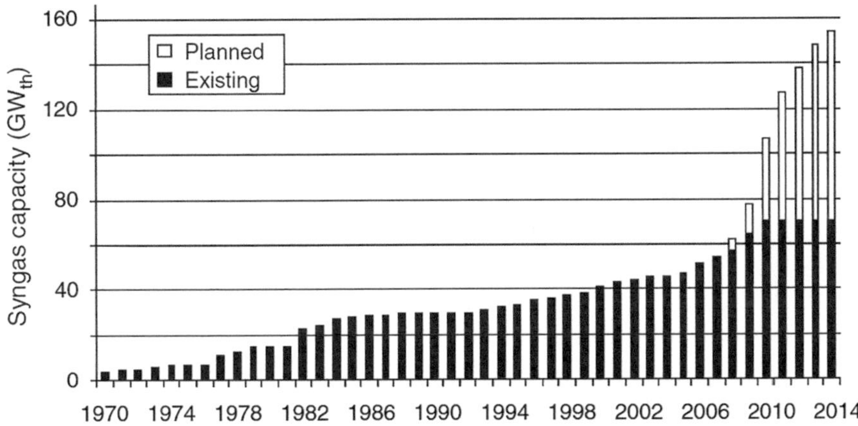

Fig. 5.1 Synthesis gas volumes produced up to 2007 [3]

by coal gasification. The production of synthesis gas from renewable organic raw materials is considered to be almost CO2-neutral and represents a particularly interesting possibility for saving CO2, especially from the point of view of climate technology. However, the process engineering effort required for biomass gasification plants is very high. This refers, among other things, to the inferiority of the biomass (high O2 content), a very complex gas purification of the synthesis gas produced, and the complex processing of the biomass due to its great inhomogeneity.

Figure 5.1 shows the synthesis gas volumes produced worldwide by gasification up to 2007 and a forecast up to 2014. Coal, heavy oil fractions, gas, petroleum coke and biomass were considered as feedstocks.

5.3 Thermodynamic Fundamentals

The aim of thermochemical gasification or reforming is to convert the feedstock as completely as possible into a gaseous product—"synthesis gas". The basic prerequisite for gasification is the supply of an oxygen-containing gasification medium such as air, oxygen, steam or mixtures of these components. In addition, the gasification medium CO2 can be used for the gasification of solid or liquid feedstocks. The oxygen can convert the carbon present in the feedstock into CO. Table 5.2 shows the basic reaction equations of the gasification of solid and gaseous feedstocks, divided into heterogeneous gas-solid reactions and homogeneous gas-phase reactions. For simplicity, reactions involving hydrocarbons are shown using methane. However, similar equations can also be set up for other hydrocarbons.

If O2 is made available by the gasification medium during the gasification of solids or liquids, partial oxidation and complete oxidation (5.1, 5.2) of the carbon can take place. Both reactions are exothermic, meaning that heat is released during

Table 5.2 Basic reactions of gasification and reforming [4]

Reaction equation	Enthalpy [kJ/mol], 25 °C	Equations
Heterogeneous gas-solid reactions		
$C + O_2 \rightarrow CO_2$	-393	5.1
$C + \frac{1}{2}\,O_2 \rightarrow CO$	-110	5.2
$C + H_2O \rightarrow CO + H_2$	131	5.3
$C + CO_2 \leftrightarrows 2\,CO$	172	5.4
$C + 2\,H_2 \leftrightarrows CH_4$	-74	5.5
Homogeneous gas phase reactions		
$CO + H_2O \leftrightarrows CO_2 + H_2$	-41	5.6
$CH_4 + H_2O \leftrightarrows CO + 3\,H_2$	206	5.7
$CH_4 + CO_2 \leftrightarrows 2\,CO + 2\,H_2$	247	5.8
$CH_4 + 2\,O_2 \rightarrow CO_2 + 2\,H_2O$	-808	5.9
$CH_4 + \frac{1}{2}\,O_2 \rightarrow CO + 2\,H_2$	-36	5.10
$CO + \frac{1}{2}\,O_2 \leftrightarrows CO_2$	-283	5.11
$H_2 + \frac{1}{2}\,O_2 \rightarrow H_2O$	-242	5.12

the reaction. Furthermore, the gases CO2 and water vapor react to form CO and H2 on the solid carbon according to the water-gas reaction (5.3) and the Boudouard reaction (5.4). The CO2 and water vapor may be present as a gasification medium or as a reaction product. Furthermore, the H2 can react at the solid carbon to form CH4 (5.5) [4].

In the homogeneous gas phase, the gases formed during the gas-solid reactions react with each other. Especially when steam is used as the gasification medium, the water-gas shift reaction (5.6) and the steam reforming (4.7) of CH4, among others, take place. The two reactions are explained in the next section under reforming.

The basic reaction equations of syngas production by reforming gaseous feedstocks are explained below. If steam is used as the gasification medium -called steam reforming- CH4 is partially converted to CO and H2 by the reaction eq. (5.7). The reaction is strongly endothermic, which means that heat must be added. In addition, the water gas shift reaction (5.6) takes place, i.e., the conversion of CO to H2 and CO2 with the addition of water vapor.

The proportion of CO in the synthesis gas is further influenced by the reaction (5.8). Here, CH4 reacts with CO2 to form CO and H2. The Boudouard reaction (5.4) and methane conversion (5.5) can lead to soot formation [3].

Since the reforming occurs with the addition of stoichiometrically insufficient oxygen (oxygen deficiency), it is also referred to as partial oxidation (POX), which is exothermic. The partial oxidation of CH4 produces CO and H2 (5.10). In part, the complete oxidation of CH4 to CO2 and water (5.9) also takes place. In addition, oxidation of previously formed CO (5.11) and H2 (5.12) can occur. The overall energy balance of partial oxidation is exothermic. The reactions that occur during gasification or reforming and how advanced they are depend on the gasification medium, residence time, and reaction conditions [4].

The energy balance of the gasification or reforming reactions is highly dependent on the reaction conditions and can be exothermic or endothermic. In the case where the required heat is provided directly by the partial oxidation of the feedstock, it is called autothermal gasification. If the required heat is supplied indirectly by external heat sources, it is referred to as allothermic gasification. As a rule, the ratio between oxygen and gasification medium is adjusted so that energy neither has to be removed nor supplied [3, 5].

5.4 Gasification Technology

5.4.1 Solid Matter

In addition to the reactor type, the available gasification technologies can also be differentiated on the basis of the type of heat supply (allothermic, autothermic), the gasification medium (air, steam, O2), the pressure (atmospheric, pressurized) and the number of reaction stages (single-stage, multi-stage). With regard to the reactor type for solid gasification, a basic subdivision into fixed-bed, fluidized-bed and entrained-flow gasifiers has become widely accepted in the literature [4]. In addition to the aforementioned gasifiers, however, there are other reactor types and special forms of more or less technical relevance, which are also briefly mentioned below.

5.4.1.1 Fixed-Bed Gasifier

In fixed-bed gasifiers, the feedstock to be gasified migrates through the reactor as a gas-permeable packed bed driven by gravity and, depending on the type of gasifier, is flowed through by the gasification medium in co-current or counter-current. On the basis of the direction of flow of the gasification medium and due to the major influence this has on the gasification process, this type of reactor can be further subdivided into co-current and counter-current fixed-bed gasifiers. Figure 5.2 compares these two main fixed-bed designs:

Depending on the type of gasifier, zones of different stages of thermochemical conversion of the feedstock are formed within the reactor, which means that they are largely spatially separated from each other. These zones are not spatially fixed and can shift vertically and overlap depending on the operating mode of the reactor [6].

In the co-current gasifier, the feedstock and gasification medium move in the same direction, i.e., downward along gravity. The feedstock introduced at the reactor head is first heated and dried before its pyrolytic decomposition takes place. Pyrolysis gas and pyrolysis coke then enter the hot oxidation zone, where the gasification medium is typically added. In the oxidation zone, the mostly long-chain hydrocarbons (tars) of the pyrolysis gas are thermally broken up ("cracked") and converted into short-chain hydrocarbon molecules. To ensure that the pyrolysis gas flows through the hot coke layer and tars are thermally broken up, co-current fixed-bed gasifiers are usually somewhat narrowed in the oxidation zone [7]. In the subsequent reduction zone, the solid and gas phases can react further, with part of the solid passing into the gas phase (see Table 5.2, Eqs. 5.3 to 5.5). The still raw,

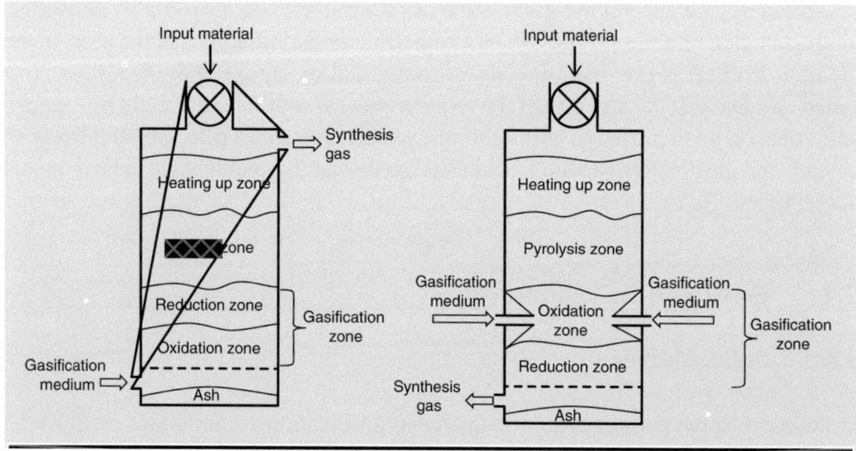

Fig. 5.2 Direct-current fixed-bed gasifier (left) and countercurrent fixed-bed gasifier (right)

unpurified synthesis gas then leaves the reactor at the bottom, while inert components of the feedstock and any ash formed remain in the lower reactor area and have to be removed regularly.

An advantage of the co-current fixed-bed gasifier is the lower tar content in the syngas (<100 mg/Nm^3) compared to the countercurrent fixed-bed gasifier [6, 8], which is ensured by the fact that the gas phase generated in the pyrolysis and reduction zone is passed through the hot oxidation zone and reformed [4].

On the other hand, the disadvantages of the co-current fixed bed gasifier are the limitation in terms of reactor size (up-scale) and the high feedstock quality requirements (especially particle size and water content). Both are due to the fact that uniformly high temperatures must be present across the reactor cross section in the individual zones in order to ensure consistently good synthesis gas quality. Areas with low temperatures here can lead to a greater proportion of uncracked, long-chain hydrocarbons in the synthesis gas [4, 9].

Another disadvantage is the high reactor temperature (500–900 °C) [10] and the associated high thermal energy of the syngas at the outlet of the reactor. Insofar as this heat energy cannot be recovered or can only be recovered incompletely, the overall efficiency of the process decreases [6].

The greatest challenge is to set a uniformly high temperature in the packed bed in order to thermally decompose as many tars as possible contained in the pyrolysis gas and to ensure good synthesis quality.

In the countercurrent fixed-bed gasifier, the feedstock is also introduced at the top of the reactor head and passes through the reactor driven by gravity. In contrast to the co-current fixed bed, however, the gasification medium is added at the bottom of the reactor and flows through the moving bed in countercurrent, so that the synthesis gas produced exits at the reactor head. As a result, the sequence in which the gas flow passes through the zones of thermal conversion formed in the reactor differs (see

Fig. 5.2). The hot gas generated in the oxidation zone releases a large part of its thermal energy as it flows through the zones located above it, thus supplying the energy required for the reactions.

One advantage of the counterflow fixed-bed gasifier is its high efficiency, which results from the low gas outlet temperature (200–300 °C [10]). In addition, it is much more flexible and robust in terms of feedstock selection [10]. On the one hand, good preheating and drying of the feedstock is ensured due to the good heat transfer from the gas to the solid phase, so that feedstocks with comparatively high water contents are also suitable [6]. On the other hand, this type of reactor is less susceptible to large particle sizes, since there is no narrowing of the reactor cross-section compared to the co-current fixed bed gasifier.

On the other hand, a major disadvantage is that the tar-containing pyrolysis gases generated in the pyrolysis zone do not pass through the hot reduction and oxidation zone, but exit at the top of the reactor head. As a result, the tar content contained in the syngas is several times higher than in the case of gasification in the co-current fixed-bed gasifier, which in turn requires more elaborate gas purification techniques [7].

Countercurrent gasifiers are used in the field of biomass gasification in the fuel power range from 100 kW to 10 MWth. In particular, several gasifiers ("Bioneer" gasifiers) in the thermal power range of 5–10 MWth were installed in Finland and Sweden in the 1980s [6]. Also worth mentioning is the Lurgi gasifier, which was developed by the Lurgi company as early as the 1930s and is still the most common type of gasifier in use worldwide in the coal-to-liquid or coal-to-gas sector [11].

In addition to the two fixed-bed gasifier types mentioned above, there are also several variants in which the gasification medium is fed to the packed bed in two or more stages. This is to try to combine the advantages of both co-current and fixed-bed gasifiers. An example of this is the double-fired gasifier [4].

Also to be counted among the fixed-bed reactor type is plasma gasification, in which a gas stream is electrically heated up to 5000 °C and the organic part of the feedstock is practically completely thermally decomposed. Although high cold gas efficiencies can be achieved with this process, complex mechanical preparation of the feedstock [4] and high electrical energy [12] are required to operate the gasifier.

5.4.1.2 Fluidized Bed Gasifier

In fluidized bed gasification, the feedstock is introduced into a fluidized bed of inert (e.g. quartz sand) or catalytically active (e.g. olivine, limestone) bed material, which is fluidized by the gasification medium introduced in the lower reactor section. Fluidized beds exhibit largely homogeneous reaction conditions in the entire fluidized bed area if the feedstock and fluidized bed are well mixed, i.e., unlike fixed-bed gasifiers, no distinct temperature and reaction zones are formed. Therefore, the gasification reactions for this type of reactor also take place in parallel throughout the reactor and are not spatially separated as in the fixed-bed reactor. The good temperature controllability and thus the very precise adjustability of the reaction conditions are major advantages of fluidized bed technology. In principle, fluidized bed reactors can be divided into (a) stationary and (b) circulating (see Fig. 5.3).

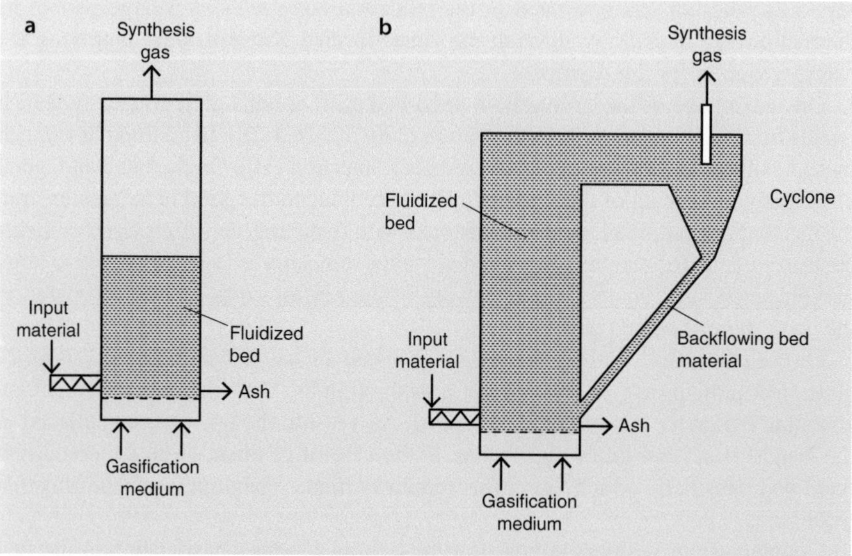

Fig. 5.3 (**a**) Stationary fluidized bed; (**b**) Circulating fluidized bed

In the case of the stationary fluidized bed, the flow velocity of the gas is only so great that the packed bed inside the reactor is placed in a state of suspension, but no significant quantities of solids are discharged from the top of the reactor. A gas velocity of approximately 5–15 times the loosening velocity of the packed bed has proven effective in this regard [4]. The synthesis gases generated in the fluidized bed exit upwards into the empty reactor space ("freeboard"), where they continue to react due to the prevailing high temperatures. Heterogeneous and homogeneous gasification and reforming reactions take place, more or less approaching chemical equilibrium depending on the residence time of the gas.

The good controllability of the process parameters and the high homogeneity of the reaction conditions are major advantages of the stationary fluidized bed. For this reason, this type of reactor is easily scalable upwards, and significantly higher thermal outputs can be achieved than is possible for the fixed-bed gasifiers already described. A disadvantage, however, is that the outlet temperature from the reactor chamber—similar to that for the countercurrent fixed-bed gasifier—necessitates costly heat recovery or otherwise only low efficiencies can be achieved [4].

Fluidized bed gasifiers were originally designed for large-scale gasification of coal (Winkler gasifier, 1926) and have been increasingly used for biomass gasification in recent years [6]. The Winkler process was further developed by the Rheinbraun and Uhde companies into the high-temperature Winkler (HTW) process, in which the feedstock is gasified in a stationary fluidized bed at temperatures below the ash softening temperature and tars still present in the syngas are thermally decomposed by an additional injection of a gasification medium above the fluidized bed [13]. Atmospherically operated stationary fluidized bed gasifiers can be considered state of the art today, whereas successful market maturity is still pending for pressurized operation [4].

The operation of the circulating fluidized bed is comparable to the stationary variant described above. However, there is a significant difference with regard to the gas velocity, which for the circulating fluidized bed is above the suspension velocity of the individual particles, so that the bed material is discharged from the reactor. As a result, there is no clearly defined bed surface for this type of reactor, but there is a descending bulk density gradient along the reactor height. The discharged bed material (and coke particles) is (are) removed from the synthesis gas stream in downstream cyclones and returned to the fluidized bed reactor, so that there is a closed circulation of the bed material.

Similar to stationary fluidized beds, up-scaling to large plant capacities is also easier for the circulating fluidized bed than for fixed bed gasifiers. In addition, the circulating fluidized bed has higher cross-sectional loadings compared to the stationary version, i.e., the same thermal output can be realized with smaller reactor sizes. However, this comes at the cost of more expensive plant engineering [4]. Circulating fluidized beds are a successfully commercially proven process and are considered state of the art.

In addition to the two designs of fluidized bed gasifier described above, there is also the possibility of combining several fluidized beds. In the two-bed fluidized bed, a steam gasification fluidized bed is typically combined with a combustion fluidized bed, the latter providing the thermal energy required for gasification. A detailed explanation of this type of gasifier can be found in [4].

5.4.1.3 Entrained Flow Gasifier

Entrained-flow gasifiers (Fig. 5.4) are usually designed as elongated tube/shaft reactors and can be used for gasification of solids, liquids, slurries (solid-liquid mixtures), and gases. Solid feedstocks must be finely ground (<100 µm) before entering the reactor and are mostly injected co-currently with the gasification medium in the head of the reactor. Using oxygen as the gasification medium and at temperatures ranging from 1250 to 2000 °C [10], the feedstock is almost completely thermally decomposed in very short residence times (≈ 1 s) and at high pressures (25–60 bar). Due to the high temperatures, the ash contained in the feedstock is liquefied and must be removed as slag at the bottom of the reactor.

For solid fuels, entrained-flow gasifiers are usually operated autothermally, i.e., the thermal energy required for gasification is provided by oxidation of part of the feedstock.

A major advantage of the entrained-flow gasifier is its high power density, which makes it possible to achieve high thermal plant outputs. Furthermore, the high gasification temperature almost completely decomposes tars and pollutants, so that the synthesis gas produced is of very high quality [14] and well suited for subsequent synthesis gas steps.

A particular disadvantage is the high technical and equipment costs, which only permit economical operation of entrained-flow gasifiers for very high plant capacities of several hundred MWth [10]. While entrained-flow gasifiers for the

Fig. 5.4 Entrained-flow gasifier

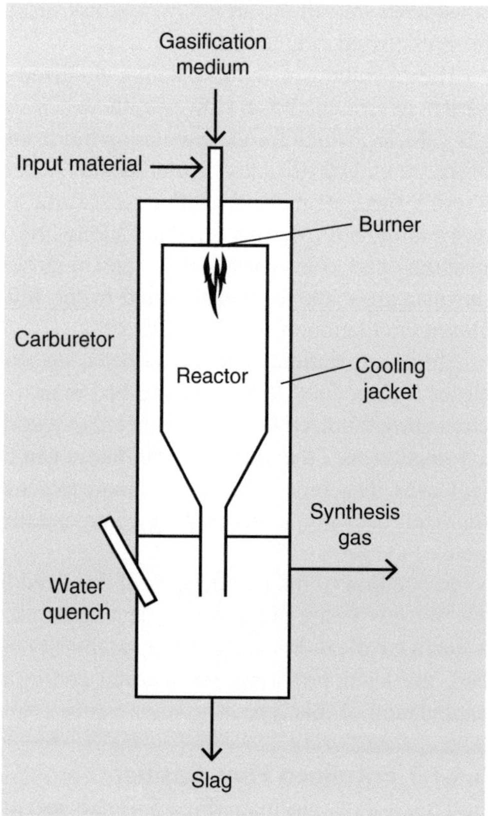

gasification of coal have already been commercially tested, this reactor type is currently of secondary importance in the field of biomass gasification [4].

5.4.1.4 Multi-stage Gasifiers

The basic idea behind multistage gasifiers is the spatial separation of different phases of the thermochemical conversion process in order to be able to set and optimize the process conditions for these phases separately. Potentially, higher efficiencies and a lower-tar synthesis gas can be achieved in this way. According to initial findings, multi-stage processes are the only ones that promise the possibility of reducing tar contents to such an extent that subsequent synthesis gas purification is no longer necessary [4].

In principle, a combination of different process principles and reactor types is possible, and numerous multistage gasifier concepts have been developed in recent years [4, 6]. While in the lower power range (<1 MWth) multistage biomass gasifiers have been realized and successfully operated in recent years [15], commercial success is still lacking, especially for the higher power range. Here, the staged reforming process (Chap. 19) promises great potential.

5.4.2 Production of Synthesis Gases from Gaseous and Liquid Feedstocks

Compared to the often very complex and slow heterogeneous reactions, the process control for homogeneous reactions is usually much simpler. In Sect. 5.4.1.1, the plant concepts of synthesis gas production based on the heterogeneous reactions of solid feedstocks (coal, biomass and similar materials) are presented. Similar concepts are also used for high-viscosity, long-chain liquid hydrocarbons. In contrast, other processes and reactors have been developed for gaseous and short-chain liquid hydrocarbons [16]. Plants based on these feedstocks can be constructed at much lower cost. As a rough guide, the investment costs of a plant for synthesis gas production based on coal as feedstock will be about three times higher than for plants using natural gas as feedstock [17].

The conversion of natural gas and short-chain hydrocarbons to synthesis gas is called reforming. The following processes are commonly understood by this term:

Steam reforming (SR)
Partial oxidation (POX)
Autothermal reforming (ATR)

In reforming, catalysts are generally used to accelerate the reactions. Only in partial oxidation (POX) are catalysts not normally used. However, processes have also been developed here that use catalysts—for example, to lower the reaction temperatures. To distinguish them from purely thermal partial oxidation (TPOX), catalytic processes are often referred to as CPOX.

5.4.2.1 Steam Reforming (SR)

Steam reforming is mainly used to produce a hydrogen-rich synthesis gas—for example, when the synthesis gas is to be used to produce ammonia. In recent years, interest in steam reforming has increased significantly as the search for a carbon-free energy supply has focused more attention on hydrogen as an energy carrier. Thus, steam reforming is not only important for large plants producing more than 100,000 m^3 of hydrogen per hour (e.g., [18]), but also for small plants producing hydrogen for fuel cells, for example (e.g., [19]).

In Fig. 5.5, the main steps of steam reforming are shown. The feedstock for steam reforming—in this case, for example, natural gas - usually has to be pretreated before being fed into the reformer. In particular, sulfur compounds must be removed, since sulfur is a strong catalyst poison for the nickel-based catalysts typically used. In this process, traces of sulfur are sufficient to create a deactivating layer on the catalyst. In general, several processes are available for desulfurization. In many cases,

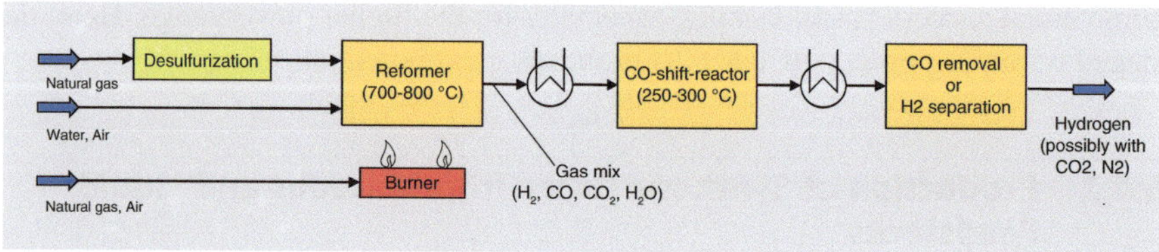

Fig. 5.5 Stages of steam reforming [20]

desulfurization is carried out by reacting the sulfur components with ZnO, since this allows the lowest residual sulfur content in the natural gas to be achieved.

In the reformer, the reaction of CH4 with water vapor proceeds according to Table 5.2, Eq. (5.7). Similar equations apply to the conversion of longer-chain hydrocarbons to H2 and CO. The reforming reaction is strongly endothermic. It follows that energy must be supplied to the reformer from outside. In order to shift the equilibrium of reaction 5.7 to the product side, it is advantageous to set high temperatures and, for gas phase reactions, comparatively low pressures in the reactor.

The feed product - here, for example, natural gas again—is mixed with steam and preheated. The mixture is then passed over the catalyst in the reformer. Typically, the reformer is a tubular reactor filled with a catalyst containing nickel inside the tubes (Fig. 5.6). Outside the tubes, in the shell space, gas or oil burners provide the required thermal energy for the endothermic reactions in the tubes. Usual operating conditions for reforming are temperatures of 700–900 °C and pressures of 20–30 bar. An overview of the catalytic reforming of CH4 is given, for example, [22].

After reforming, the syngas contains residual methane contents of 1–10%, depending on the operating pressure, temperature, and steam/carbon ratio [21]. In addition, the gas still contains appreciable amounts of CO. This CO can be used in a water-gas shift reactor according to reaction 5.6, Table 5.2 to increase the hydrogen content. On the one hand, high temperatures would be favorable for the course of the shift reaction in order to achieve high reaction rates; on the other hand, the water gas shift reaction is exothermic. Thus, the equilibrium shifts to the reactant side at high temperatures. For this reason, water gas shift reactions are often used in two stages at different temperatures. The first stage, the high-temperature shift (HTS) stage, is carried out in the temperature range of 320–450 °C and by using Fe3O4/Cr2O3 catalysts. This stage reduces the CO concentration to about 3%. In the second stage, the low-temperature shift (LTS) stage, the CO concentration is reduced to about 500 ppm at a temperature of about 180–250 °C and by using Cu/ZnO/Al2O3 catalysts [23]. In some cases, three-stage water gas shift reactors are also used. In this case, a medium temperature stage (temperature range from 220 to 270 °C) is added.

High-temperature conversion is used as standard in most H2 plants.

After the shift stage, it may still be necessary to separate H2 or remove CO (or other harmful components) from the synthesis gas. Since, for example, CO must

Fig. 5.6 Schematic diagram
of a reforming reactor [21]

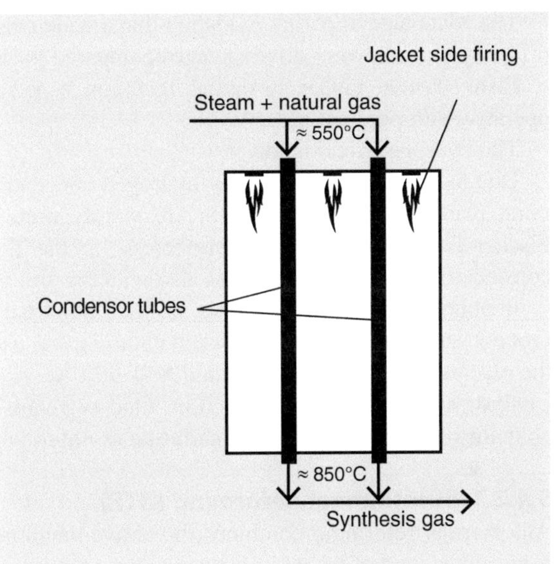

not be present in the synthesis gas during ammonia synthesis and CO concentrations cannot be sufficiently reduced by the shift reactions, post-purification of the synthesis gas is necessary. CO can be sufficiently reduced by methanation (reversal of reaction 5.7 in Table 5.2) catalytically at temperatures around 100 °C.

5.4.2.2 Partial Oxidation (POX)

In partial oxidation, the feedstocks (here: gaseous or liquid hydrocarbons) are reacted with oxygen in an exothermic process. The oxygen can be supplied to the reactor either as atmospheric oxygen, as pure oxygen or as oxygen-enriched air. When pure oxygen is used, it is avoided that the synthesis gas produced contains significant amounts of nitrogen.

In the case of partial oxidation, the oxygen supplied is not sufficient to completely burn the feedstock, i.e., the combustion proceeds substoichiometrically. Due to the lack of oxygen, the desired products CO and H_2 are formed in addition to the usual flue gas components, such as CO_2 and H_2O. Even if a large number of gaseous or liquid feedstocks are suitable for partial oxidation, mainly highly viscous, long-chain hydrocarbons with high boiling temperatures (above 240 °C) are used. Steam reforming is not suitable for these feedstocks. Partial oxidation is comparable to gasification of solids - both processes are based on the same or very similar reactions.

Due to the high reaction temperatures, partial oxidation is carried out in lined reactors, e.g. in shaft reactors. The reaction partners are usually converted to synthesis gas at pressures of 3 to 6 MPa and temperatures of 1200–1600 °C.

The advantage of partial oxidation that a wide range of feedstocks can be used is offset by the following disadvantages compared to steam reforming:

Plant costs are higher for partial oxidation, especially if partial oxidation is to be operated with pure oxygen.

The hydrogen yield is lower.

Due to the comparatively low hydrogen concentration and thus the low H2/CO ratio, plants for partial oxidation are mainly used to generate synthesis gas for Fischer-Tropsch synthesis. Nevertheless, in the field of small plants there are considerations to use this process also with the aim of hydrogen production.

In almost all cases, no catalysts are used in partial oxidation. However, there are process variants—for example, when natural gas is used—that use catalysts to lower the reaction temperatures to around 800–900 °C. A prerequisite for this is that the catalyst is not used in the reaction. One prerequisite here is that the feedstock contains only small amounts of sulfur so as not to poison the catalyst used.

5.4.2.3 Autothermal Reforming (ATR)

Autothermal reforming combines the above-mentioned processes. In this process, for example, part of the gas is burned to raise the temperature in the catalyst layer for steam reforming. The overall reaction is slightly exothermic. The advantage of autothermal gasification over steam reforming is that, due to direct heating, the thermal efficiency of the process is higher. The higher temperatures in autothermal reforming ensure higher reaction rates and thus greater throughput. The limiting factors, however, are the application ranges of the materials, which determine the maximum reaction temperatures. It should also be noted that pure oxygen is required if no nitrogen is to be present in the synthesis gas.

Another problem is that soot can be formed during partial combustion due to parallel reactions (according to the equilibrium reaction 5.3 and 5.4 from Table 5.2 in the reverse direction). This can be avoided if a high steam concentration (a high steam-to-carbon [S/C] ratio) is set. However, high S/C ratios are associated with high costs, so work on optimizing steam addition has been ongoing for some time [24].

5.4.2.4 Small Reformers

In the past, the main interest in syngas production was on the development and optimization of large-scale plants. Due to the increasing importance of fuel cells in mobile systems or decentralized hydrogen supply as a carbon-free energy carrier, small reformers are becoming more and more in the focus of interest. All the synthesis gas production processes presented here are also suitable for small reformers. In some cases, it is even being considered to use reformers in automobiles to produce hydrogen from fuels in the vehicle. The hydrogen is used in fuel cells to generate electricity, with the aim of operating auxiliary systems, so-called APUs (auxilliary power units) [25].

5.5 Synthesis Gas Composition and Utilization

The synthesis gas produced in the gasification process is a mixture of desirable high calorific value gas components, inert gases, undesirable pollutant gases, and possibly solid particles. The specific composition depends mainly on the following factors [4]:

Feedstock (chemical composition, particle size, impurities).
Type and quantity of the gasification medium (air, water vapor, CO2, gas mixtures)
Type of gasifier (Sect. 5.4)
Temperature, residence time and pressure

A selection of the gasification concept that covers all the above factors must take into account the use of the synthesis gas produced, since the requirements for the synthesis gas composition depend significantly on the utilization technique. In the following subchapter, the dependencies of the synthesis gas composition on the above factors are first described, and then the requirements of subsequent utilization techniques are addressed.

5.5.1 Factors Influencing the Synthesis Gas Composition

5.5.1.1 Gasification Medium

Basically, a distinction must be made between nitrogen-containing (air) and nitrogen-free (steam, CO2, O2) gasification media. N2 as an inert gas does not participate in any chemical reaction and has a calorific value of 0 MJ/m^3. Therefore, it causes an undesired dilution of the synthesis gas and may have to be removed before further synthesis gas utilization in order to enable economic utilization. Nevertheless, air is the most commonly used gasification medium, especially in the lower power range, due to its ease of use and low supply costs [4]. For example, syngas produced in an air gasification has a calorific value in the range of 3 to 6.5 MJ/m^3 and is thus considered a lean gas [4]. In engine combustion, the lower power density would result in lower engine power than would be possible for a gas with a higher calorific value (e.g., natural gas). In comparison, a syngas from a steam gasification typically has calorific values in the range of 10 to 16 MJ/m^3 [4].

5.5.1.2 Gasifier Type

As already described in Sect. 5.4, the gasifier design and process control have a significant influence on the tar and particle content of the synthesis gas. Table 5.3 shows typical values for the gasifier types described in the previous chapters:

The highest tar contents by far are recorded for the countercurrent fixed-bed gasifier. For co-current fixed-bed gasifiers—as described above—the stock flow is much more favorable with regard to tar decomposition. Therefore, tar concentrations are low. Fluidized bed gasifiers, on the other hand, lie between co-current and countercurrent fixed bed gasifiers in terms of tar content in the synthesis gas.

Table 5.3 Typical tar and particle contents for different gasifier types [4]

	Tar content [g/m^3]	Particle content [g/m^3]
Direct current fixed bed	50	1
Countercurrent fixed bed	0.5	1
Stationary fluidized bed	12	4
Circulating fluidized bed	8	20
Second bed fluidized bed	2	20
Airstream	<0.1	<0.05
Multilevel	<0.1	No data

However, the tar content can vary considerably depending on the process temperature, the residence time, and the selected bed material (catalytic or non-catalytic). entrained-flow gasifiers produce virtually tar-free synthesis gas, which is due in particular to the high temperatures prevailing in the reactor. The situation is similar for multistage gasifiers, which were developed explicitly to reduce the tar content in the synthesis gas, among other things. Here, the pyrolysis gas with a high tar content produced in the pyrolysis stage is fed to a subsequent reforming stage in which tars are decomposed under high temperatures.

In terms of particle content in the synthesis gas, fluidized bed gasifiers are at a distinct disadvantage compared with other gasifier designs due to high flow velocities and the use of a fine-grained bed material. The dusts contained in the syngas are sometimes very fine and often cannot be separated by a simple and inexpensive cyclone [4].

5.5.1.3 Temperature, Residence Time and Pressure

The gasification reactions described in Sect. 5.3 (Table 5.2) require a certain time to reach chemical equilibrium. Typically, chemical equilibrium is not reached in gasifiers [26]. However, the reaction rate is largely determined by the prevailing temperature, i.e., the higher the temperature, the faster the chemical reactions proceed and the faster chemical equilibrium is approached. Moreover, according to Le Chatelier's rule, high gasification temperatures favor endothermic reactions, which increases the fraction of CO and H2 in the syngas at higher temperatures according to the reaction equations described in Sect. 5.3 [27]. The temperature prevailing in the gasification reactor thus represents one of the most important process parameters and one that significantly determines the final synthesis gas composition [28].

The position of the chemical equilibrium between reactants and products of the above-mentioned gasification reactions is also determined by the prevailing pressure in the gasification reactor. In this context, the H2 and CO contents decrease with increasing pressure if an increase in volume is associated with the underlying reactions [27]. This also follows Le Chatelier's "principle of least constraint", according to which pressure increases favor volume-reducing reactions, i.e., the equilibrium shifts to the side of the chemical reaction equation that has the smaller number of molecules.

5.5.1.4 Feedstock

The chemical composition of the feedstock influences the content of undesirable pollutant gases and impurities in the synthesis gas, with particular reference to sulfur, nitrogen and halogen compounds, as well as alkalis and heavy metals. Such impurities, e.g. as catalyst poisons, can impede a material utilization of the synthesis gas in further synthesis steps or an energetic utilization by combustion in gas engines or gas turbines and must therefore, depending on the further utilization technology, be removed in a downstream synthesis gas purification.

5.5.2 Requirements of Synthesis Gas Utilization Technology

Energy or material utilization places requirements on the composition and, in particular, the impurities in the synthesis gas, so that complex and cost-intensive gas purification measures may be necessary. However, as described in Sect. 5.5.1, the synthesis gas composition is already significantly influenced by upstream factors. Therefore, it can be concluded that different gasifier types, feedstocks, and gasification conditions are better or worse suited for downstream utilization techniques and require different gas purification techniques.

5.5.2.1 Energetic Utilization

As shown in Sect. 5.4, tar-containing components that pass into the gas phase during feedstock pyrolysis are thermally broken up and split into short-chain molecules in the gasification reactor. In practice, this process is never completely finished, so that a certain tar content is always found in the synthesis gas. The quantity as well as the composition of the tars are determined to a large extent by the type of gasifier and the process conditions [29].

Regardless of the specific utilization technique, tars must be prevented from condensing out in the piping and equipment and causing sticking or blockage. On the other hand, provided that condensation is avoided, tars can be combusted without problems as high calorific value components of the synthesis gas in gas engines and, in particular, in gas turbines.

The particle content of the synthesis gas can also be problematic in downstream piping and apparatus due to erosion. This is especially true for gas turbines, which allow much lower particle concentrations in the syngas than gas engines [4].

For gas engines, there is also a risk of high-temperature corrosion, e.g. as a result of sulfur and halogen compounds. High-temperature corrosion leads to erosion of the metallic material surface due to the formation of corrosion products during the reaction of the ambient medium (in this case synthesis gas) and the material, and thus to long-term damage and possibly failure of the component. Halogens and also alkalis are also problematic for the engine oil in the form of acid gases, e.g. (H2S) and ammonia (NH3), due to their life-shortening effect.

In general, gas turbines place higher demands on the particle and alkali content in the synthesis gas than gas engines. Due to the significantly higher flow velocities of the gas in the turbine, the risk of erosion is significantly increased here. With regard to alkalis, condensation of alkali sulfates in the gas turbine and thus sulfate-induced high-temperature corrosion must be avoided [30].

5.5.2.2 Material Utilization

The requirements for the synthesis gas composition in the case of use for the production of liquid and/or gaseous fuels are significantly higher compared to energetic use in gas engines or gas turbines [4]. A comprehensive overview of the requirements of essential synthesis processes on the synthesis gas composition can be found in [14]. Because of these stringent requirements, low-temperature scrubbing, such as rectisol scrubbing, is typically used in large-scale commercial syngas production plants.

Also, the lowest possible nitrogen content of the synthesis gas must be maintained in order to reduce the required compressor power on the one hand and not to hinder the reactions on the other. High nitrogen contents also lead to a reduction in the partial pressures of the reactants, which would impair the subsequent catalytic synthesis [4]. For this reason, air is ruled out as a gasification medium for such downstream utilization techniques, and steam or steam/O2 gasifiers are used.

The selection of the gasification medium is also important with respect to the H2/CO ratio in the syngas. There are optimum ranges for the respective synthesis processes [14], which should ideally be maintained by the synthesis gas composition. A water gas shift reactor interposed between the gasifier and the synthesis can readjust the H2/CO ratio, if necessary, but involves additional effort and cost.

References

1. Baerns M, Behr A, Brehm A, Gmehling J, Hofmann H, Onken U, Renken A (2013) Technische Chemie. Wiley-VCH Verlag, Weinheim
2. Cerbe G, Lendt B (2004) Grundlagen Der Gastechnik. 6. Auflage. Carl Hanser Verlag, München
3. Moulijn JA, Makkee M, Van Diepen AE (2013) Chemical process technology. 2. Auflage. Wiley, Hoboken, NJ
4. Kaltschmitt M, Streicher W (2009) Regenerative Energien in Österreich – Energie aus Biomasse. Springer, Heidelberg
5. Vogel A, Bolhàr-Nordenkampf M, Kaltschmitt M, Hofbauer H (2006) Schriftenreihe "Nachwachsende Rohstoffe", Bd. 29. Analyse und Evaluierung der thermo-chemischen Vergasung von Biomasse. Fachagentur Nachwachsende Rohstoffe
6. Knoef HAM (2005) Handbook Biomass Gasification. 2. Auflage. BTG Biomass Technology Group BV, Enschede
7. Campareda SC (2011). Biomass energy conversion. In: Nayeripour M, Kheshti M (Eds.) Sustainable growth and applications in renewable energy sources. Intech Open, London, S 209–226
8. Sulc J, Stojdl J, Richter M, Popelka J, Svoboda K, Smetana J, Vacek J, Skoblja S, Buryan P (2012) Biomass waste gasification – Can be the two stage process suitable for tar reduction and power generation? Waste Manag 32(4):692–700
9. Balat M, Balat M, Kirtay E, Balat H (2009) Main routes for the thermo-conversion of biomass into fuels and chemicals. Part 2: Gasification systems. Energy Convers Manag 50(12): 3158–3168. https://doi.org/10.1016/j.enconman.2009.08.013

10. Nagel F-P (2008) Electricity from wood through the combination of gasification and solid oxide fuel cells – systems analysis and proof-of-concept. Dissertation, ETH Zürich
11. Weiss M-M, Turna O (2009). Lurgi's FBDB gasification – recent developments and project up-dates. IEC Gasification Conference Publication. Lurgi GmbH
12. Williams RB, Jenkins BM, Nguyen D (2003) Solid waste conversion: a review and database of current and emerging technologies. Bericht erstellt für das California Integrated Waste Management Board, Davis, CA
13. Herdel P, Ströhle J, Epple B, Kolmorgen B, Keller D (2014) Aufbau einer Pilotanlage zur HTW-Vergasung von Kohle und Biomasse. DGMK-Fachbereichstagung Konversion von Biomassen
14. E4tech (2009) Review of technologies for gasification of biomass and wastes. Biomass: 125. http://www.nnfcc.co.uk/tools/review-of-technologies-for-gasification-of-biomass-and-wastes-nnfcc-09-008
15. Larcher B (2017) Erfahrungen aus dem Betrieb einer SynCraft Anlage. 9. Internationale Anwenderkonferenz Biomassevergasung
16. Liu K, Song C, Subramani V (2009) Hydrogen and syngas production and purification technologies. American Institute of Chemical Engineers, New York
17. Nielsen JR, Christiansen J, Lars J (2011) Concepts in syngas manufacture. Catal Sci Series 10: 379
18. Wolf J (2003) Die Neuen Entwicklungen der Technik/Elemente der Wasserstoff-Infrastruktur von der Herstellung bis zum Tank. Director:1–11
19. Thormann J (2009) Diesel-Dampfreformierung in Mikrostrukturreaktoren. Dissertation, Technische Universität Clausthal
20. Aicher T, Blum L, Specht M (2004) Wasserstoffgewinnung aus Erdgas – Anlagenentwicklung und Systemtechnik. FVS Themen 2004:60–64
21. Gellert S (2013) Thermochemische Herstellung von Wasserstoff aus Biomasse unter besonderer Berücksichtigung der Rohgasreformierung. Dissertation, Universität Hamburg
22. van Beurden P (2004) On the catalytic aspects of steam reforming methane – a literature survey. ECN
23. Kalamaras CM, Efstathiou AM (2013) Hydrogen production technologies: current state and future developments. Conference Papers in Energy 2013:690627. https://doi.org/10.1155/2013/690627
24. Aasberg-Petersen K, Christensen TS, Stub Nielsen C, Dybkjær I (2003) Recent developments in autothermal reforming and pre-reforming for sythesis gas production in GTL applications. Fuel Process Technol 83(1):253–261
25. Tschöke H (2014) Die Elektrifizierung des Antriebsstrangs. Springer Vieweg, Wiesbaden
26. Schuster G, Löffler G, Weigl K, Hofbauer H (2001) Biomass steam gasification - an extensive parametric modeling study. Bioresour Technol 77(1):71–79
27. Kienberger T (2010) Methanierung bioogener Synthesegase mit Hinblick auf die direkte Umsetzuung von höheren Kohlenwasserstoffen. Dissertation, Technische Universität Graz
28. Kumar A, Jones DD, Hanna MA (2009) Thermochemical biomass gasification: a review of the current status of the technology. Energies 2(3):556–581
29. van Paasen SVB, Kiel JHA (2004) Tar formation in a Fluidised-Bed Gasifier: impact of fuel properties and operating conditions. Kardiol Pol 67(6):58
30. Müller M (2009) Freisetzung und Einbindungen von Alkalimetallverbindungen in kohlebefeuerten Kombikraftwerken. Dissertation, Forschungszentrum Jülich

Chemical-Catalytic Conversion of CO_2 and CO

6

Robert Schlögl

Abstract

Recycling CO_2 from combustion processes offers the option of reducing CO_2 emissions from these processes. Possible products are formic acid, methane, methanol and products of the Fischer-Tropsch reaction class (alkanes, olefins, alcohols). Only Fischer-Tropsch catalysis has been established on a large scale to date. Although considerable amounts of energy are required to convert CO_2, its utilization is fundamentally viable. The processes convert electricity into material energy carriers that act as energy storage for excess electricity from volatile sources. Therefore, recycling CO_2 contributes to the system efficiency of an energy system based on renewable energy.

Keywords

CO_2, formic acid, methane, methanol, Fischer-Tropsch reaction, electricity, renewable energy

6.1 Evaluation of Possibilities for the Reduction of CO_2

If carbonaceous substances are oxidized to the thermodynamic minimum of energy, carbonate is obtained. In this form, by far the most carbon is stored in rocks on earth. The gas CO_2, on the other hand, is still considerably reactive and can therefore be oxidized to organic carbonates or carboxylates as well as reduced to hydrocarbons. The first reduction product, CO, plays a central role in this process. This was recognized early on, and research began to address the material use of CO_2 in the chemical industry long before "energy research". In particular, the reaction of CO_2

R. Schlögl (✉)

Max Planck Institute for Chemical Energy Conversion, Mülheim an der Ruhr, Germany

e-mail: robert.schloegl@cec.mpg.de

© The Author(s), under exclusive license to Springer Nature Switzerland AG 2023

M. Kircher, T. Schwarz (eds.), *CO2 and CO as Feedstock*, Circular Economy and Sustainability, https://doi.org/10.1007/978-3-031-27811-2_6

as a "ligand" in organometallic systems [1, 2] was intensively studied. Furthermore, reactions in which CO_2 is integrated into a molecule as a building block have been studied. Organic carbonates and their polymers are obvious candidates for such reactions, taking an example from the large-scale synthesis of urea from ammonia

$$CO_2 : CO_2 + 2NH_3 \rightleftharpoons NH_2CONH_2 + H_2O \tag{6.1}$$

The heterogeneously catalyzed formation of dimethyl carbonate can also be understood as the incorporation of methanol into CO_2:

$$CO_2 + 2CH_3OH \rightleftharpoons CO(OCH_3)_2 + H_2O \tag{6.2}$$

Another type of reaction is the reaction of CO_2 with hydrogen. This produces formic acid, CO or methanol.

$$CO_2 + H_2 \rightleftharpoons HCOOH \tag{6.3}$$

$$CO_2 + H_2 \rightarrow CO + H_2O \tag{6.4}$$

$$CO + H_2O \rightarrow CO_2 + H_2 \tag{6.5}$$

$$CO_2 + 3H_2 \rightleftharpoons CH_3OH + H_2O \tag{6.6}$$

The reactions Eq. 6.4 and Eq. 6.5 form the important water-gas equilibrium, which plays a relevant role in all reactions of CO_2 with hydrogen and thus limits the achievable yields of other molecules.

A common property of the kinetically fast reactions with CO_2 is the formation of the stable product water. Its high stability forms a kinetic driving force for the reaction, whose desired further product is usually much higher in energy than CO_2. However, to ensure a useful conversion rate, two hydrogen atoms must be "sacrificed", which represent a significant portion of the losses in the chemical activation of CO_2. Reactions without kinetic driving force, on the other hand, proceed much more slowly and require either highly active coreactants and/or catalysts.

In Table 6.1, some energetic ratios are compiled that provide a framework for the possibilities of CO_2 activation. When comparing the numerical values across such different classes of substances, it should be borne in mind that a whole number of effects beyond the molecular structure of the carbon fragment determine the energetic position.

gas : gaseous; solid : solid; aq : liquid.

It is obvious that considerable amounts of energy must be expended if CO_2 is to be used as a molecule either in chemical synthesis or as an energy carrier. To a large extent, this is desired, which gives the activation of CO_2 the status of a "chemical battery". However, and unlike the electrochemical battery (accumulator), which

Table 6.1 Standard enthalpies of formation of carbon-containing molecules

Species	State	Enthalpy of formation H_{0f} [kJ/mol]
C_6H_6	gas	82.6
C_2H_4	gas	52.5
CH_4	gas	−74.4
C	solid	0.0
CO	gas	−110.53
C_2H_5OH	gas	−235.1
CO_2	gas	−393.51
CO_3^{2-}	aq	−675.23
CH_3COONa	solid	−708.8
$CaCO_3$	solid	−1100

largely reversibly releases its energy and thus stores electrical energy with little loss, this is not the case with the chemical battery. The necessity of breaking and forming covalent bonds on carbon requires considerable additional kinetic energy over and above the energy content of the bond, which occurs as "losses". The amounts of these losses [3–6] reach the amounts of energy to be stored. They depend very much on the processes and their embedding in other technologies [7, 8]. Basically, these amounts can be reduced [9] but not eliminated by advances in catalysis and chemical technology.

This quite complex situation leads to the fact that the family of CCU technologies, which we are concerned with here, experiences a controversial assessment in the discussion of energy science. Although the term "CCU"(carbon capture and utilization) is relatively new [10, 11], the considerations and laboratory experiments on the chemical conversion of CO_2 are much older and already abundantly worked on. In the literature, assessments range from "monstrous thermodynamic crime" [12] to "CO_2 chemistry wonderland" [13]. In this situation, it would be meritorious to have a rational guide for assessing the possibilities of CO_2 reduction. This assessment is essential because it may involve misguiding the transformation of energy systems, as suggested in the literature [14]. Moreover, the costs associated with the introduction of such technologies are enormous, both in research and later in a technological development. This is a mega-technology [15] if it were to have a noticeable impact on the CO_2 emissions balance. To date, no chemical technology has been realized that uses CO_2 to limit its emission. These arguments reinforce the desire for early assessment before investing significant resources, potentially raising unwarranted expectations, and certainly losing valuable time that is already in short supply to address the greenhouse gas problem [16].

A more rational approach has been attempted by combining assessments of chemical-catalytic achievements with an energy balance via synthesis [17] of the reaction products. This results in a ranking for products and technologies of their production, which initially gives the impression that this solves the assessment problem. The authors' argumentation also goes in this direction and thus has a selective effect on chemical technologies for CO_2 activation. As positive as it is to have a uniform view of the different possibilities, both methodologically and in

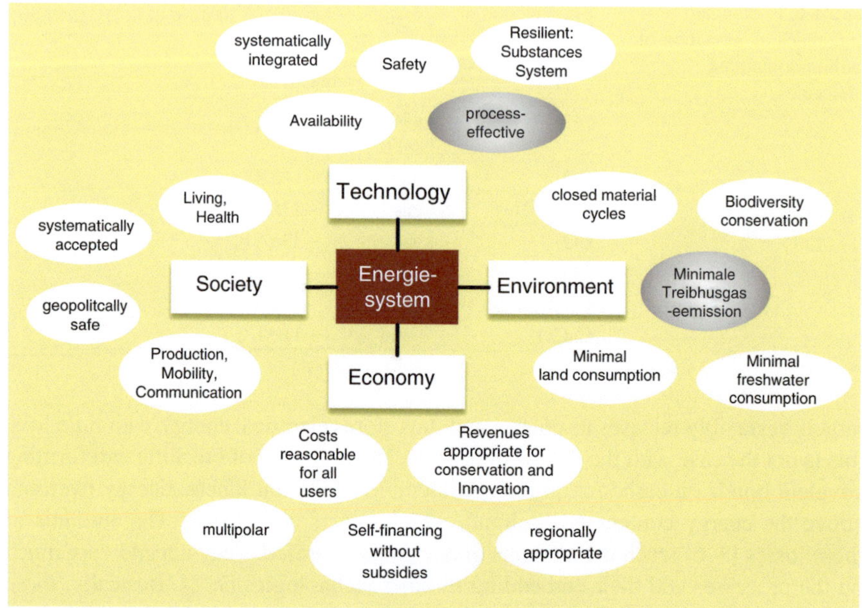

Fig. 6.1 Essential criteria for assessing measures in the transformation of an energy system. The four rectangles symbolize four levels of criteria. Important elements in these levels are grouped in the oval boxes around the level labels. This representation is still very simple in its complexity. The relationships between the elements are not listed for clarity

terms of system boundaries, this narrows down the view of the actual requirement of an assessment.

The assessment of CO_2 reduction possibilities can only be meaningful if it takes into account all the essential factors that determine an energy system. Among them, the reduction potential of CO_2 is an important factor, but certainly not the only relevant one. Since there are regionally very different energy systems on earth with different framework conditions, an assessment would have to be made in each case under regional and temporal aspects. Figure 6.1 gives an impression of the complexity that has to be considered in the assessment and how much the "technological approach" narrows the view.

Figure 6.1 expresses with the four levels of elements of an energy system the complexity of the criteria for assessing the benefits of an individual measure. In [17], the reduction of the carbon footprint of a given reaction and its synthesis technology is used as the central criterion. It is indeed extremely helpful to determine such a measure according to uniform criteria. However, assumptions about framework conditions play an important role. These have been disclosed in the paper. Many additional arguments arising from the elements of Fig. 6.1 have not been considered.

Thus, the conclusion remains that, because of the difficulty of an adequate rational evaluation of technology options, science and its subsequent technological development should be pursued in a maximally technology-open manner. In the overall context of future energy systems, it makes only limited sense to restrict the scope of scientific and technological options for the use of CO_2. Rather, the possibilities should be tested in the sense of a "toolbox" to such an extent that demonstrators in system-relevant dimensions are developed and operated for a sufficiently long time. This requires an overarching cooperation of all stakeholders addressed by the elements in Fig. 6.1. Too narrow a scientific fixation is just as damaging as a naïve neglect of technical-economic interrelationships. The latter must always be reassessed with caution because they change faster on the relevant time scales of infrastructure projects than the introduction of a new technology can occur. Therefore, the basic systemic significance of the use of CO_2 by chemical processes is discussed first in the following, followed by an overview of individual processes. It should be noted that such a comprehensive presentation is given in the overview by Leitner [2] that a similar presentation is deliberately omitted here.

6.2 The Importance of Chemical Energy Conversion

The current status of the energy turnaround in Germany and in Europe makes it clear that the greenhouse gas reduction targets agreed in the Agreement of Paris can in no way be achieved with the efforts initiated to date. These measures are essentially based on the widespread introduction of renewable energies for power generation. Separately, there are numerous efforts in the area of heat supply to reduce the use of fossil energy sources. These efforts involve both thermal insulation and the use of non-fossil energy sources such as biomass and, for example, heat pumps. In the transport sector, few efforts are currently successful, despite the introduction of bio-based fuels. So there is no lack of initiatives and good ideas, nevertheless the sum of the efforts is not sufficient.

Figure 6.2 illustrates this for the European energy area (EU 28). There, about the same amount of energy is consumed today as 30 years ago, which is mainly due to the restructuring of the economy (globalization, relocation of production). The import dependency of supply has increased significantly, which does not involve the import of nuclear fuel. Renewables now contribute about as much to energy supply as nuclear, which can be seen as a success story of a special kind. However, the growth of renewables expected today through 2050, based on governmental reports, is in no way sufficient. Many influences are responsible for the inhibition of the growth, of which the volatility of energy production is a significant factor. Another factor is the lack of significant capacity to store energy beyond what is needed to keep the power grid functioning in the short term. Finally, there are numerous regulatory hurdles in various countries, and some countries are barely participating in the renewable expansion program anyway, for very different reasons.

A key obstacle to the expansion of renewables is thus their current exclusive use as electric power to be used immediately. To make this possible with substantial generation capacities of over 30% of the total load, conventional power plants must be operated dynamically and throttle their output when many renewables are

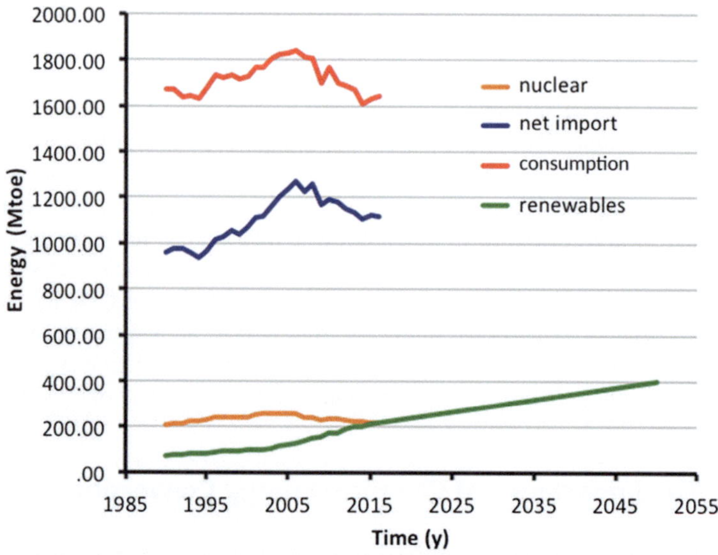

Fig. 6.2 Selected key figures of the European energy system. Data were compiled from country fact sheets from EUROSTAT (2018). The extrapolation of renewable sources is from a 2017 estimate by the European Environmental Agency (EEA)

available. This is currently done only to a limited extent for technical and regulatory reasons and leads to underutilization of generation potential. In addition, this inhibits the expansion of renewables because their economic effectiveness is increasingly diminishing.

If renewables were converted into material energy sources and thus stored and transported, the picture [18] would change drastically. One could store surplus renewables locally and move them to parts of the energy system that cannot be easily electrified (parts of mobility as the largest users, heat generation for decentralized and industrial applications). Even more significant would be the import of renewables from regions with structural surplus to regions with consumption focal points, quite analogous to today's import of fossil fuels. Looking at the figures in Fig. 6.2, it becomes clear that it is very unlikely that the EU (and other world regions) can do without importing energy from outside its territory. If renewables cannot be made transportable on a very large scale, then truly meaningful progress beyond the projection in Fig. 6.2 is likely to be difficult.

Thus, large-scale conversion of electricity into material energy carriers is a key prerequisite for the success of the energy transition. For this, chemical reactions between primarily generated hydrogen from renewables ("green") and CO_2 to form energy carriers that resemble today's fossil carriers in their chemical structures (liquid hydrocarbons) are probably central building blocks. Such an approach has the enormous advantage that today's infrastructure for transport, distribution and use, which is based on fossil carriers, can largely be reused, which will reduce the

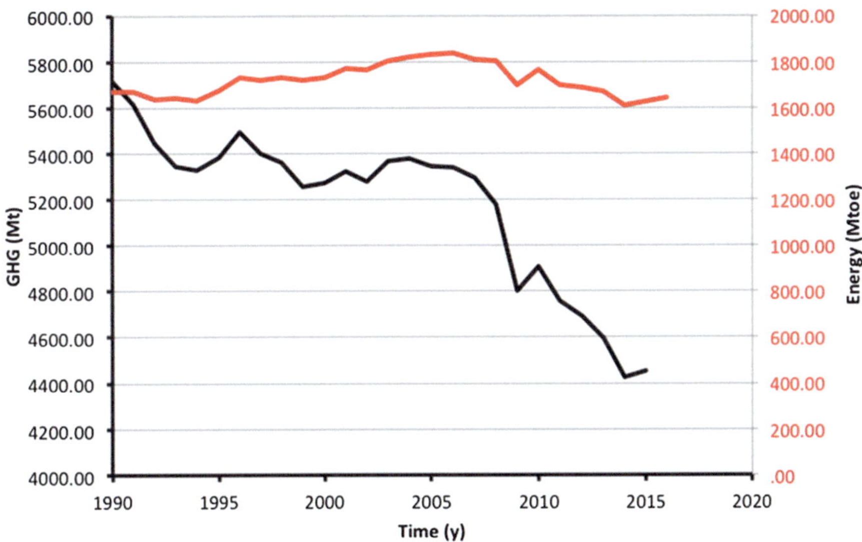

Fig. 6.3 Comparison of energy-related greenhouse gas emissions with gross energy consumption in the EU 28. For data sources, see text and signature for Fig. 6.2

costs of such an energy transition (Fig. 6.1) very significantly, both economically and socially (e.g. with regard to acceptance).

However, the idea that one should then dispense with the local production of renewables after all and replace fossil fuels with renewables from deserted areas of the earth with the help of transports does not make sense for systemic reasons. On the contrary, the chemical conversion of energy and the conversion back into useful energy are associated with very considerable losses. This is also the case with fossil fuels, as can be seen when comparing the primary energy consumption of Europe, for example, with the final energy consumption. Across all energy applications, the loss is 24%, and for power generation and mobility it is more than twice as high. Added to this are the considerable conversion losses for chemical energy conversion, which account for around 50%. Thus, up to 80% of the original renewables are lost through a transport and reuse process.

The advantage of using renewables as directly as possible in an energy system can again be demonstrated by the example of the EU 28. In Fig. 6.3, energy consumption and CO_2 emissions are compared in trend.

It is evident that the change in primary energy mix away from coal and toward renewables has enabled significant CO_2 savings without affecting consumption. However, the data from Fig. 6.2 suggest that the favorable trend of the past decade will not continue easily. Rather, significant structural efforts are needed to make the deployment of renewables more effective. Essentially, this includes exploiting the options arising from the chemical use of CO_2.

6.3 Chemical Reduction of CO_2 Is a Case of CCU (Carbon Capture and Utilization)

The above-mentioned dispute about the value of CCU as a measure in the energy transition overshadows the assessment of the development potential of chemical reduction of CO_2. This hinders the construction of practical demonstrators of relevant size, their integration into existing energy systems, and the gathering of operational as well as economic experience. Only when all this has been successfully carried out for a number of options can a reliable, overall assessment of the potentials and necessary boundary conditions be made. This conservative approach is warranted given the enormity of the deployment of CCU technologies if they are to act as climate-relevant options.

To help clarify the issues, it seems useful [19] to conceptualize the role of CCU and the resulting consumption of renewables. For this purpose, a model of an energy system with two processes is used, each of which provides equal amounts of energy by burning material energy sources. In Fig. 6.4, three operating options are represented by different colored connections of the elements.

In conventional operation (black), both processes use fossil carriers and emit equal amounts of CO_2. In linear CCU operation (blue), the emissions from process 1 are collected and processed into an energy carrier for process 2 using hydrogen

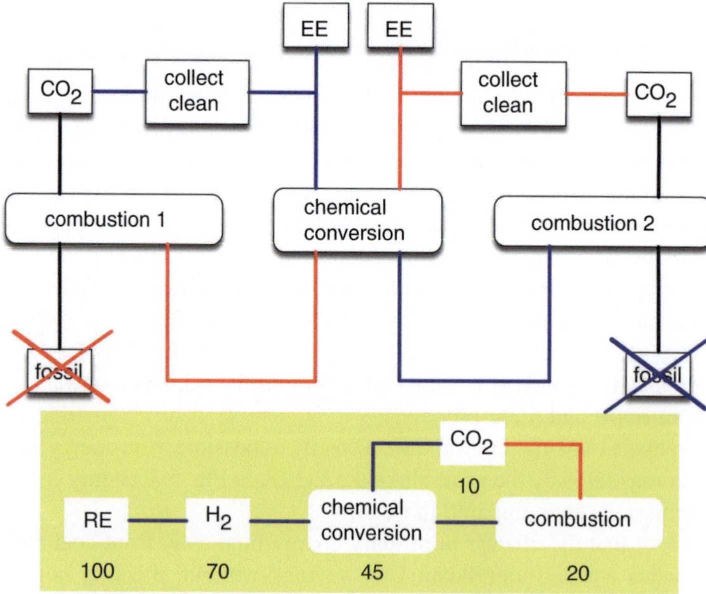

Fig. 6.4 A very simple model system for energy supply. The different colored connections refer to different operating options discussed in the text. The yellow highlighted box gives an estimate of the process efficiencies. These are rough approximations, in [20] and [21] one can find considerable variations of the figures, which are significantly caused by the assumptions of the considered process boundaries

obtained from renewable energies in a chemical conversion process. This therefore does not require fossil fuels. However, the CO_2 from process 1 is then released in process 2 and not saved. In relation to the entire system, 50% of the occurring CO_2 is thus saved in linear operation, but only if only renewable energies are used for all sub-steps of the collection, purification and processing of the CO_2 and not, for example, the current electricity mix in the EU with approximately 50% fossil sources. In the cyclic CCU process (red), the CO_2 from combustion 2 is also collected and processed, and the energy carrier obtained is made available to process 1. Then the use of fossil energy sources is unnecessary. The storage and transport options of chemical conversion eliminate the volatility of renewables and mitigate the need to obtain renewables locally in processes 1 and 2. However, and here is the problem with chemical reduction of CO_2, this requires a significant additional input of renewable energies. From the roughly estimated efficiencies of the necessary process steps given in Fig. 6.4, it can be seen that, compared to a hypothetical electrification of combustion processes 1 and 2, ten times the energy input or, in the case of linear CCU, five times the energy input must be provided for a 50% CO_2 reduction.

6.4   Some Principles of Chemical Energy Conversion

The enormous demand for renewable energy for the chemical use of CO_2 brings considerable criticism to the concept of energy conversion. There is a preference for a solution using fossil energy and the CCS (carbon capture and storage) process to achieve the targets of the Paris Agreement at a much lower financial and technical cost [22]. Initially, it remains to state (Fig. 6.4) that a process of collecting renewables and storing/transporting them and converting them back to useful energy requires about five times the energy input as direct use of renewable electricity in an area that can be accessed by power lines. This rough estimate omits transmission losses from the transmission line system.

Thus, energy systems should consist of a combination of as much as possible local extraction of renewables and their immediate use in electrical applications, and as little as possible remote extraction and storage/transport of materially converted renewables. A sense of the relationship between these two sustainable forms of energy is provided by the regional balance of renewables. This determines how many full load hours of the energy system can be met with locally extractable renewables. For Europe, with a few exceptions on the coasts, this value is about 4000 full load hours [23]. This results in the need to store or transport about 50% of the energy consumption. In the current scenarios, this is made possible by imports of gas and biomass; so far, these considerations exclude the import of renewables. It should be clearly noted here that there are significant differences over the numerical values, as the scenarios significantly downscale these values by including across-the-board efficiency gains and energy savings [24]. However, it remains to be noted that material energy sources from renewables can account for a share of 20–50% of the energy system. This gives one an order of magnitude for the importance of chemical energy conversion.

This order of magnitude implies some principles, which should be defined before the following considerations.

1. Climate relevance: Chemical energy conversion is often presented as a method for reducing CO_2. This is only partially true for the following reasons. First, a CCU process of current design (linear) saves only 50% of the converted CO_2. Secondly, the possibility of absorbing CO_2 in climate-relevant quantities [15] (well over 100 Mt. per year) in the area of basic and intermediate chemical products is not sufficiently large. Only mobility, heat and balancing power plants for electricity reach the necessary dimensions. In addition to technical arguments, the design of markets for renewable energies (a CCU process cannot be an "end user", for example, in the sense of the German Renewable Energies Act (EEG)) and the products (price erosion due to oversupply) must also be viewed critically here (see Fig. 6.1).
2. Energy losses: The large loss of renewables associated with CCU clearly dictates that priority be given to serving all direct applications in which fossil fuels are replaced by renewable electricity. This rule finds its limits where there is significant balance surplus energy (in Germany, for example, where renewables already account for close to 40% of the electricity system) and in cases where insufficient renewables can be generated locally and long-distance transmission is necessary.
3. Process energies: The conversion chain of renewables into material energy carriers requires hydrogen for the reduction of CO_2 as well as considerable amounts [21] of process energy, a small part of which ends up in the energy carriers and a larger part as losses. These process energies must not come from fossil fuels (electricity mix in the EU, for example), otherwise the intended reduction in CO_2 emissions system-wide will fall well below the 50% level.
4. Systemic relevance: There will always be industrial processes (see point 1 and basic industries such as steel, cement, glass) that emit CO_2 in significant quantities. Systemic sustainable energy concepts require integrating these sources into a carbon cycle. Again, the seemingly simpler CCS (carbon capture and storage) option is not sustainable because the carbon cycle is not closed. For these unavoidable sources of CO_2, there will be CCU processes that save 50% CO_2, at least in a transitional period, and have the potential to be expanded into cycles in the long term. In the current critical discussion, it is often overlooked that favorable synergies can already arise locally and within an industry that make CO_2 hydrogenation with subsequent upgrading steps advantageous [25, 26].

## 6.5	High-Carbon Gases

This term refers to waste gases from combustion processes, roasting gases and gases from fermentation processes that use carbon-containing feedstocks. This does not include CO_2 from ambient air, as it is too dilute at 0.04% for direct chemical processing. Depending on the process of origin, these gases are mixtures of CO_2 with water and the residual components of the process air (oxygen, nitrogen, inert gases). The air components must be removed as unwanted volumes. This is done either by separating the air components or by washing out the CO_2 from the carrier gas. Which of the technically well-developed methods [18] is used in each case is determined primarily by the relative quantity of the gas constituents.

Furthermore, the gases contain a whole series of critical constituents in trace amounts. In this context, nanoparticles from the pressed materials should be mentioned first, which interfere with chemical processing and must be carefully removed. Furthermore, CO, hydrogen and methane are found in varying quantities, which must be included in the chemical processing as reducing agents. As catalyst poisons, classes of trace substances with sulfur, nitrogen, phosphorus, halogens and organometallic components must be removed as thoroughly as possible. Limit concentrations of less than 0.0001% are required in order not to endanger catalytic processes in the subsequent steps by premature deactivation and not to prematurely wear out valves of the factories. Finally, families of hydrocarbons are included that may be harmful or tolerable depending on the process [27].

Gas composition problems vary greatly with feedstocks. The cleanest feedstock is natural gas. This is followed by distillates of petroleum, with heavy boiling fractions already carrying significant impurities of all the classes of substances mentioned. Highly complex are then processes with coals and biomass. Cement plants, steel production and waste incineration also produce complex gas matrices that require a sequence of purification steps. Therefore, thermal utilization in a power plant is often preferred for such carbon-rich gases [28, 29] and chemical reduction has been avoided so far.

It is obvious that such gas mixtures have to be treated in complex processes [29, 30] before one can think of chemical reduction. Even static loading with the above-mentioned families of pollutants poses a considerable challenge. In many real-world situations, temporal variability of loading is then added, due to the variability of feedstocks and the peculiarities of process control. This makes the design of gas purification systems particularly complex, since chemical catalysts often respond to even short-term breakthroughs of trace amounts of contaminants with irreversible deactivation, and the chemical process must be shut down. A "pathological" molecule here is carbonyl sulfide, which has a very damaging effect on metal catalysts and may not only be present as a concentration peak in the gas stream, but also forms under unfavorable conditions with the cooperation of the purification carbons (activated carbon) in the purification process, making it difficult to detect in gas samples.

Another boundary condition of gas treatment is the cost. CO_2 is often planned as a free resource. In the economic calculation of a CCU process, then, only minimal approaches to collection and purification are provided, which exclude sufficiently safe dimensioned and sophisticated gas treatment systems and therefore lead to unrealistic estimates of operating results of a CCU process.

Gas scrubbing, filtration, and adsorptive purification generate complex and often toxic residuals that must be considered significant cost factors. Reliable data and methods for the treatment of these residues in the context of CCU concepts are still lacking. In principle, such problems are known, for example, from the technology of combustion processes in power plants.

In summary, it can be stated that the recovery of chemically usable gas streams of carbon-rich gases from energetic or industrial processes is a complex challenge. One has numerous process building blocks and much experience with such processing tasks, but is currently not confident enough in the design and operation of such plant components. In any case, this aspect should not be neglected when evaluating CCU processes. General assumptions about qualities and their temporal variability should be regarded with caution, since even trace substances that occur rarely in time can result in significant and not easily diagnosable errors in subsequent chemical process chains.

6.6   Conversion Processes of Carbon Dioxide

There are numerous extensive literature reviews on the reaction possibilities of carbon dioxide. Among others, the Aresta and Leitner groups have dealt extensively with this topic, with their emphasis on the use of CO_2 as a synthesis building block in chemistry. Both small-scale syntheses of intermediate and final chemical products and large-scale syntheses of fuels are addressed. The opposite perspective is taken by review articles from the Centi group. There, possible uses of CO_2 from unavoidable sources are discussed. Further, the direct photochemical conversion of CO_2 by photoelectrocatalysis is investigated with the aim of realizing a technical carbon cycle analogous to the natural carbon cycle with an arrangement called an "artificial leaf". This approach is also being followed by major U.S. research collaborations. Despite considerable scientific and conceptual progress, these processes are still so far away from a practically relevant application that they will not be considered further in the following. This is due exclusively to the low level of technological maturity and in no way to the concept or the results achieved.

For the evaluation of such approaches, it is important to precisely define the goal of the chemical use of CO_2. In chemistry, the initial assumption is to effectively use CO_2 as a synthesis building block, thus opening up synthesis pathways that are shorter or can be pursued with less waste or less effort than current methods to the same target products. This results in clear evaluation criteria [31], but they do not take into account possible consequences or benefits outside chemistry. Conversely, one can ask what climate effect the use of (unavoidable) CO_2 might have in an integrated energy system [19, 32]. Then the first question is the volume of use. This question is controversial because it has a different dimension on a decentralized and local scale than on a global scale considering the goals of the Paris Climate Agreement. The author strongly argues that such assessments should be avoided at a time when almost all of these discussions are hypothetical. We need all processes to make CO_2 usable as a feedstock. The sole assessment should first be whether this will cause a systemic increase in the production of CO_2 or other waste materials that are not recycled. This assessment must therefore include how the processes are to be

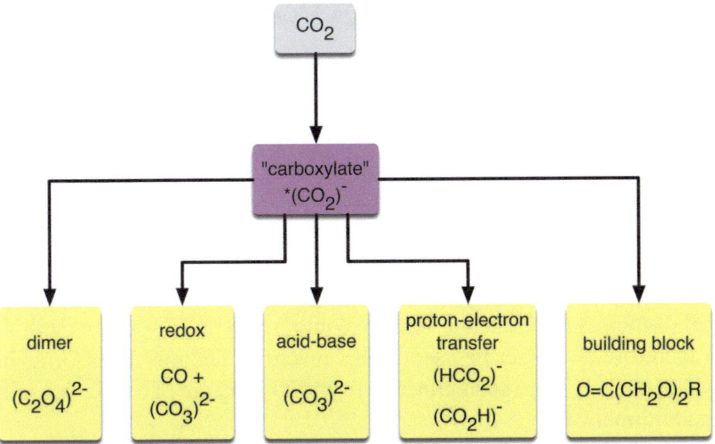

Fig. 6.5 Possibilities of chemical activation of CO$_2$

managed and supplied with energy, and where the necessary reactants are to come from. If these questions can be answered in such a way that no deterioration of the overall systemic emissions is associated with the process, the research should first be brought to a result state that experimental data on the effect of this process in an energy system are available. This will require demonstrators at relevant size scales (thousands of tons per year). Only then may one discuss the fit of the appropriate process into an energy system on a local or global scale and the possible competitive situations for the amounts of renewable energy to be invested in a CO$_2$ utilization process.

From the chemical point of view, there remain five basic approaches for the utilization of CO$_2$, which are shown in Fig. 6.5.

First, the CO$_2$ molecule must be deflected from its stable molecular constitution. This is usually done via the highly reactive [33] "carboxylate". This is an angled particle found on catalyst surfaces (including electrocatalysts).

This reaction, like all processes involving ion formation, is favored by solvation—which is the stabilization of the carboxylate with other neutral CO$_2$ molecules—or by donor molecules, such as water in electrochemistry. Solvation with CO$_2$ results in two reactions between carboxylate and CO$_2$, which are called "dimerization" and "redox" in Fig. 6.5. Once, the stable oxalate anion is formed, which decomposes at elevated temperatures or catalytically into the products of the redox process CO and carbonate.

Three other types of reactions require another molecule in addition to CO$_2$. By far the most important reaction when it comes to quantities is the reaction of the weak acid H2CO3 (carbonic acid), which results from reaction with water, with strong bases to form carbonates.

$$CO_2 + M(OH)_2 \rightleftharpoons MCO_3 + H_2O \tag{6.7}$$

The mountain-forming rocks MgCO3 and CaCO3 may serve as illustration for the fact, already discussed with Table 6.1, that CO_2 is by far not a reaction-bearing molecule. Even at room temperature, carbonate formations occur as central reactions in construction chemistry during "setting." Hence the idea of binding large amounts of CO_2 by "mineralization" [34–36] and simultaneously putting it to a use that utilizes the entire molecule rather than the carbon atom in CO_2 alone. The challenge with this concept is to find the strong bases in the form of oxides that do not require energetically expensive pretreatments to become reactive against CO_2. Naturally, almost all of these substances are already combined with CO_2 at the earth's surface and can therefore no longer be used in this reaction. The processes "lime burning" and "cement burning" free such minerals from CO_2, but are of course not suitable for binding additional CO_2.

A technically particularly important reaction from this class is the formation of urea from CO_2 and ammonia, which is carried out on the scale of 125 Mt. per year.

$$CO_2 + 2NH_3 \rightleftharpoons CO(NH_2)_2 + H_2O \tag{6.8}$$

Further, there is a wide range of reactions in which CO_2 reacts as a synthesis building block with highly reactive molecules such as ethylene oxide to produce a variety of molecular products of chemistry [37]. Prominent among these, due to the maturity of the application, is the synthesis of functional polymers of the class of polyurethanes. The general possibility is the reaction of the weak acid CO_2 with bases other than metal hydroxides. This reaction type, prototypically illustrated by urea synthesis (ammonia as a strong base), leads to many organic molecules.

The last and here most important reaction type is the reaction of the carboxylate with activated hydrogen, which is conceived as a proton-coupled electron transfer (PCET) reaction. This process leads to two intermediates containing hydrogen bound either to the carbon or to the oxygen. This results in either water-gas equilibrium to produce synthesis gas with the CO molecule as the actual carbon source, or methanol synthesis. This equilibrium thus combines the CO, CO_2 and methanol molecules. This is made possible by the structures of the common molecular intermediates shown in Fig. 6.6 inside the "carbon triangle".

The reaction network of CO_2 with hydrogen knows the two central intermediates methanol and CO, which can be converted into further important chemical products. In Fig. 6.7, the important reaction chains are shown.

The simplest storage molecule for energy results from the reaction of carboxylate with hydrogen to formic acid. This reaction is easily accomplished catalytically [38], as is the splitting back [39, 40] into CO_2 and hydrogen. In particular, the synthesis of formic acid is readily accomplished by homogeneous catalysis and also does not require noble metal catalysts. Possibly, the low gravimetric storage density and the relatively high reactivity of formic acid are reasons for the little attention this reaction has received in energy chemistry so far.

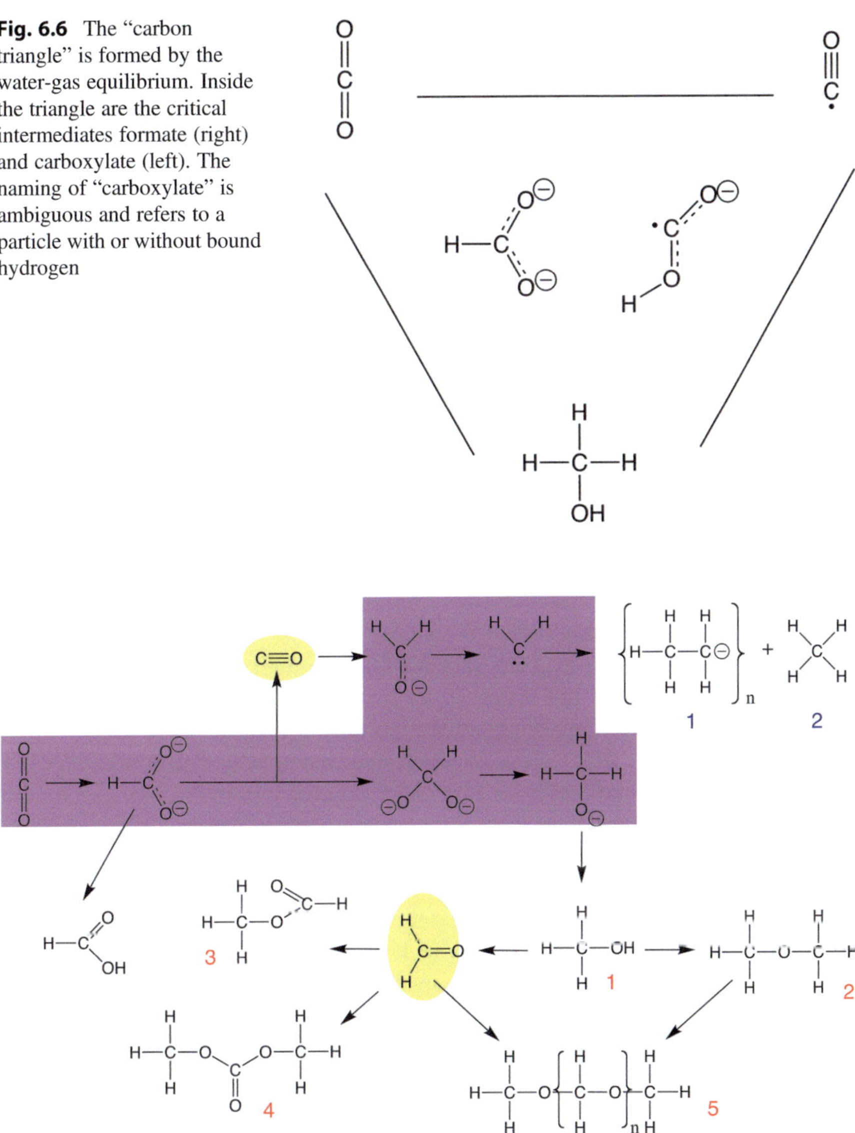

Fig. 6.6 The "carbon triangle" is formed by the water-gas equilibrium. Inside the triangle are the critical intermediates formate (right) and carboxylate (left). The naming of "carboxylate" is ambiguous and refers to a particle with or without bound hydrogen

Fig. 6.7 Section of the reaction network for the reaction of CO_2 with hydrogen. Reactive intermediates are shown in the part highlighted in lavender. The molecules CO and formaldehyde highlighted in yellow are chemical intermediates. The other products can be used for energy storage and as "solar fuels" in addition to applications in chemistry. The following can be used as fuels: (blue) 1 Fischer-Tropsch alkanes, 2 Sabatier methane; (red) 1 methanol, 2 dimethyl ether (DME), 3 methyl formate (MF), 4 dimethyl carbonate (DMC), 5 oxymethylene ether (OME)

The intermediate CO is followed by the products of the Fischer-Tropsch reaction class [41, 42]. This reaction class refers to a series of condensation reactions in which, with formation of water, the CO is first hydrogenated to CH2 fragments that are statistically likely to assemble into longer alkane chains [43], forming a mixture of products. Depending on the choice of catalyst [44, 45], this produces mainly alkanes (Co), olefins (Fe), or even higher alcohols if some undecomposed CO molecules are incorporated (Cu plus promoters). Unfortunately, the control of the reaction by catalysis is far from perfection [42, 44, 46]: complex mixtures are always formed, which require complex separations. Another problem is the difficult-to-suppress full hydrogenation of the -CH2 fragment to methane [47], which is an undesirable product here.

Direct generation of methane [48] under the term power to gas (p2g) is a technically well feasible but controversial option [49] for coupling renewable energy into the thermal subsystem of a power system. The application as "autogas" is also technically fully developed and well available, but is currently not well accepted in the market for non-technical reasons (at least in Europe, where there is a very good distribution system for gas). Chemically, the "Sabatier reaction" is a well studied reaction for a long time [47, 50, 51]. It runs very well over noble metal catalysts and offers no insurmountable difficulties [52], revolving around issues of reaction completeness to avoid subsequent product purification. The avoidance of noble metals by using Ni as a catalyst is more difficult. There are problems of selectivity (only partial reaction to CO) and stability (formation of carbon and destruction of the catalyst). Given the currently very low commercial value of methane (the hydrogen needed to produce it is much more valuable than the market price of the product), stable methanation that is not disturbed by impurities from CO_2 (from biological processes or flue gas) would be valuable especially if it could serve decentralized processing and storage of local excess renewable energy [49]. The conceivable use of methane as a universal feedstock for further chemistry from the consideration of Fig. 6.7 is not feasible because there is no way to activate methane except to burn it. Despite extensive studies on this, there is currently little hope [53] of developing a new methodology here.

The Fischer-Tropsch reaction is the only reaction pathway that has made it to technological maturity in energy supply and that already has technical applications today on a world scale [54–56]. In this special variant for the production of high-quality gasoline, the Fischer-Tropsch reaction is used to produce a mixture of the longest possible chain alkanes (wax), which is then partially cracked again and isomerized (linear to branched chains) using the established methods of refinery chemistry. In principle, it is conceivable to also use biomass [57] and "green hydrogen" as feedstocks, which would provide a renewable fuel for mobility (for example, for aircraft). At present, it is not so much technical-chemical as commercial issues and business models that are hindering such a development.

Nonetheless, a wide variety of scientific questions exist in this complex of issues, ranging from definitive analysis of the means of controlling the condensation reaction, to clever production of higher alcohols as desirable fuels [58], to combining the Fischer-Tropsch reaction with other functionalization reactions to produce molecularly unitary fuels from the oxygenate family. The issue of methane production would be largely relaxed in a hypothetical integrated plant producing fuel for heat generation in addition to fuel. Again, issues of economics, operating models, and business models are more likely to be obstacles than the chemical technology, which would also be sufficient for a first generation of such solutions today. The reader notes the caveat that such concepts require the availability of economically favorable hydrogen with a cost threshold (not price!) of below 3000 €/t, which is realistic in some areas of the world today, but not in Central Europe.

From Fig. 6.7 it can be seen that a large number of relevant molecules can be obtained from methanol, which is accessible from CO$_2$ via formate (formal!) and methoxy, and a subsequent reaction. Methanol can therefore be described as a platform molecule that makes a number of products accessible. In Fig. 6.7, only the small class of molecules that do not contain C–C bonds is considered. Numerous more complex molecules with applications in polymer chemistry or, again, as fuels, are readily accessible via the downstream technical processes MTO (methanol to olefin) and MTG (methanol to gasoline), each involving acid-catalyzed dehydrations followed by condensation steps. Therefore, the synthesis of methanol as the first chemical platform product of the hydrogenation of CO$_2$ appears to be a very attractive option for coupling renewable energy into a variety of applications using a single chemical building block.

The synthesis of methanol, like the Fischer-Tropsch reaction, is also a world-scale reaction with one of the largest growth rates in its application (source: IGP Energy. com). This is traditionally justified by the platform character of methanol in chemistry (which still largely sustains today's development) and, more recently, by the possibility of using large coal reserves on the one hand and contributing to the use of CO$_2$ produced elsewhere in large quantities on the other, via the "methanol economy" [59], which was preconceived in Germany [60] but is now being implemented in China [61]. Currently, about 65 Mt. of methanol is produced per year in the world; this amount will almost double in the next decade, considering only projects announced or under construction. This projection does not yet include projects dedicated to the utilization of CO$_2$. The application market for methanol is currently expanding dramatically due to the possibility of using methanol as a fuel [62, 63] or as an intermediate compound for the production of fuels. The chemical applications of methanol alone could not support such momentum or even the prospect of using methanol to achieve climate-relevant uses of CO$_2$.

In Fig. 6.7, the class of C1 oxygenates that can be used as fuels is indicated by the red numbers. These molecules burn in diesel engines (DME, OME) or gasoline engines, which require minor engineering and software modifications [64, 65] in their fuel delivery systems. Estimates of manufacturing costs [62, 66] are very dependent on the electricity prices for the hydrogen; it is believed that today's fuel prices are achievable at about a hydrogen cost of 2000 €/t. For research purposes, prices of 5 €/kg are discussed.

The use of oxygenates in general as fuel, besides the possibilities of sector coupling and the use of CO$_2$, holds an option that arises from the single-variety molecular structure of these molecules. No local emissions [58, 67–69] of carbon particles and NOx occur; in combination with a simple exhaust aftertreatment system against organic intermediates, a "zero emission" [70, 71] of regulated pollutants can thus be achieved even with internal combustion engines. This proven effect is an additional benefit whose value lies in the continued use of existing technologies for mobility and its supply. Thus, there is an argument to provide regulatory support for the introduction of synthetic fuels that exhibit this property, since both CO_2 savings and pollutant emission reductions would result quickly (within 10 years) at minimal cost compared to alternative solutions.

Regarding methanol synthesis, it should be noted that there is currently a discussion in the literature [72, 73] about the necessary catalysts. All technical methanol processes use a mixture of CO and CO_2 as the carbon source, which is chosen, if possible, so that the product water resulting from the hydrogenation of CO_2 (the molecular source of all methanol [74]) is converted by the water-gas equilibrium and its processes in the catalyst, and thus a very low-water methanol and, if possible, no CO gas appears in the product mixture of the reactor. If pure CO_2 is used as the synthesis gas, both water and CO will appear in the reactor output stream. This is thought to affect the functionality of current catalysts [73], citing that the measured productivity of methanol in pure CO_2 is much lower than in a syngas mixture containing CO and CO_2. It is further shown that the CO content can be significantly reduced when Pd/Zn [75] is used instead of copper as the active material. Such observations follow from the different reaction conditions applied in each case and an insufficient test for the dependence of the observations on the reaction conditions and the location of the measured reaction rates with respect to the equilibria from Fig. 6.6. It can well be shown [76, 77] that technical catalysts are suitable for methanol synthesis with pure CO_2 and that there is no deterioration in their performance compared with a CO-containing synthesis gas. What is new, however, is the finite selectivity to CO, which cannot be suppressed at high production rates of methanol. It is not clear at present how much the settings of the equilibria from Fig. 6.6 can be kinetically separated. This may succeed [78] if there are materially independent reaction centers for methanol synthesis and water gas synthesis that differ in terms of their binding effect of intermediates.

As a summary, important characteristics of the major processes for the utilization of CO_2-rich gases are compared again. Table 6.2 summarizes some characteristics. The details have already been discussed in the text.

From a chemical point of view, all processes can be realized on a "climate-relevant" scale of over 100 Mt. per year. Their biggest disadvantage is the hydrogen requirement. Part of this is stored in the products as a "chemical battery", and another part is lost as reaction water. All processes except urea synthesis still require considerable research and development in order to achieve optimum plant efficiency in later generations. The optimization goals here are free choice of plant size, catalyst service life, simplification of the process sequence and, ultimately, specific costs. However, these requirements in no way prevent the benefits of these processes for the sector coupling of energy systems and for faster defossilization from being exploited in a first generation. One can think of large, optimally economical centralized units (at large point sources, for example) that conventional experience suggests can be operational within 10 years. But one can also think of small decentralized units located at renewable electricity generation sites to meet local demands or to store and transport "tankable" renewable energy. For this, the

Table 6.2 Some properties of large processes of CO_2 utilization

Property	Methane (Sabatier)	Hydrocarbons (FT)	Methanol	Urea
Carbon source	CO	CO	CO_2	CO_2
Hydrogen demand	4	2n + 1	3	3 (in NH_3)
Water[a]	2	n	1	1
Stored molecules hydrogen	2	n + 1	2	2
Catayst	Noble metal, Ni	Co, Fe	Cu	–
Platform	No	Conditional	Yes	No
Infrastructure	Yes	Yes	Conditional	Yes
Challenges	Soot formation	Product selectivity	CO Co-product	none
Complexity	Low	Very high	Low	Low
Realized on a large scale	No (for gas purification only)	Yes	Yes	Yes

[a]Molecules per product molecule

economic hurdles are lower and the time scales shorter, but there is a lack of much design and operational experience with decentralized chemical plants, which will have to be operated differently than large plants. The author strongly advocates the simultaneous pursuit of both approaches, as they both have their systemic justification and both can significantly support the transformation of energy systems.

6.7   Concluding Remarks

Chemical utilization of carbon-rich gases for renewable energy storage and energy sector coupling is an indispensable element of future energy strategies. In this context, chemistry offers the possibility of converting the primary chemical energy carrier in such a way that it can be easily stored, transported and used in existing technology systems. The "price" to be paid for this of low process efficiency of the chemical energy carriers is offset by their system efficiency. Advances in fundamental and applied understanding of chemical reaction control can significantly reduce the disadvantage of low process efficiency and contribute substantially to exceeding economically important cost thresholds. Here, an organization of research efforts based on the division of labor would be highly desirable, at least in the process-related part of the work. The prerequisite and systemic, central challenge remains the provision of hydrogen at systemically justifiable costs.

The introduction of the numerous preconceived and laboratory-tested conversion processes does not fail because of the chemistry or process technology, although there is still considerable potential for optimization here for possible later generations of processes. Entry is currently failing due to the lack of business models and/or regulatory framework conditions. In particular, there is a lack of an efficient industry to drive the introduction of the technologies and the marketing of the products. This, in turn, is likely due to the unfavorable framework conditions in Europe (high energy prices, complex regulation), which can be mitigated or eliminated by government intervention (treating renewables for chemical processes as feedstock rather than energy). Entry into government-organized valuation of CO_2 (emissions trading scheme ETS 2.0 as an example) would be another possible way to facilitate chemical use of CO_2. In the economic regions of the Middle and Far East, which operate under different political frameworks, the emergence of corresponding structures can already be observed today, and clear plans exist for the large-scale use of hydrogenation of carbon-rich gas streams.

The biggest single obstacle to the further development of chemical energy storage in carbonaceous molecules is the controversial discussions [14] on the subject. They are characterized by inaccurate expectations, lack of knowledge about technical and scientific developments, and a less than creative approach to the non-technical design possibilities of the environment surrounding these technologies. As a result, uncertainty and inaction arise, preventing economic advantage through technological maturity. Moreover, these uncertainties make it difficult to finance appropriate projects. A climate of hesitation is created, which others exploit in order to become not only suppliers of appropriate technology, but also to supply us with the products manufactured with it, thus further increasing energy dependence, at least in Europe. A breakthrough in the debate on carbon-based energy storage is seen as the single most important building block for the development of this field.

References

1. Aresta M, Dibenedetto A (2007) Utilisation of CO2 as a chemical feedstock: opportunities and challenges. Dalton Trans 28:2975–2992
2. Leitner W (1996) The coordination chemistry of carbon dioxide and its relevance for catalysis: a critical survey. Coord Chem Rev 153:257–284
3. Zhang XJ, Bauer C, Mutel CL, Volkart K (2017) Life cycle assessment of power-to-gas: approaches, system variations and their environmental implications. Appl Energ 190:326–338
4. Naraharisetti PK, Yeo TY, Bu J (2017) Factors influencing CO2 and energy penalties of CO2 mineralization processes. ChemPhysChem 18(22):3189–3202
5. Wilson G, Trusler M, Yao J, Lee JSM, Graham R, Mac Dowell N, Cuellar-Franca R, Dowson G, Fennell P, Styring P, Gibbins J, Mazzotti M, Brandani S, Muller C, Hubble R (2016) End use and disposal of CO2 – storage or utilisation? General discussion. Faraday Discuss 192:561–579
6. Smit B (2016) Carbon capture and storage: introductory lecture. Faraday Discuss 192:9–25
7. Perez-Fortes M, Schoneberger JC, Boulamanti A, Tzimas E (2016) Methanol synthesis using captured CO2 as raw material: techno-economic and environmental assessment. Appl Energ 161:718–732
8. Jadhav SG, Vaidya PD, Bhanage BM, Joshi JB (2014) Catalytic carbon dioxide hydrogenation to methanol: a review of recent studies. Chem Eng Res Des 92(11):2557–2567
9. Seh ZW, Kibsgaard J, Dickens CF, Chorkendorff IB, Nørskov JK, Jaramillo TF (2017) Combining theory and experiment in electrocatalysis: insights into materials design. Science 355(6321):146

10. Bruhn T, Naims H, Olfe-Krautlein B (2016) Separating the debate on CO2 utilisation from carbon capture and storage. Environ Sci Pol 60:38–43
11. Markewitz P, Kuckshinrichs W, Leitner W, Linssen J, Zapp P, Bongartz R, Schreiber A, Mueller TE (2012) Worldwide innovations in the development of carbon capture technologies and the utilization of CO2. Energy Environ Sci 5(6):7281–7305
12. Lee JSM, Rochelle G, Styring P, Fennell P, Wilson G, Trusler M, Clough P, Blamey J, Dunstan M, MacDowell N, Lyth S, Yao J, Hills T, Gazzani M, Brandl P, Anantharaman R, Brandani S, Stolaroff J, Mazzotti M, Maitland G, Muller C, Dowson G, Gibbins J, Ocone R, Campbell KS, Erans M, Zheng LY, Sutter D, Armutlulu A, Smit B (2016) CCS – a technology for now: general discussion. Faraday Discuss 192:125–151
13. Aresta M (2017) My journey in the CO2-chemistry wonderland. Coord Chem Rev 334:150–183
14. Mac Dowell N, Fennell PS, Shah N, Maitland GC (2017) The role of CO2 capture and utilization in mitigating climate change. Nat Clim Chang 7(4):243–249
15. Mikkelsen M, Jorgensen M, Krebs FC (2010) The teraton challenge. A review of fixation and transformation of carbon dioxide. Energ Environ Sci 3(1):43–81
16. Rockstrom J, Gaffney O, Rogelj J, Meinshausen M, Nakicenovic N, Schellnhuber HJ (2017) A roadmap for rapid decarbonization. Science 355(6331):1269–1271
17. Artz J, Muller TE, Thenert K, Kleinekorte J, Meys R, Sternberg A, Bardow A, Leitner W (2018) Sustainable conversion of carbon dioxide: an integrated review of catalysis and life cycle assessment. Chem Rev 118(2):434–504
18. Koytsoumpa EI, Bergins C, Kakaras E (2018) The CO2 economy: review of CO2 capture and reuse technologies. J Supercrit Fluid 132:3–16
19. Abanades JC, Rubin ES, Mazzotti M, Herzog HJ (2017) On the climate change mitigation potential of CO2 conversion to fuels. Energy Environ Sci 10(12):2491–2499
20. Valente A, Iribarren D, Dufour J (2017) Life cycle assessment of hydrogen energy systems: a review of methodological choices. Int J Life Cycle Assess 22(3):346–363
21. Cuellar-Franca RM, Azapagic A (2015) Carbon capture, storage and utilisation technologies: a critical analysis and comparison of their life cycle environmental impacts. J CO2 Util 9:82–102
22. Haszeldine RS (2009) Carbon capture and storage: how green can black be? Science 325(5948): 1647–1652
23. Philibert C (2017) Renewable energy for industry. International Energy Agency, Paris, S 65
24. Palzer A, Henning HM (2014) A comprehensive model for the German electricity and heat sector in a future energy system with a dominant contribution from renewable energy technologies – part II: results. Renew Sust Energ Rev 30:1019–1034
25. Navarrete A, Centi G, Bogaerts A, Martin A, York A, Stefanidis GD (2017) Harvesting renewable energy for carbon dioxide catalysis. Energ Technol 5(6):796–811
26. Perathoner S, Gross S, Hensen EJM, Wessel H, Chraye H, Centi G (2017) Looking at the future of chemical production through the European roadmap on science and technology of catalysis the EU effort for a long-term vision. ChemCatChem 9(6):904–909
27. Bukthiyarova M, Lunkenbein T, Kähler K, Schlögl R (2017) Methanol synthesis from industrial CO2 sources: a contribution to chemical energy conversion. Catal Lett 147(2):416–427
28. Maruoka N, Akiyama T (2006) Exergy recovery from steelmaking off-gas by latent heat storage for methanol production. Energy 31(10–11):1632–1642
29. Yoon YI, Kim YE, Nam SC, Park SY, Chun IS, Lee SD, Kim HS (2017) Economic analysis of CCU in the Korean cement industry: CO2 capture using KIERSOL & PCC conversion. Energy Procedia 114:6240–6245
30. Bridgewater AV (1995) The technical and economic feasibility of biomass gasification for power generation. Fuel 74(5):631–653
31. Boot-Handford ME, Abanades JC, Anthony EJ, Blunt MJ, Brandani S, Mac Dowell N, Fernandez JR, Ferrari MC, Gross R, Hallett JP, Haszeldine RS, Heptonstall P, Lyngfelt A, Makuch Z, Mangano E, Porter RTJ, Pourkashanian M, Rochelle GT, Shah N, Yao JG, Fennell PS (2014) Carbon capture and storage update. Energy Environ Sci 7(1):130–189

32. Freund HJ, Roberts MW (1996) Surface chemistry of carbon dioxide. Surf Sci Rep 25(8): 225–273
33. Grasa G, Martinez I, Diego ME, Abanades JC (2014) Determination of CaO carbonation kinetics under recarbonation conditions. Energy Fuel 28(6):4033–4042
34. Hariharan S, Mazzotti M (2017) Kinetics of flue gas CO_2 mineralization processes using partially dehydroxylated lizardite. Chem Eng J 324:397–413
35. Prigiobbe V, Mazzotti M (2013) Precipitation of Mg-carbonates at elevated temperature and partial pressure of CO_2. Chem Eng J 223:755–763
36. Klankermayer J, Wesselbaum S, Beydoun K, Leitner W (2016) Selective catalytic synthesis using the combination of carbon dioxide and hydrogen: catalytic chess at the interface of energy and chemistry. Angew Chem Int Ed 55(26):7296–7343
37. Federsel C, Ziebart C, Jackstell R, Baumann W, Beller M (2012) Catalytic hydrogenation of carbon dioxide and bicarbonates with a well-defined cobalt dihydrogen complex. Chem Eur J 18(1):72–75
38. Zhang S, Shao Y, Yin G, Lin Y (2010) Facile synthesis of PtAu alloy nanoparticles with high activity for formic acid oxidation. J Power Sources 195(4):1103–1106
39. Rienacker G, Muller H (1968) Catalytic studies in alloys. 31. Decomposition of formic acid vapor on silver-palladium alloys. Z Anorg Allg Chem 357(4–6):255
40. Göthel H (1996) Fischer-Tropsch-Synthese und verwandte Verfahren. In: Zerbe C (Ed.) Mineralöle und verwandte Produkte. Springer, Berlin/Heidelberg, S 572–580
41. Van der Laan GP, Beenackers A (1999) Kinetics and selectivity of the Fischer-Tropsch synthesis: a literature review. Catal Rev Sci Eng 41(3–4):255–318
42. Anderson RB, Hofer LJE, Storch HH (1958) Der Reaktionsmechanismus der Fischer-Tropsch-Synthese. Chem Ing Tech 9:560–566
43. Iglesia E, Reyes SC, Madon RJ, Soled SL (1993) Selectivity control and catalyst design in the Fischer-Tropsch synthesis – sites, pellets, and reactors. Adv Catal 39:221–302
44. Biloen P, Sachtler WMH (1981) Mechanism of hydrocarbon synthesis over Fischer-Tropsch catalysts. Adv Catal 30:165–216
45. Dry ME (2002) The Fischer-Tropsch process: 1950–2000. Catal Today 71(3–4):227–241
46. Ponec V (1978) Some aspects of mechanism of methanation and Fischer-Tropsch synthesis. Catal Rev Si Eng 18(1):151–171
47. Lorenzi G, Lanzini A, Santarelli M, Martin A (2017) Exergo-economic analysis of a direct biogas upgrading process to synthetic natural gas via integrated high-temperature electrolysis and methanation. Energy 141:1524–1537
48. Ronsch S, Schneider J, Matthischke S, Schluter M, Gotz M, Lefebvre J, Prabhakaran P, Bajohr S (2016) Review on methanation – from fundamentals to current projects. Fuel 166:276–296
49. Ichikawa S (1989) Reactive chemisorption and methanation of carbon-dioxide on rhodium particles approaching atomic dispersion. J Mol Catal 53(1):53–65
50. Lunde PJ, Kester FL (1974) Carbon-dioxide methanation on a ruthemium catalyst. Ind Eng Chem Process Des Dev 13(1):27–33
51. Ronsch S, Kochermann J, Schneider J, Matthischke S (2016) Global reaction kinetics of CO and CO_2 methanation for dynamic process modeling. Chem Eng Technol 39(2):208–218
52. Schwarz H (2011) Chemistry with methane: concepts rather than recipes. Angew Chem Int Ed 50(43):10096–10115
53. Khodakov AY, Chu W, Fongarland P (2007) Advances in the development of novel cobalt Fischer-Tropsch catalysts for synthesis of long-chain hydrocarbons and clean fuels. Chem Rev 107(5):1692–1744
54. Roh HS, Koo KY, Joshi UD, Yoon WL (2008) Combined H_2O and CO_2 reforming of methane over Ni-Ce-ZrO_2 catalysts for gas to liquids (GTL). Catal Lett 125(3–4):283–288
55. Hao X, Djatmiko ME, Xu YY, Wang YN, Chang J, Li YW (2008) Simulation analysis of a GTL process using ASPEN plus. Chem Eng Technol 31(2):188–196
56. Ail SS, Dasappa S (2016) Biomass to liquid transportation fuel via Fischer Tropsch synthesis – technology review and current scenario. Renew Sust Energ Rev 58:267–286

57. Leitner W, Klankermayer J, Pischinger S, Pitsch H, Kohse-Hoinghaus K (2017) Advanced biofuels and beyond: chemistry solutions for propulsion and production. Angew Chem Int Ed 56(20):5412–5452
58. Olah GA (2005) Beyond oil and gas: the methanol economy. Angew Chem Int Ed 44(18): 2636–2639
59. Asinger F (1986) Methanol, Chemie- und Energierohstoff. Springer, Berlin, S 407
60. Xu XY, Liu Y, Zhang F, Di W, Zhang YL (2007) Clean coal technologies in China based on methanol platform. Catal Today 298:61–68
61. Schmitz N, Burger J, Strofer E, Hasse H (2016) From methanol to the oxygenated diesel fuel poly(oxymethylene) dimethyl ether: an assessment of the production costs. Fuel 185:67–72
62. Zhen XD, Wang Y (2015) An overview of methanol as an internal combustion engine fuel. Renew Sust Energ Rev 52:477–493
63. Maus W, Jacob E (2015) Future-safe combustion-engined drives – the role of sustainable fuels. International Emgione Congress, Baden, pp 283–284
64. Hartl M, Seidenspinner P, Jacob E, Wachtmeister G (2015) Oxygenate screening on a heavy-duty diesel engine and emission characteristics of highly oxygenated oxymethylene ether fuel OME1. Fuel 153:328–335
65. Lautenschutz L, Oestreich D, Haltenort P, Arnold U, Dinjus E, Sauer J (2017) Efficient synthesis of oxymethylene dimethyl ethers (OME) from dimethoxymethane and trioxane over zeolites. Fuel Process Technol 165:27–33
66. Deutsch D, Oestreich D, Lautenschutz L, Haltenort P, Arnold U, Sauer J (2017) High purity oligomeric oxymethylene ethers as diesel fuels. Chem Ing Tech 89(4):486–489
67. Haertl M, Seidenspinner P, Jacob E, Wachtmeister G (2015) Oxygenate screening on a heavy-duty diesel engine and emission characteristics of highly oxygenated oxymethylene ether fuel OME1. Fuel 153:328–335
68. Deutz S, Bongartz D, Heuser B, Katelhon A, Langenhorst LS, Omari A, Walters M, Klankermayer J, Leitner W, Mitsos A, Pischinger S, Bardow A (2018) Cleaner production of cleaner fuels: wind-to-wheel – environmental assessment of CO2-based oxymethylene ether as a drop-in fuel. Energy Environ Sci 11(2):331–343
69. Peter A, Fehr SM, Dybbert V, Himmel D, Lindner I, Jacob E, Ouda M, Schaadt A, White RJ, Scherer H, Krossing I (2018) Towards a sustainable synthesis of oxymethylene dimethyl ether by homogeneous catalysis and uptake of molecular formaldehyde. Angew Chem Int Ed 57(30): 9461–9464
70. Omari A, Heuser B, Pischinger S (2017) Potential of oxymethylenether-diesel blends for ultra-low emission engines. Fuel 209:232–237
71. Tursunov O, Kustov L, Kustov A (2017) A brief review of carbon dioxide hydrogenation to methanol over copper and iron based catalysts. Oil Gas Sci Technol 72(5):30
72. Frei MS, Capdevila-Cortada M, Garcia-Muelas R, Mondelli C, Lopez N, Stewart JA, Ferre DC, Perez-Ramirez J (2018) Mechanism and microkinetics of methanol synthesis via CO2 hydrogenation on indium oxide. J Catal 361:313–321
73. Studt F, Behrens M, Kunkes EL, Thomas N, Zander S, Tarasov A, Schumann J, Frei E, Varley JB, Abild-Pedersen F, Nørskov JK, Schlögl R (2015) The mechanism of CO and CO2 hydrogenation to methanol over Cu-based catalysts. ChemCatChem 7(7):1105–1111
74. Bahruji H, Esquius JR, Bowker M, Hutchings G, Armstrong RD, Jones W (2018) Solvent free synthesis of PdZn/TiO2 catalysts for the hydrogenation of CO2 to methanol. Top Catal 61(3–4): 144–153
75. Zurbel A, Kraft M, Kavurucu-Schubert S, Bertau M (2018) Methanol synthesis by CO2 hydrogenation over Cu/ZnO/Al2O3 catalysts under fluctuating conditions. Chem Ing Tech 90(5):721–724

76. Bukhtiyarova M, Lunkenbein T, Kähler K, Schlögl R (2017) Methanol synthesis from industrial CO_2 sources: a contribution to chemical energy conversion. Catal Lett 147(2):416–427
77. Kunkes EL, Studt F, Abild-Pedersen F, Schlögl R, Behrens M (2015) Hydrogenation of CO_2 to methanol and CO on $Cu/ZnO/Al_2O_3$: Is there a common intermediate or not? J Catal 328:43–48
78. Li M, Rao AD, Brouwer J, Samuelsen GS (2010) Design of highly efficient coal-based integrated gasification fuel cell power plants. J Power Sources 195(17):5707–5718

Microbial Processes: Biocatalytic Conversion

7

Peter Dürre and Frank R. Bengelsdorf

Abstract

So-called C1 gases are utilized by acetogenic bacteria, aerobic methanotrophic bacteria, cyanobacteria, and photoautotrophic microalgae. The C1 metabolic pathways of all the aforementioned groups are understood and thus, in principle, amenable to genetic optimization for industrial application. For example, acetogenic bacteria for the utilization of carbon monoxide to ethanol are used on an industrial production scale since 2018. The stage of development for the production of biomass with a high nutritional value by anaerobic methanotrophic bacteria based on methane has also passed the demonstration stage, and a commercial plant is under construction. Already established is the use of photoautotrophic microalgae and cyanobacteria for the production of vitamins, antioxidants, pigments, and special fatty acids.

Keywords

Acetogenic bacteria · Aerobic methanotrophic bacteria · Cyanobacteria · Photoautotrophic microalgae

7.1 C1 Gases as Microbiological Carbon Sources

A variety of compounds containing only one carbon atom are gaseous at atmospheric pressures of 1015 hPa and temperatures below 100 °C. These include carbon dioxide (CO_2), carbon monoxide (CO), methane (CH_4), and halogenated derivatives of methane. Most of these compounds are highly toxic to humans and animals even at very low concentrations. Microorganisms can fix these gaseous C1 compounds by

P. Dürre (✉) · F. R. Bengelsdorf
Institute for Molecular Biology and Biotechnology of Prokaryotes, Ulm University, Ulm, Germany
e-mail: peter.duerre@uni-ulm.de; frank.bengelsdorf@uni-ulm.de

different metabolic pathways and use them as a carbon source for their growth. The focus of this chapter is on microorganisms that are able to fix CO_2, CO, and CH_4. The ability of living organisms to produce organic carbon compounds only from inorganic C1 compounds is called autotrophy and the organisms are called autotrophs. Carbon dioxide is assimilated by autotrophic organisms in a variety of ways. Plants and cyanobacteria use the Calvin(-Benson-Bassham, CBB) cycle for this purpose. Some anaerobic bacteria, called acetogens, fix carbon monoxide or carbon dioxide using hydrogen via the Wood-Ljungdahl (reductive acetyl-CoA) pathway. Other pathways of carbon assimilation include the reductive citrate cycle, the 3-hydroxypropionate cycle, the 3-hydroxypropionate/4-hydroxybutyrate cycle, and the dicarboxylate/4-hydroxybutyrate cycle. Methane is used as a carbon source by methanotrophic bacteria. These bacteria require oxygen to activate methane. Cell substance is then formed via the serine pathway in some species and the ribulose-monophosphate cycle in others. The fundamentals of selected C1 metabolic pathways are described in Sect. 7.3.

The combination of CO_2 assimilation via the CBB cycle and light as an energy source is realized in many photoautotrophic organisms. These include plants and some phototrophic bacteria such as cyanobacteria and Rhodospirillaceae. Other phototrophic bacteria use the other CO_2 fixation pathways mentioned above. The necessary energy, in the form of adenosine triphosphate (ATP), and nicotinic acid amide adenine dinucleotide phosphate (NADPH) to reduce CO_2 are provided by the reactions of the photosynthetic apparatus (localized in the cytoplasmic membrane in bacteria and in the chloroplasts of the eukaryotic organisms). In this process, three ATP and two NADPH are numerically required per fixed molecule of CO_2.

The Wood-Ljungdahl pathway is the oldest pathway for CO_2 fixation; it is used by anaerobic microorganisms. The energy for CO_2 reduction is provided by oxidation of hydrogen (on the one hand, formation of an ion gradient using reduced ferredoxin and membrane protein complexes to synthesize ATP, on the other hand, provision of reduced ferredoxin and NADH by hydrogenase). Computationally, 0.5 reduced ferredoxin, 0.5 ATP, and 2 NADH are required to reduce a CO_2 molecule to the stage of acetyl-CoA. The reduction of CO by some acetogens is also possible by this route; in fact, the bacteria may even forgo hydrogen entirely.

Although anaerobic methane oxidation also occurs, the use of this gas as a carbon source is typically associated with aerobic methanotrophs. The representatives of the so-called type I include, for example, the genera Methylomonas, Methylobacter, Methylococcus, and Methylomicrobium, and assimilate carbon (as formaldehyde) via the ribulose monophosphate cycle. In contrast, type II organisms (e.g., Methylocystis and Methylosinus) use the serine pathway for C1 fixation. Reduction equivalents are derived from the oxidation of methane, and ATP is formed in large quantities via the respiratory chain.

7.2 Overview of Relevant Microorganisms

7.2.1 Acetogenic Bacteria

Anaerobic bacterial gas fermentation attracts wide interest in various scientific and industrial fields. This microbial process is carried out by a specific group of anaerobic bacteria called acetogens. These organisms use the Wood-Ljungdahl pathway to fix CO_2, but belong to different taxonomic groups within the phyla Acidobacteria, Firmicutes, and Spirochaetes. Only one bacterial strain has been described in each of Acidobacteria and Spirochaetes, whereas approximately 60 different strains have been described in Firmicutes [1]. All acetogenic bacteria use the Wood-Ljungdahl pathway to produce the central intermediate acetyl-CoA and subsequently alcohols and acids. These include biofuels such as ethanol, butanol, hexanol (Sect. 7.3.1), and 2,3-butanediol on the one hand, and acetate, butyrate, caproate, and lactate on the other. All these biobased basic chemicals can thus be produced from gases as carbon and energy sources by acetogenic bacteria. The majority of known acetogenic species produce acetic acid as their sole metabolic product. A few organisms additionally produce butyrate. Namely, these are *Acetonema longum* [2], "*Butyribacterium methylotrophicum*" [3], *Clostridium carboxidivorans*, *Clostridium drakei* [4], *Clostridium magnum* [5] *Clostridium scatologenes* [6], *Clostridium methoxybenzovorans* [7], *Eubacterium aggregans* [5], *Eubacterium limosum* [8, 9], and *Oxobacter pfennigii* [10]. The bacterial strains "*B. methylotrophicum*" [3], *C. carboxidivorans*, and *C. drakei* [4] are also known for their ability to form ethanol and butanol in addition to acetate and butyrate as metabolic end products. In addition, strains *C. carboxidivorans* [11, 12] and *E. limosum* [8, 9] can produce hexanoate, with *C. carboxidivorans* also producing hexanol [11, 12].

7.2.1.1 *Acetobacterium woodii* DSM 1030^T

A. woodii was isolated from an estuary in Woods Hole, MA, USA and named after Harland G. Wood [13]. The type strain WB1 stains Gram positive, grows chemolithoautotrophically using a $H_2 + CO_2$ gas mixture and heterotrophically with fructose, glucose, lactate, glycerate, and formate. Acetate is normally the only metabolic end product, only under phosphate-limiting conditions does the strain also produce ethanol [14]. *A. woodii* is one of the two model organisms used to study acetogenic metabolism in detail. The other model organism is *Clostridium ljungdahlii*. *A. woodii* is dependent on Na^+ ions in the medium [15] and has a Na^+-dependent ATPase that produces ATP [16]. Müller and coworkers used cells from *A. woodii* to demonstrate that the Rnf complex translocates Na^+ across the cytoplasmic membrane [17]. The energy required for this is provided by electron-bifurcating hydrogenases. Both, Rnf and the hydrogenases are involved in the mechanism of energy conservation (Sect. 7.3.1). *A. woodii* was first genetically accessed in 1994 by Strätz et al. [18]. Twenty years later, this was revisited by Straub et al. (2014) [19] and used for selective overproduction of acetate in a continuously $H_2 + CO_2$-gassed and fully mixed reactor. Furthermore, the metabolism of *A. woodii* was genetically modified such that the recombinant strains were able to produce acetone fermentatively from a H_2 + CO_2 gas mixture [20].

7.2.1.2 *"Butyribacterium methylotrophicum"* DSM 3468

"*B. methylotrophicum*" is a Gram-positive, motile, catabolically versatile, mesophilic, spore-forming anaerobic bacterium isolated from a wastewater digester in Marburg, Germany [21]. This so-called Marburg strain can reduce H_2 + CO_2 as well as CO + CO_2 and forms acetate, butyrate, ethanol, and butanol [22]. The ratio of acids and alcohols can be changed in favor of alcohols by reducing the fermentation pH [3]. Various sugars and the C1 compound methanol are fermented under heterotrophic growth conditions. Lactate is another possible fermentation product of "*B. methylotrophicum*" [23]. "*B. methylotrophicum*" is very similar to *E. limosum* ATCC 8486^T [24]. The metabolic characteristics of both strains as well as phylogenetic analyses of the corresponding 16S rRNA gene sequences indicate a close relationship between the two strains [25, 26].

"*B. methylotrophicum*" was provisionally classified as a risk group 2 organism (biological safety level 2) due to its close relationship with *E. limosum*. However, a comparative genome analysis of the strains "*B. methylotrophicum*" DSM 3468 and *E. limosum* ATCC 8486^T [27] revealed an average nucleotide identity of only 94.6%. The threshold for species differentiation is between 95 and 96% [28], suggesting that "*B. methylotrophicum*" DSM 3468 might be a different species compared to *E. limosum* ATCC 8486^T.

7.2.1.3 *Clostridium aceticum* DSM 1496^T

C. aceticum was the first described bacterial strain that grew autotrophically under anaerobic conditions using H_2 + CO_2 and formed acetate from this gas mixture. This endospore-forming organism is equally capable of using CO as a carbon source [29]. The strain was isolated from the mud of a canal in Wageningen in 1936 [30, 31]. After World War II, it was assumed that the strain had been lost. However, in 1981, a spore preparation of the original *C. aceticum* was rediscovered in a laboratory culture collection in California [32]. At the same time, the organism was also successfully re-isolated in Wageningen using "50 g dry ditch mud" [33]. Meanwhile, the complete genome sequence of *C. aceticum* (4.2 Mbp) and its native plasmid (5.7 kbp) is public [34]. Although cytochromes have been detected [32], the bacterial genome does not contain genes whose gene products are required for quinone synthesis [35]. Thus, an electron transport chain via quinones and cytochromes is not possible. *C. aceticum* contains an Rnf complex that acts as a proton pump [35] (Sect. 7.3.1). The bacterium can use fumarate as a carbon source, which is dismutated to succinate, acetate, and CO_2 [35]. Most of the fumarate is directly reduced to succinate, while the rest is converted via malate and oxaloacetate to pyruvate and further to acetate and CO_2, similar to *C. formicaceticum* [35, 36].

7.2.1.4 *Clostridium carboxidivorans* DSM 15243^T

C. carboxidivorans (strain P7^T) was isolated from an agricultural settling lagoon using CO as the sole substrate. For the first time, this bacterial strain was used for the anaerobic conversion of CO to acetic acid, butyric acid, ethanol and butanol in 2002 [37]. *C. carboxidivorans* P7^T was described as an acetogenic bacterium that can grow autotrophically with CO or H_2 + CO_2 [4]. A more detailed analysis and description of the genome sequence of *C. carboxivorans* P7^T is provided by Bruant et al. (2010; [38]). In addition to the chromosome, a native plasmid (p19) of approximately 19 kbp was described. The function of the plasmid is unknown, as most of the annotated genes encode putative "hypothetical proteins". All necessary genes encoding enzymes required for CO_2 fixation via the Wood-Ljungdahl pathway, as well as genes encoding enzymes for acetate, butyrate, ethanol, and butanol production have been described in detail [1]. In 2015, *C. carboxidivorans* P7^T was described as a hexanol- and hexanoate-producing bacterium by two different groups [11, 12].

7.2.1.5 *Clostridium ljungdahlii* DSM 13528^T and Closely Related Strains

C. ljungdahlii is an anaerobic, chemolithotrophic, Gram-positive, motile, spore-forming, rod-shaped mesophilic bacterium named after Lars G. Ljungdahl. The bacterial strain was isolated and described from chicken manure in 1993 by the research group of Ralph S. Tanner [39]. This was followed only a few months later by the description of the strain "*C. autoethanogenum*" [40]. "*C. ragsdalei*" was initially designated strain P11 and described in a patent by Huhnke et al. (2008; [41]). "*C. coskatii*" was also described in a patent [42]. In this patent, it is claimed that "*C. coskatii*" produces ethanol as a major metabolic product. The bacterial names "*C. autoethanogenum*", "*C. ragsdalei*", and "*C. coskatii*" have not yet been officially validated. Only *C. ljungdahlii* has been validly described and is officially listed [39]. Genome sequence analyses of the four strains showed high average nucleotide identity (ANI) values between *C. ljungdahlii* and "*C. autoethanogenum*" (99.3%) and "*C. coskatii*" (98.3%), respectively. *C. ljungdahlii* and "*C. ragsdalei*" showed an ANI-based similarity of 95.8%. The high similarity of the strains on genetic basis suggests that all strains belong to the same species [43] *C. ljungdahlii* (4.63 Mbp) has the largest genome, followed by "*C. coskatii*" (4.51 Mbp), "*C. ragsdalei*" (4.41 Mbp), and "*C. autoethanogenum*" (4.35 Mbp). Thus, the genome of "*C. autoethanogenum*" is 6.4% (0.28 Mbp) smaller than that of *C. ljungdahlii*. "*C. coskatii*" is the only one of the four strains that lacks genes encoding an aldehyde:ferredoxin oxidoreductase [43]. Therefore, "*C. coskatii*" is unable to form ethanol via the conversion of acetaldehyde from acetate. Ethanol formation via this metabolic pathway has been described for "*C. autoethanogenum*" and for *C. ljungdahlii* [44, 45].

C. ljungdahlii and "*C. autoethanogenum*" are genetically accessible and well-studied with respect to their conversion in syngas fermentation processes. Both share the metabolic feature that an almost complete conversion of the necessarily produced acetate into ethanol is possible. By so-called metabolic engineering of wild-type strains, the construction of recombinant strains is possible. These genetically engineered strains are then capable of producing substances such as isopropanol, butyrate, butanol, and 3-hydroxybutyrate from gas-containing substrates [46]. The bacterial strain "*C. autoethanogenum*" is currently already used industrially by LanzaTech for fermentative ethanol production from CO-rich steel mill waste gases.

7.2.1.6 *Moorella thermoacetica* (DSM 521[T] and DSM 2955[T])

M. thermoacetica (formerly *Clostridium thermoaceticum*) was isolated from horse feces [47] and was the model organism used to elucidate the enzymatic reactions of the Wood-Ljungdahl pathway. For this purpose, the two strains *M. thermoacetica* DSM 521[T] and *M. thermoacetica* ATCC 39073 were mainly used. *M. thermoacetica* is a moderately thermophilic bacterium with an optimal growth temperature of 55 ° C. The biochemistry of reduction of CO_2 or CO has been described in detail in several review articles (see, e.g., [48–50]). About 40 years after the isolation of *M. thermoacetica*, Kerby & Zeikus (1983; [51]) obtained a spore preparation of the originally isolated strain from Elizabeth McCoy. It was shown that the cultured strain also used CO as its sole source of energy and carbon, not only H_2 and CO_2 [51]. This significant difference compared to the strain DSM 521[T] led to the decision to deposit a second type strain of the species *M. thermoacetica* with the strain number DSM 2955[T] in the German Collection of Microorganisms and Cell Cultures GmbH (DSMZ). The first complete genome sequence of strain *M. thermoacetica* ATCC 39073 was published by Pierce et al. (2008; [52]). The genome sequences of the two type strains DSM 521[T] and DSM 2955[T], respectively, have been published separately [53, 54]. Compared to mesophilic acetogenic bacteria, no genes encoding the Rnf complex are found in the sequenced genomes of *M. thermoacetica* strains. Instead, genes encoding cytochromes, quinones and an energy-conserving hydrogenase complex (Ech) were detected. The latter enzyme complex serves to generate a proton gradient across the cell membrane and thus conserve energy via ATPase [50, 52].

7.2.1.7 *Thermoanaerobacter kivui* DSM 2030[T]

T. kivui (formerly *Acetogenium kivui*) was isolated from the sediment of Lake Kivu in Africa [55]. This acetogenic bacterial strain (LKT-1) is chemolithotrophic, thermophilic (66 °C), and immobile. The rod-shaped cells do not form spores and stain Gram-negative. The doubling time of the cells during autotrophic growth with H_2 + CO_2 is about 2 h. Thus, this strain is one of the fastest growing acetogenic bacteria. *T. kivui* can also use CO as the sole carbon and energy source, but the doubling time then increases to 33 h [56]. Autotrophic growth (H_2 + CO_2) or acetate formation is not dependent on Na^+ ions present in the medium; instead, a protonophore in resting cells inhibits acetate formation, suggesting H^+-dependent energy conservation. Two distinct gene clusters have been detected in the *T. kivui* genome, both encoding an energy-conserving hydrogenase complex [57]. The strain exhibits a natural competence to incorporate foreign genetic material into its own genome via homologous recombination [58]. Thus, the implementation of gene cassettes encoding metabolic pathways for industrially relevant products into the genome of *T. kivui* is realistic. Appropriate recombinant strains offer the advantage that process heat is utilized by the thermophilic bacterium during appropriate fermentation, thus reducing cooling costs. In addition, metabolic and diffusion rates under thermophilic conditions are higher than under mesophilic conditions [59].

7.2.2 Photoautotrophic Microalgae and Cyanobacteria

Microalgae are commonly referred to as the prokaryotic cyanobacteria and the microscopic eukaryotic algae. The most common classes of microalgae are the cyanobacteria (*Cyanophyceae*), the green algae (*Chlorophyceae*), the *Bacillariophyceae* (including the diatoms), and the *Chrysophyceae* (including the golden algae). Microalgae have been used by mankind for centuries as a food source or as a dietary supplement. In this regard, García et al. (2017; [60]) mention, for example, the Aztecs, who harvested cyanobacteria of the genus Arthrospira (formerly *Spirulina* sp.) from lakes and prepared the biomass as a dry cake, which they called "tecuitlatl." Other examples where different cyanobacteria are processed into food include "Dihé" from Chad, "Fa cai" and "Laceplum" from South America, or "Suizenji-nori" from Japan. Some microalgae are of increasing biotechnological interest because they can produce so-called nutraceutical compounds. These molecules can be defined as nutrients from food products that not only supplement the diet, but also facilitate the prevention or treatment of a disease and/or disorder.

The ability to use sunlight as an energy source and CO_2 from the air as a carbon source under aerobic conditions naturally attracted economic interest to these organisms since ancient times. With these undoubted advantages, however, come some disadvantages. The efficiency of photosynthesis (efficiency of conversion of sunlight into energy) is about 3–7%, well below that of photovoltaics [61]. A number of factors are responsible for this, such as the limited radiation spectrum that photosynthetically active organisms can use, mutual shading of cells (or cell assemblies), temperature variations, and suboptimal CO_2 concentration. For large-scale use in simple, open systems (raceway ponds), a large surface area must be maintained at shallow water depths to keep shading effects low. In addition, there are the constant nutrient supply and contamination problems. In closed systems (photobioreactors), sufficient light supply is the biggest challenge (wall growth, diameter of tube systems).

The CO_2 uptake rate is significantly lower for photoautotrophic microalgae and cyanobacteria than for acetogenic bacteria (Table 7.1). However, it must be taken into account that the cultivation conditions of the organisms showed significant differences (e.g., different CO_2 concentrations, gas transfer to the liquid phase, energy source).

7.2.2.1 *Arthrospira* sp. (Formerly *Spirulina* sp.)

Microorganisms of the genus *Arthrospira* are cyanobacteria (*Cyanophyceae*), i.e. not eukaryotic algae, but prokaryotic bacteria. Cells of this genus cluster together to form free-floating, filamentous, helical, cylindrical trichomes characterized by an open left-hand helix. Based on their morphological similarity, these photosynthetic organisms were initially referred to as blue-green algae before classification into the

Table 7.1 Cell dry weights, CO_2 uptake rates, growth rates of acetogenic bacteria and phototrophic microorganisms

Microorganism	Cell dry weight [g L^{-1}]	CO_2-uptake rate [mmol L^{-1} d^{-1}]	Growth rate μ_{max} [d^{-1}]	References
Acetobacterium woodii	1.15	780	1.9	[5]
Acetobacterium wieringae	1.36	530.9	1.76	[5]
Clostridium magnum	0.35	132	0.53	[5]
Clostridium aceticum	1.0	336	1.69	[62]
Sporomusa acidovorans	0.43	228	0.59	[5]
Sporomusa ovata	1.25	631	2.9	[5]
Chlorella vulgaris	1.31	3.66	0.31	[63, 64]
Dunaliella tertiolecta	3.42	7.11	n.r.	[65, 66]
Haematococcus pluvialis	4.09	3.25	0.8	[67, 68]
Phaeodactylum tricornutum	13	6.36	0.15	[69–71]
Scenedesmus almeriensis	n.r.	18	0.34	[72]
Scenedesmus obtusiusculus	n.r.	22	0.43	[71, 73]
Scenedesmus obliquus	2.12	0.26	0.33	[63]
Spirulina sp.	4.13	0.413	0.42	[63]
Spirulina platensis	n.r.	6.36	0.65	[71]
Spirulina maxima	n.r.	6.36	0.6	[71]

n.r. not reported

kingdom of bacteria was proposed in 1962 [74] and then validated in 1974 in Bergey's Manual of Determinative Bacteriology [75]. The two species *A. maxima* and *A. platensis* were once classified in the genus *Spirulina*, and although both genera (*Arthrospira* and *Spirulina*) are generally accepted, there have been many disputes in the past regarding the classification of the two species; the resulting taxonomic confusion is enormous.

7.2.2.2 *Chlorella* sp.

Chlorella vulgaris is a unicellular green alga (taxon *Chlorophyceae*) and has been used as a model organism in research. Thus, the biochemical mechanism of photosynthesis has been elucidated using this alga. Currently, 24 Chlorella species are taxonomically validated. The different Chlorella species form spherical cells with a diameter of 2–10 μm and are green in color. The cells contain a single chloroplast and mitochondria scattered in the cytoplasm. Reproduction is apparently exclusively asexual; no gamete formation has yet been demonstrated.

7.2.2.3 *Dunaliella* sp.

Dunaliella species are green algae and belong to the genus *Chlorophyceae*. *D. salina* lives in hypersaline environments and is reddish in color due to a high content of β-carotene. This trait is used commercially in cultivation facilities in Australia, Israel, and the United States. *D. tertiolecta* is a marine green alga with a size of about 10 to 12 µm. Due to the high content of lipids (approx. 37%) this organism is interesting for the production of biodiesel.

7.2.2.4 *Haematococcus pluvialis*

H. pluvialis, the blood rain alga, is a freshwater species of green algae (*Chlorophyceae*). This species is known for producing astaxanthin, which acts as an antioxidant. This is a reddish to purple natural pigment responsible, for example, for the red coloration of crustaceans that eat astaxanthin-containing algae. Astaxanthin accumulates in cells, especially in cysts (permanent stages), when conditions are unfavorable for growth. Examples of such conditions include bright light, high salinity, and low nutrient availability [76]. Astaxanthin presumably protects cysts from the harmful effects of UV radiation when exposed to direct sunlight [77]. In 2011, an economic evaluation of astaxanthin production by large-scale cultivation of *H. pluvialis* concluded that the estimated costs could be lower compared to chemically synthesized astaxanthin [78]. This contrasts with the assessment of Nguyen (2013; [79]) stating that astaxanthin production by algae was almost four times more expensive than chemical synthesis.

7.2.2.5 *Nannochloropsis* sp.

Microalgae of the genus *Nannochloropsis* (class *Eustigmatophyceae*) are single-celled organisms with a cell diameter of 2–5 µm [80]. They are found in both marine and freshwater habitats and use only chlorophyll a and violaxanthin as the main pigment for photosynthesis.

The major biotechnological interest is due to the considerable amounts of lipids produced by these microalgae. On average, the lipid content of *Nannochloropsis* sp. ranges from 25 to 45%. Therefore, biomass from these microalgae have great potential as feedstock for biofuel production [81]. Furthermore, these microalgae also contain significant amounts of omega-3 fatty acids, mostly in the form of eicosapentaenoic acid (EPA, 20:5), which has been shown to provide many health benefits to humans [82]. A biorefinery approach for *Nannochloropsis* was proposed by Chua and Schenk (2017; [81]). It starts with the induction of lipid and EPA production by nitrogen-limited cultivation conditions. Subsequently, cell sedimentation can be induced by flocculants and food-grade biomass harvested. This biomass can then be concentrated by centrifugation and dried. The omega-3 fatty acids are extracted, and the remaining biomass can be further processed either as an animal feed ingredient or even as a high quality protein dietary supplement. Omega-3 fatty acids are currently extracted from the fat of fish or krill, which is not sustainable in the long term [82].

7.2.3   Aerobic Methanotrophic Bacteria

Although the focus of this book is on the utilization of the oxidized C1 gases CO_2 and CO, the reduced compound methane (CH_4) will also be briefly discussed. The conversion of this gas plays a very crucial role in the Earth's climate balance, as methane is a very potent greenhouse gas. Utilization as a carbon source can occur under both anaerobic and aerobic conditions. In the anaerobic range, Candidatus *Methylomirabilis oxyfera* is currently the only known bacterium that can oxidize methane under anaerobic conditions and with nitrite as cosubstrate. This apparently involves the intermediate formation of oxygen, which can be used directly by a methane monooxygenase (the key enzyme in aerobic methane oxidation). Candidatus *M. oxyfera* is not yet available as a pure culture. In addition, some archaea are known (also not available in pure culture) that can reverse the known pathway of methanogenesis by coupling with sulfate or nitrate reduction and oxidize methane anaerobically in this way (Candidatus *Methanoperedens nitroreducens*) [83, 84].

Under aerobic conditions, a whole range of bacteria can oxidize methane. In the phylum Proteobacteria they belong to the classes Gammaproteobacteria, Alphaproteobacteria, and Verrucomicrobia. In Gammaproteobacteria, methanotrophs are organized in the family *Methylococcaceae*. These are primarily organisms referred to as type I methylotrophic bacteria. Characteristics of these are the fixation of formaldehyde via the ribulose-monophosphate cycle and the organization of intracellular membrane systems in the form of stacks with integrated methane monooxygenase (particulate MMO, pMMO). Typical representatives are the genera *Methylobacter* and *Methylomonas*. A subgroup is called type X methylotrophic bacteria. This is the genus *Methylococcus*, which also has a soluble methane monooxygenase (soluble MMO, sMMO) and elements of the serine pathway for formaldehyde fixation. Alphaproteobacteria include the type II methylotrophic bacteria, which form their intracellular membrane systems as rings at the cell periphery, fix formaldehyde via the serine pathway, and possess pMMO and sMMO. They are organized in the family *Methylocystaceae*; typical representatives are the genera *Methylocystis* and *Methylosinus*. Among Verrucomicrobia, representatives of the order Methylacidiphilales are capable of methane oxidation.

7.2.3.1 *Methylococcus capsulatus*

M. capsulatus is an immobile, Gram-negative, methanotrophic bacterium that grows optimally under mesophilic conditions. In addition to methane, methanol can also be utilized. Growth is still possible even under oxygen limitation. The genome sequence has been determined [85]. The organism is now also accessible by molecular biology. In 2012, the construction of a mutant was described in which the gene *hpnB*, whose gene product catalyzes hopanoid methylation at position C-3, was deleted [86].

7.2.3.2 *Methylocystis parvus*

M. parvus is the type strain of the genus *Methylocystis*. The organism is immotile, Gram-negative, and forms small rods. *M. parvus* can also use methanol as a carbon source in addition to methane. Optimal growth occurs between 25 and 35 °C. The genome sequence has been determined [87]. *M. parvus* can form large amounts of poly-3-hydroxybutyrate [88].

7.2.3.3 *Methylosinus trichosporium*

M. trichosporium is a Gram-negative rod that oxidizes methane under mesophilic conditions. The bacterium has become the model organism for the study of both methane monooxygenases (pMMO and sMMO). The genome sequence has been determined [89]. *Methylosinus* species form exospores.

7.3 Fundamentals of C1 Metabolism in Selected Classes of Microorganisms

7.3.1 Acetogenic Bacteria

Acetogenic bacteria use the Wood-Ljungdahl metabolic pathway (reductive acetyl-CoA pathway) to synthesize acetyl-CoA from H_2 + CO_2 or CO-rich gases as a carbon and energy source. Acetyl-CoA is the key intermediate that subsequently serves for the production of biofuels, such as ethanol, butanol or hexanol, as well as biological feedstocks, such as acetate, lactate, butyrate, hexanoate, 2,3-butanediol, isopropanol, or acetone.

The current definition of an acetogenic bacterium is [90]: an anaerobic bacterium that uses the reductive acetyl-CoA pathway as (1) a mechanism for reductive synthesis of acetyl-CoA from CO_2, (2) an energy-conserving process, and (3) a mechanism for fixation (assimilation) of CO_2 to form cell mass. Accordingly, the formation of acetate is irrelevant; only the process of acetyl-CoA production is important. The terms reductive acetyl-CoA pathway and Wood-Ljungdahl pathway are used interchangeably. The corresponding biochemistry of the pathway has been presented in a number of review articles [1, 49, 50, 91–95]. The Wood-Ljungdahl pathway consists of a methyl and a carbonyl branch, in which CO_2 is reduced by means of enzymatic reactions and ultimately the central intermediate acetyl-CoA is formed (Fig. 7.1). The methyl branch is used for the stepwise reduction of CO_2 to a methyl group. Another molecule of CO_2 is reduced to CO by the CO-DH/acetyl-CoA synthase complex (CO-DH/Acs) to enter the carbonyl branch. When acetogenic bacteria use CO as their sole carbon source, one molecule enters the carbonyl branch directly, while another CO is oxidized to CO_2 by carbon monoxide dehydrogenase (CO-DH) and reduced ferredoxin is formed simultaneously [96]. In the methyl branch, CO_2 is reduced to formate by a formate dehydrogenase (Fdh).

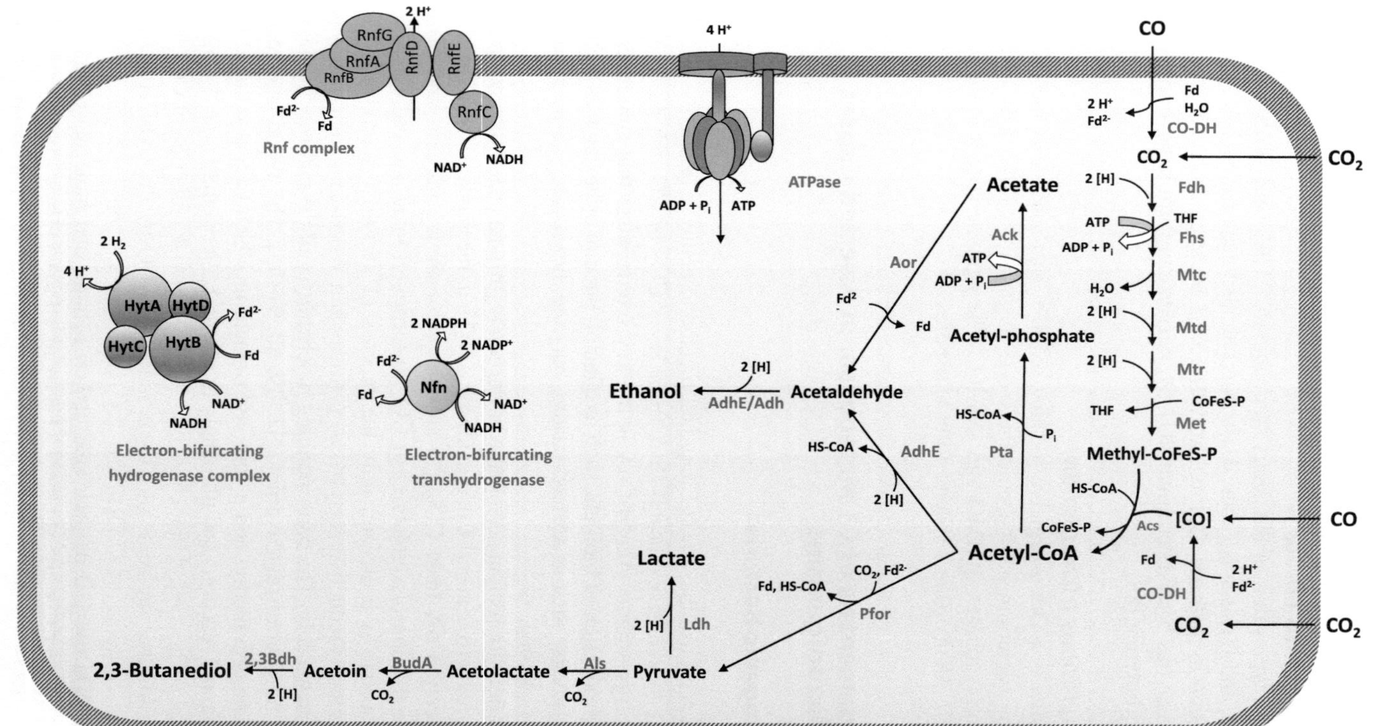

Fig. 7.1 Wood-Ljungdahl pathway and native metabolites of *Clostridium ljungdahlii*. 2,3-Bdh 2,3-butanediol dehydrogenase, Ack acetate kinase, Acs acetyl-CoA synthase, AdhE aldehyde/alcohol dehydrogenase, Als acetolactate synthase, Aor aldehyde:ferredoxin oxidoreductase, BudA acetoin decarboxylase, CO-DH CO dehydrogenase, Fdh formate dehydrogenase, Fhs formyl-THF synthetase, Ldh lactate dehydrogenase, Met methyltransferase, Mtc methenyl-THF cyclohydrolase, Mtd methylene-THF dehydrogenase, Mtr methylene-THF reductase, Pfor pyruvate:ferredoxin oxidoreductase, Pta phosphotransacetylase

Formate and tetrahydrofolate (THF) are fused to formyl-THF, and formyl-THF synthetase (Fhs) consumes an adenosine triphosphate (ATP). The stepwise reduction of formyl-THF to methyl-THF is catalyzed by the enzymes methenyl-THF cyclohydrolase (Mtc), methylene-THF dehydrogenase (Mtd), and methylene-THF reductase (Mtr).

In the next step of the methyl branch, the methyl group is transferred from methyl-THF to a corrinoid iron-sulfur protein. The CO-DH/Acs enzyme complex finally fuses the methyl group, the carbonyl group, and coenzyme A, which together form the key intermediate acetyl-CoA. This CO-DH/Acs complex is an $\alpha2\beta2$-tetramer in which the α-subunit catalyzes the formation of acetyl-CoA and the β-subunit reduces CO_2 to CO [94, 97–99]. Figure 7.1 shows the metabolism described using *C. ljungdahlii* as an example.

Acetate is the major product of most acetogenic bacteria, and in many cases it is the only product. Acetate is formed via the enzymes phosphotransacetylase (Pta) and acetate kinase (Ack) by converting acetyl-CoA to acetyl phosphate and the latter to acetate. In the latter reaction, ATP is obtained. This ATP is needed to activate formate to formyl-THF. Thus, during autotrophic growth of acetogenic bacteria, there is no net ATP gain from substrate chain phosphorylation. Acetogenic bacteria use different mechanisms for energy conservation (ATP formation). Schuchmann and Müller (2014; [50]) classified acetogenic bacteria bioenergetically as "Rnf-acetogenic bacteria" and "Ech-acetogenic bacteria". In both groups, the Wood-Ljungdahl pathway is coupled to a transmembrane ion gradient that drives ATP synthesis. The ions used by both complexes (and their respective ATP synthases) are either protons (H^+) or sodium ions (Na^+), resulting in two subclasses. Each class contains H^+-dependent and Na^+-dependent organisms. In *A. woodii*, Na^+ ions are used to form a chemiosmotic ion gradient via the Rnf complex (name stands for Rhodobacter nitrogen fixation) integrated into the cell membrane. A sodium-dependent ATPase then pumps the Na^+ ions back into the cell, thereby generating ATP [100]. In addition to pumping function, the Rnf complex also acts as a ferredoxin:NAD^+ oxidoreductase. In contrast, *C. ljungdahlii* and "*C. autoethanogenum*" possess the H^+-dependent Rnf complex, which establishes an H^+ gradient across the cell membrane, which in turn is used by an H^+-dependent ATPase to generate ATP [101]. In *M. thermoacetica*, such an Rnf complex is absent. Energy conservation in this organism may occur via cytochromes that generate an H^+ gradient [102]. In addition, *M. thermoacetica* contains a membrane-bound energy-conserving hydrogenase (Ech) complex. Compared to the Rnf complex, the Ech complex releases less energy [50]. In the class of "Ech-acetogenic bacteria", only "H^+-dependent organisms" are known so far.

7.3.2 Photoautotrophic Microalgae and Cyanobacteria

The conversion of light into biologically usable energy (ATP) can in principle take place in two different ways. In some archaea and bacteria, bacterio- or proteorhodopsin ensures that a proton is translocated across the membrane by conformational change after light irradiation. The proton gradient can then be used for ATP synthesis via an ATPase. Photosynthesis in phototrophic bacteria and plants is completely different. In principle, light energy knocks an electron out of a (bacterio)chlorophyll molecule and transports it to a strongly negative redox potential. From there, it is transported back to the (bacterio)chlorophyll via a cyclic electron transport chain, whereby protons are translocated across the membrane by the quinone cycle used in the process (quite analogous to the respiratory chain), which are then used for ATP synthesis via an ATPase. This form of photosynthesis is found in various developmental stages in bacteria and plants. Bacteria of the family Rhodospirillaceae carry out anoxygenic photosynthesis under anaerobic conditions using a simple electron transport system as described above. Via light-harvesting complexes (with antenna pigments such as carotenoids and bacteriochlorophyll), photons are concentrated onto the reaction system, where release and potential shift of electrons from a special pair of bacteriochlorophyll occurs. Electrons are returned to the reaction center-bacteriochlorophyll via bacteriopheophytin (bacteriochlorophyll without a central atom), the quinone cycle, and the cytochrome bc1 and c2 complexes. ATP is generated via proton translocation in the quinone cycle and ATPase, and NADH is generated via reverse electron transport with the aid of ATP. CO_2 assimilation occurs via the CBB cycle. Similar photosystems are used by the Chromatiaceae and the Chloroflexaceae. The Chromatiaceae can also oxidize sulfur compounds very efficiently and use the electrons to synthesize NADH via reverse electron transport. The Chloroflexaceae have their antennal pigments organized in so-called chlorosomes on the inside of the cytoplasmic membrane. They fix CO_2 via the 3-hydroxypropionate cycle. Photosynthesis proceeds somewhat differently in Chlorobiaceae and Heliobacteriaceae. In these families, the excited electrons are transferred via an iron sulfur protein to either quinone or ferredoxin, which can then be used directly for NADH formation. The electron gap that inevitably occurs in the formation of NADH is filled by electrons from the oxidation of sulfur compounds, which are returned to the reaction center-bacteriochlophyll via a cytochrome c (i.e., either ATP or NADH synthesis). Chlorobiaceae also have chlorosomes and use the reductive citrate cycle for CO_2 assimilation. In cyanobacteria and plants, photosynthesis is accomplished in two photosystems. One photosystem (PS I) is similar to that of Chlorobiaceae and Heliobacteriaceae and allows the formation of NADH. The other provides electrons to replenish PS I, passing through the quinone cycle and thus contributing to ATP formation. Electrons to replenish PS II originate from the oxidation of H_2O, thus realizing oxygenic photosynthesis. Some cyanobacteria can also still use PS II to oxidize H_2S, illustrating the evolutionary developmental process. The use of two photosystems allows the organisms to produce ATP and NADH simultaneously. CO_2 fixation proceeds in cyanobacteria and plants via the CBB cycle.

The key enzymes of the CBB cycle are ribulokinase and ribulosebisphosphate carboxylase/oxygenase (RuBisCO). The former forms 1,5-ribulose bisphosphate, which is then converted by RuBisCO with CO_2 to two molecules of 3-phosphoglycerate, the first tangible fixation product. Phosphorylation with ATP and reduction yields glyceraldehyde 3-phosphate, which is used to build cell substance on the one hand and serves to regenerate the acceptor molecule 1,5-ribulose bisphosphate on the other (Fig. 7.2). The ATP requirement is very high in the CBB

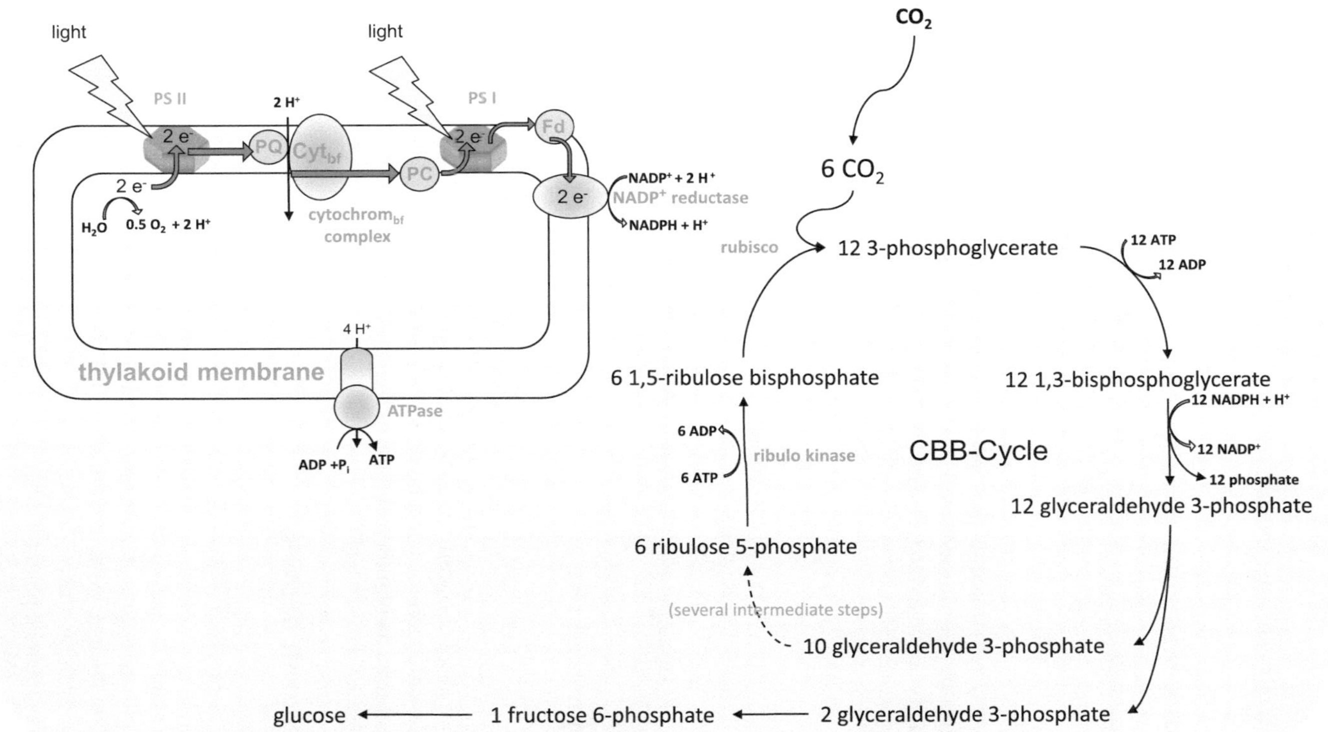

Fig. 7.2 Photosystems (PS) in microalgae and cyanobacteria and Calvin(-Benson-Bassham, CBB) cycle. Cyt bf cytochrome bf complex, Fd ferredoxin, PQ plastoquinone, PC plastocyanin

cycle (more than 7 ATP for the synthesis of 1 molecule of pyruvate) because the RuBisCO also carries out an oxygenase reaction and thus virtually wastes part of the acceptor molecules (conversion of 1,5-ribulose bisphosphate into 3-phosphoglycerate and 2-phosphoglycolate).

7.3.3   Aerobic Methanotrophic Bacteria

Methanotrophic bacteria are aerobic and use the respiratory chain to produce ATP. The corresponding reduction equivalents come from the oxidation of methane. This is first oxidized to methanol by methane monooxygenase with the help of NADH and oxygen. Further reaction products are NAD^+ and H_2O. CH_3OH is then oxidized to formaldehyde by pyrroloquinoline quinone-dependent alcohol dehydrogenases. CH_2O is converted into formate and NADH with the aid of NAD^+ and water. HCOOH is finally converted to CO_2 and NADH with the help of NAD^+. The overall balance is thus:

$$CH_4 + O_2 \rightarrow CO_2 + 2 \times 2[H] \tag{7.1}$$

The fixation of carbon in cell substance occurs predominantly at the formaldehyde stage. In type I methanotrophic bacteria, the ribulose monophosphate cycle is used for this purpose. Type II species use either the serine pathway or, if they lack the key enzyme isocitrate lyase, the ethylmalonyl-CoA pathway for C assimilation.

7.4   Commercial Use

The low price of C1 gases and their global availability make them attractive carbon and energy sources for biotechnological use. Indeed, there are now industrial applications for all of the above groups of organisms. Another positive aspect is the sustainable reduction of greenhouse gases, provided that the fermentation products are not used as biofuel and thus immediately burned again.

7.4.1   Acetogenic Bacteria

In the industrial use of acetogenic bacteria, the company LanzaTech Inc. (Skokie, USA) is currently the world market leader. Typically, process waste gases that correspond to synthesis gas in their composition (high CO content) are converted to ethanol. Such waste gases are mainly found in steel production, but also in the gasification of municipal waste and biomass. Ethanol can be further converted to aircraft fuel. In addition, the production of various basic and specialty chemicals is envisaged. A commercial plant for ethanol production from steel mill waste gas was commissioned in 2018 in Caofeidian (China) together with the Shougang Group (capacity: 48,000 t ethanol/year) and a second plant in 2021. Another plant is scheduled to open in 2022 at an ArcelorMittal steel mill in Ghent, Belgium (capacity: 62,000 t ethanol/year). Plants in South Africa (capacity: 52,000 t ethanol/year), India (capacity: 34,000 t ethanol/year) and California, USA (capacity: 35,000 t ethanol/year) are then to be added by 2022/2023.

7.4.2 Photoautotrophic Microalgae and Cyanobacteria

The current market with these organisms and their products is limited (about 10,000 t *Arthrospira/Spirulina* and about 4000 t *Chlorella* per year). The two main lines include, on the one hand, dried microalgae (*Chlorella, Spirulina*) with high nutritional content (vitamins C, D_2 and B_{12}). On the other hand, it is about products extracted from microalgae such as antioxidants (e.g. β-carotene), fatty acids (e.g. omega-3 fatty acids, docosahexaenoic acid, eicosapentaenoic acid), pigments (e.g. astaxanthin) and proteins (e.g. phycocyanin). Currently, there are a number of companies worldwide that cultivate microalgae and manufacture or sell corresponding products. These include, for example, Algatechnologies Ltd. (Israel), BASF (Germany), Cyanotech Co. (USA, Hawaii), Earthrise Nutritionals, LLC (USA, California), ecoduna GmbH (Austria), Hainan-DIC Microalgae Co. Ltd. (China), and Roquette Klötze GmbH & Co. KG (Germany). BASF SE (Nutrition & Health Division) operates the largest open pond system for the cultivation of microalgae, covering approximately 740 hectares. The company Roquette Klötze GmbH & Co. KG cultivates chlorella algae in a 500 km long glass tube system, where the organisms are supplied with sunlight as an energy source. Product categories offered include Chlorella-containing powder, feed and food, and tablets. Another company offering "Chlorella dry powder" is eparella GmbH. The biomass of this green alga is also used for the production of food, food supplements, and also cosmetics. The blood rain alga *H. pluvialis* is used by various companies (e.g. Algatechnologies Ltd. and Cyanotech Co.) to obtain astaxanthin. Cyanotech Co. cultivates *H. pluvialis* in large, open freshwater tanks in Hawaii. In these, the microalgae's natural environment is mimicked, while taking advantage of Hawaii's long hours of sunshine. Another company, Algatechnologies Ltd, uses a patented technology to cultivate the blood rain algae in photobioreactors. The company ecoduna GmbH offers "100% NANNOCHLOROPSIS" and "100% EPA-STRAIN OMEGA-3" as quality products. The cyanobacteria *A. platensis* and *A. maxima* are used to produce a food supplement called "Spirulina" [103]. Cyantech Co. produces the finished product Hawaiian Spirulina® using a large open cultivation system. Comparable processes are used by Earthrise Nutritionals, LLC and Hainan-DIC Microalgae Co. Ltd. (both DIC Corporation), which achieve a total production capacity of about 800 tons per year in complex *Spirulina* farms.

7.4.3   Aerobic Methanotrophic Bacteria

Commercial use is also already taking place with aerobic methanotrophic bacteria [104]. Calysta Inc. operates a small plant for the conversion of natural gas into biomass in Teesside, England (capacity about 100 t/year). The organism used is *Methylococcus capsulatus*; the harvested, pressed, and dried biomass is marketed as fish feed with high nutritional value. A large-scale plant is currently planned to be commissioned in 2022 in China via Calysseo, a joint venture of Calysta and Adisseo (final capacity: 20,000-60,000 t/year). UniBio in Kalundborg, Denmark, operates a very similar process. Approx. 80 tons of *M. capsulatus* biomass/year have been produced there since 2016.

References

1. Bengelsdorf FR, Beck BH, Erz C, Hoffmeister S, Karl MM, Riegler P, Wirth S, Poehlein A, Weuster-Botz D, Dürre P (2018) Bacterial anaerobic synthesis gas (syngas) and CO_2 + H_2 fermentation. Adv Appl Microbiol 103:143–221
2. Kane MD, Breznak JA (1991) *Acetonema longum* gen. nov. sp. nov., an H_2/CO_2 acetogenic bacterium from the termite, Pterotermes occidentis. Arch Microbiol 156:91–98. https://doi.org/10.1007/bf00290979
3. Worden R, Grethlein A, Jain M, Datta R (1991) Production of butanol and ethanol from synthesis gas via fermentation. Fuel 70:615–619. https://doi.org/10.1016/0016-2361(91)90175-a
4. Liou JSC, Balkwill DL, Drake GR, Tanner RS (2005) *Clostridium carboxidivorans* sp. nov., a solvent-producing *Clostridium* isolated from an agricultural settling lagoon, and reclassification of the acetogen *Clostridium scatologenes* strain SL1 as *Clostridium drakei* sp. nov. Int J Syst Evol Microbiol 55:2085–2091. https://doi.org/10.1099/ijs.0.63482-0
5. Groher A, Weuster-Botz D (2016) Comparative reaction engineering analysis of different acetogenic bacteria for gas fermentation. J Biotechnol 228:82–94. https://doi.org/10.1016/j.jbiotec.2016.04.032
6. Küsel K, Dorsch T, Acker G, Stackebrandt E, Drake HL (2000) *Clostridium scatologenes* strain SL1 isolated as an acetogenic bacterium from acidic sediments. Int J Syst Evol Microbiol 50:537–546. https://doi.org/10.1099/00207713-50-2-537
7. Mechichi T, Labat M, Patel BK, Woo TH, Thomas P, Garcia J (1999) Clostridium methoxybenzovorans sp. nov., a new aromatic o-demethylating homoacetogen from an olive mill wastewater treatment digester. Int J Syst Bacteriol 49:1201–1209. https://doi.org/10.1099/00207713-49-3-1201
8. Genthner BR, Davis CL, Bryant MP (1981) Features of rumen and sewage sludge strains of *Eubacterium limosum*, a methanol- and H_2-CO_2-utilizing species. Appl Environ Microbiol 42:12–19
9. Lindley ND, Loubiere P, Pacaud S, Mariotto C, Goma G (1987) Novel products of the acidogenic fermentation of methanol during growth of *Eubacterium limosum* in the presence of high concentrations of organic acids. Microbiology 133:3557–3563. https://doi.org/10.1099/00221287-133-12-3557
10. Krumholz LR, Bryant MP (1985) *Clostridium pfennigii* sp. nov. uses methoxyl groups of monobenzenoids and produces butyrate. Int J Syst Bacteriol 35:454–456. https://doi.org/10.1099/00207713-35-4-454
11. Phillips JR, Atiyeh HK, Tanner RS, Torres JR, Saxena J, Wilkins MR, Huhnke RL (2015) Butanol and hexanol production in *Clostridium carboxidivorans* syngas fermentation: medium

development and culture techniques. Bioresour Technol 190:114–121. https://doi.org/10.1016/j.biortech.2015.04.043

12. Ramió-Pujol S, Ganigué R, Bañeras L, Colprim J (2015) Incubation at 25 °C prevents acid crash and enhances alcohol production in Clostridium carboxidivorans P7. Bioresour Technol 192:296–303. https://doi.org/10.1016/j.biortech.2015.05.077

13. Balch WE, Schoberth S, Tanner RS, Wolfe RS (1977) *Acetobacterium*, a new genus of hydrogen-oxidizing, carbon dioxide-reducing, anaerobic bacteria. Int J Syst Bacteriol 27:355–361. https://doi.org/10.1099/00207713-27-4-355

14. Buschhorn H, Dürre P, Gottschalk G (1989) Production and utilization of ethanol by the homoacetogen *Acetobacterium woodii*. Appl Environ Microbiol 55:1835–1840

15. Heise R, Müller V, Gottschalk G (1989) Sodium dependence of acetate formation by the acetogenic bacterium *Acetobacterium woodii*. J Bacteriol 171:5473–5478. https://doi.org/10.1128/jb.171.10.5473-5478.1989

16. Müller V, Aufurth S, Rahlfs S (2001) The Na^+ cycle in *Acetobacterium woodii*: identification and characterization of a Na^+ translocating F_1F_0-ATPase with a mixed oligomer of 8 and 16 kDa proteolipids. Biochim Biophys Acta 1505:108–120. https://doi.org/10.1016/s0005-2728(00)00281-4

17. Biegel E, Müller V (2010) Bacterial Na^+-translocating ferredoxin:NAD^+oxidoreductase. Proc Natl Acad Sci U S A 107:18138–18142. https://doi.org/10.1073/pnas.1010318107

18. Strätz M, Sauer U, Kuhn A, Dürre P (1994) Plasmid transfer into the homoacetogen *Acetobacterium woodii* by electroporation and conjugation. Appl Environ Microbiol 60:1033–1037

19. Straub M, Demler M, Weuster-Botz D, Dürre P (2014) Selective enhancement of autotrophic acetate production with genetically modified *Acetobacterium woodii*. J Biotechnol 178:67–72. https://doi.org/10.1016/j.jbiotec.2014.03.005

20. Hoffmeister S, Gerdom M, Bengelsdorf FR, Linder S, Flüchter S, Öztürk H, Blümke W, May A, Fischer R-J, Bahl H, Dürre P (2016) Acetone production with metabolically engineered strains of *Acetobacterium woodii*. Met Eng 36:37–47. https://doi.org/10.1016/j.ymben.2016.03.001

21. Zeikus JG, Lynd LH, Thompson TE, Krzycki JA, Weimer PJ, Hegge PW (1980) Isolation and characterization of a new, methylotrophic, acidogenic anaerobe, the Marburg strain. Curr Microbiol 3:381–386. https://doi.org/10.1007/bf02601907

22. Grethlein AJ, Worden R, Jain MK, Datta R (1991) Evidence for production of n-butanol from carbon monoxide by *Butyribacterium methylotrophicum*. J Ferment Bioeng 72:58–60. https://doi.org/10.1016/0922-338x(91)90147-9

23. Heiskanen H, Virkajärvi I, Viikari L (2007) The effect of syngas composition on the growth and product formation of *Butyribacterium methylotrophicum*. Enzym Microb Technol 41:362–367. https://doi.org/10.1016/j.enzmictec.2007.03.004

24. Lynd LH, Zeikus JG (1983) Metabolism of H_2-CO_2, methanol, and glucose by *Butyribacterium methylotrophicum*. J Bacteriol 153:1415–1423

25. Jansen M, Hansen TA (2001) Non-growth-associated demethylation of dimethylsulfoniopropionate by (homo)acetogenic bacteria. Appl Environ Microbiol 67:300–306. https://doi.org/10.1128/aem.67.1.300-306.2001

26. Bengelsdorf FR, Poehlein A, Schiel-Bengelsdorf B, Daniel R, Dürre P (2016) Genome sequence of the acetogenic bacterium *Butyribacterium methylotrophicum* DSM 3468[T]. Genome Announc 4:e01338–e01316. https://doi.org/10.1128/genomea.01338-16

27. Song Y, Cho B (2015) Draft genome sequence of chemolithoautotrophic acetogenic butanol-producing *Eubacterium limosum* ATCC 8486. Genome Announc 3:e01564–e01514. https://doi.org/10.1128/genomea.01564-14

28. Kim M, Oh H, Park S, Chun J (2014) Towards a taxonomic coherence between average nucleotide identity and 16S rRNA gene sequence similarity for species demarcation of prokaryotes. Int J Syst Evol Microbiol 64:1825. https://doi.org/10.1099/ijs.0.064931-0

29. Lux MF, Drake HL (1992) Re-examination of the metabolic potentials of the acetogens *Clostridium aceticum* and *Clostridium formicoaceticum*: chemolithoautotrophic and aromatic-dependent growth. FEMS Microbiol Lett 95:49–56. https://doi.org/10.1111/j.1574-6968.1992.tb05341.x

30. Wieringa KT (1936) Over het verdwijnen van waterstof en koolzuur onder anaerobe voorwaarden. Ant Leeuwenhoek 3:263–273. https://doi.org/10.1007/bf02059556

31. Wieringa KT (1939) The formation of acetic acid from carbon dioxide and hydrogen by anaerobic spore-forming bacteria. Ant Leeuwenhoek 6:251–262. https://doi.org/10.1007/bf02146190

32. Braun M, Mayer F, Gottschalk G (1981) *Clostridium aceticum* (Wieringa), a microorganism producing acetic acid from molecular hydrogen and carbon dioxide. Arch Microbiol 128:288–293. https://doi.org/10.1007/bf00422532

33. Adamse AD (1980) New isolation of *Clostridium aceticum* (Wieringa). Ant Leeuwenhoek 46:523–531. https://doi.org/10.1007/bf00394009

34. Poehlein A, Bengelsdorf FR, Schiel-Bengelsdorf B, Gottschalk G, Daniel R, Dürre P (2015) Complete genome sequence of Rnf- and cytochrome-containing autotrophic acetogen *Clostridium aceticum* DSM 1496. Genome Announc 3:e00786–e00715. https://doi.org/10.1128/genomea.00786-15

35. Poehlein A, Cebulla M, Ilg MM, Bengelsdorf FR, Schiel-Bengelsdorf B, Whited G et al (2015) The complete genome sequence of *Clostridium aceticum*: a missing link between Rnf- and cytochrome-containing autotrophic acetogens. MBio 6:5. https://doi.org/10.1128/mbio.01168-15

36. Dorn M, Andreesen JR, Gottschalk G (1978) Fumarate reductase of *Clostridium formicoaceticum*. Arch Microbiol 119:7–11. https://doi.org/10.1007/bf00407920

37. Rajagopalan S, Datar RP, Lewis RS (2002) Formation of ethanol from carbon monoxide via a new microbial catalyst. Biomass Bioenergy 23:487–493. https://doi.org/10.1016/s0961-9534(02)00071-5

38. Bruant G, Lévesque M, Peter C, Guiot SR, Masson L (2010) Genomic analysis of carbon monoxide utilization and butanol production by *Clostridium carboxidivorans* strain P7^T. PLoS One 5:9. https://doi.org/10.1371/journal.pone0013033

39. Tanner RS, Miller LM, Yang D (1993) *Clostridium ljungdahlii* sp. nov., an acetogenic species in clostridial rRNA homology group I. Int J Syst Bacteriol 43:232–236. https://doi.org/10.1099/00207713-43-2-232

40. Abrini J, Naveau H, Nyns E (1994) *Clostridium autoethanogenum*, sp. nov., an anaerobic bacterium that produces ethanol from carbon monoxide. Arch Microbiol 161:345–351. https://doi.org/10.1007/s002030050065

41. Huhnke RL, Lewis RS, Tanner RS (2008) Isolation and characterization of novel clostridial species. US20080057554 A1. U.S. Patent and Trademark Office, Washington, DC

42. Zahn JA, Saxena J (2012) Ethanologenic *Clostridium* species, *Clostridium coskatii*. US 8143037 B2. U.S. Patent and Trademark Office, Washington, DC

43. Bengelsdorf FR, Poehlein A, Linder S, Erz C, Hummel T, Hoffmeister S, Daniel R, Dürre P (2016) Industrial acetogenic biocatalysts: a comparative metabolic and genomic analysis. Front Microbiol 7:1036. https://doi.org/10.3389/fmicb.2016.01036

44. Marcellin E, Behrendorff JB, Nagaraju S, de Tissera S, Segovia S, Palfreyman RW et al (2016) Low carbon fuels and commodity chemicals from waste gases – systematic approach to understand energy metabolism in a model acetogen. Green Chem 18:3020–3028. https://doi.org/10.1039/c5gc02708j

45. Richter H, Molitor B, Wei H, Chen W, Aristilde L, Angenent LT (2016) Ethanol production in syngas-fermenting *Clostridium ljungdahlii* is controlled by thermodynamics rather than by enzyme expression. Energy Environ Sci 9:2392–2399. https://doi.org/10.1039/c6ee01108j

46. Liew F, Martin ME, Tappel RC, Heijstra BD, Mihalcea C, Köpke M (2016) Gas fermentation – a flexible platform for commercial scale production of low-carbon fuels and chemicals from waste and renewable feedstocks. Front Microbiol 7:694

47. Fontaine FE, Peterson WH, McCoy E, Johnson MJ, Ritter GJ (1942) A new type of glucose fermentation by *Clostridium thermoaceticum*. J Bacteriol 43:701–715

48. Drake H, Küsel K, Matthies C (2013) Acetogenic prokaryotes. In: Rosenberg E, DeLong EF, Lory S, Stackebrandt E, Thompson F (Eds) The Prokaryotes. Springer, Boston, MA, S 354–420. doi:https://doi.org/10.1007/978-3-642-30141-4_61

49. Ragsdale SW (2004) Life with carbon monoxide. Crit Rev Biochem Mol Biol 39:165–195. https://doi.org/10.1080/10409230490496577

50. Schuchmann K, Müller V (2014) Autotrophy at the thermodynamic limit of life: a model for energy conservation in acetogenic bacteria. Nat Rev Microbiol 12:809–821. https://doi.org/10.1038/nrmicro3365

51. Kerby R, Zeikus JG (1983) Growth of *Clostridium thermoaceticum* on H_2/CO_2 or CO as energy source. Curr Microbiol 8:27–30. https://doi.org/10.1007/bf01567310

52. Pierce E, Xie G, Barabote RD, Saunders E, Han CS, Detter JC et al (2008) The complete genome sequence of *Moorella thermoacetica* (f. *Clostridium thermoaceticum*). Environ Microbiol 10:2550–2573. https://doi.org/10.1111/j.1462-2920.2008.01679.x

53. Poehlein A, Bengelsdorf FR, Esser C, Schiel-Bengelsdorf B, Daniel R, Dürre P (2015c) Complete genome sequence of the type strain of the acetogenic bacterium *Moorella thermoacetica* DSM 521^T. Genome Announc 3:e01159–e01115. https://doi.org/10.1128/genomea.01159-15

54. Bengelsdorf FR, Poehlein A, Esser C, Schiel-Bengelsdorf B, Daniel R, Dürre P (2015) Complete genome sequence of the acetogenic bacterium *Moorella thermoacetica* DSM 2955^T. Genome Announc 3:e01157–e01115. https://doi.org/10.1128/genomea.01157-15

55. Daniel SL, Hsu T, Dean SI, Drake HL (1990) Characterization of the H_2- and CO-dependent chemolithotrophic potentials of the acetogens *Clostridium thermoaceticum* and *Acetogenium kivui*. J Bacteriol 172:4464–4471. https://doi.org/10.1128/jb.172.8.4464-4471.1990

56. Weghoff MC, Müller V (2016) CO metabolism in the thermophilic acetogen *Thermoanaerobacter kivui*. Appl Environ Microbiol 82:2312–2319. https://doi.org/10.1128/aem.00122-16

57. Schoelmerich MC, Müller V (2019) Energy conservation by a hydrogenase-dependent chemiosmotic mechanism in an ancient metabolic pathway. Proc Natl Acad Sci U S A 116:6329–6334. https://doi.org/10.1073/pnas.1818580116

58. Basen M, Geiger I, Henke L, Müller V (2018) A genetic system for the thermophilic acetogenic bacterium *Thermoanaerobacter kivui*. Appl Environ Microbiol 84:e02210–e02217

59. Basen M, Müller V (2016) "Hot" acetogenesis. Extremophiles 21:15–26. https://doi.org/10.1007/s00792-016 0873-3

60. García JL, Vicente M, Galan B (2017) Microalgae, old sustainable food and fashion nutraceuticals. Microb Biotechnol 10:1017–1024

61. Blankenship RE, Tiede DM, Barber J, Brudvig GW, Fleming G, Ghirardi M, Gunner MR, Junge W, Kramer DM, Melis A, Moore TA, Moser CC, Nocera DG, Nozik AJ, Ort DR, Parson WW, Prince RC, Savre RT (2011) Comparing photosynthetic and photovoltaic efficiencies and recognizing the potential for improvement. Science 332:805–809. https://doi.org/10.1126/science.1200165

62. Mayer A, Weuster-Botz D (2017) Reaction engineering analysis of the autotrophic energy metabolism of *Clostridium aceticum*. FEMS Microbiol Lett 364. https://doi.org/10.1093/femsle/fnx219

63. de Morais MG, Costa JAV (2007) Carbon dioxide fixation by *Chlorella kessleri*, *C. vulgaris*, *Scenedesmus obliquus* and *Spirulina* sp. cultivated in flasks and vertical tubular photobioreactors. Biotechnol Lett 29:1349–1352

64. Yun YS, Lee SB, Park JM, Lee CI, Yang JW (1997) Carbon dioxide fixation by algal cultivation using wastewater nutrients. J Chem Technol Biotechnol 69:451–455

65. Chang EH, Yang SS (2003) Some characteristics of microalgae isolated in Taiwan for biofixation of carbon dioxide. Bot Bull Acad Sinica 44:43–52

66. Kishimoto M, Okakura T, Nagashima H, Minowa T, Yokoyama SY, Yamaberi K (1994) CO_2 fixation and oil production using micro-algae. J Ferment Bioeng 78:479–482
67. Kaewpintong K, Shotipruk A, Powtongsook S, Pavasant P (2007) Photoautotrophic high-density cultivation of vegetative cells of *Haematococcus pluvialis* in airlift bioreactor. Bioresour Technol 98:288–295
68. Huntley M, Redjalje D (2007) Carbon dioxide mitigation and renewable oil from photosynthetic microbes: a new appraisal. Mitig Adapt Strateg Glob Chang 12:573–608
69. Meiser A, Schmid-Staiger U, Trösch W (2004) Optimization of eicosapentaenoic acid production by *Phaeodactylum tricornutumin* the flat panel airlift (FPA) reactor. J Appl Phycol 16:215–225
70. Bitaubé E, Caro I, Perèz L (2008) Kinetic model for growth of *Phaedolactynum tricornitum* in intensive culture photobioreactor. Biochem Eng J 40:520–525
71. Morales M, Sánchez L, Revah S (2018) The impact of environmental factors on carbon dioxide fixation by microalgae. FEMS Microbiol Lett 365. https://doi.org/10.1093/femsle/fnx262
72. Costache TA, Acièn FG, Morales MM, Fernández-Sevilla JM, Stamatin I, Molina E (2013) Comprehensive model of microalgae photosynthesis rate as a function of culture conditions in photobioreactor. Appl Biotechnol 97:7627–7637
73. Cabello J, Morales M, Revah S (2015) Effect of the temperature, pH and irradiance on the photosynthetic activity by *Scenedesmus obtusiusculus* under nitrogen replete and deplete conditions. Bioresour Technol 181:128–135
74. Stanier RY, van Niel CB (1962) The concept of a bacterium. Arch Mikrobiol 42:17–35
75. Sánchez M, Bernal-Castillo J, Rozo C, Rodríguez I (2003) *Spirulina* (Arthrospira): an edible microorganism: a review. Uni Sci 8:7–24
76. Boussiba S, Vonshak A (1991) Astaxanthin accumulation in the green alga *Haematococcus pluvialis*. Plant Cell Physiol 32:1077–1082
77. Dore JE, Cysewski GR (2005) *Haematococcus* algae meal as a source of natural astaxanthin for aquaculture feeds. Cyanotech Corporation, Kailua-Kona, HI. http://www.ruscom.com/cyan/web02/pdfs/naturose/nrtl09.pdf. Accessed: 29. März 2018
78. Li J, Zhu D, Niu J, Shen S, Wang G (2011) An economic assessment of astaxanthin production by large scale cultivation of *Haematococcus pluvialis*. Biotechnol Adv 29:568–574
79. Nguyen KD (2013) Astaxanthin: a comparative case of synthetic vs. natural production. Chemical and biomolecular engineering publications and other works. http://trace.tennessee.edu/utk_chembiopubs/94
80. Gwo JC, Chiu JY, Chou CC, Cheng HY (2005) Cryopreservation of a marine microalga, *Nannochloropsis oculata* (*Eustigmatophyceae*). Cryobiology 50:338–343
81. Chua ET, Schenk PM (2017) A biorefinery for Nannochloropsis: induction, harvesting, and extraction of EPA-rich oil and high-value protein. Bioresour Technol 244:1416–1424. https://doi.org/10.1016/j.biortech.2017.05.124
82. Adarme-Vega TC, Lim DK, Timmins M, Vernen F, Li Y, Schenk PM (2012) Microalgal biofactories: a promising approach towards sustainable omega-3 fatty acid production. Microb Cell Factories 11:96. https://doi.org/10.1186/1475-2859-11-96
83. Timmers PH, Welte CU, Koehorst JJ, Plugge CM, Jetten MS, Stams AJ (2017) Reverse methanogenesis and respiration in methanotrophic archaea. Archaea 2017:1654237. https://doi.org/10.1155/2017/1654
84. Welte CU, Rasigraf O, Vaksmaa A, Versantvoort W, Arshad A, Op den Camp HJ, Jetten MS, Lüke C, Reimann J (2016) Nitrate- and nitrite-dependent anaerobic oxidation of methane. Environ Microbiol Rep 8:941–955
85. Ward N, Larsen Ø, Sakwa J, Bruseth L, Khouri H, Durkin AS, Dimitrov G, Jiang L, Scanlan D, Kang KH, Lewis M, Nelson KE, Methé B, Wu M, Heidelberg JF, Paulsen IT, Fouts D, Ravel J, Tettelin H, Ren Q, Read T, DeBoy RT, Seshadri R, Salzberg SL, Jensen HB, Birkeland NK, Nelson WC, Dodson RJ, Grindhaug SH, Holt I, Eidhammer I, Jonasen I, Vanaken S, Utterback T, Feldblyum TV, Fraser CM, Lillehaug JR, Eisen JA (2004) Genomic

insights into methanotrophy: the complete genome sequence of *Methylococcus capsulatus* (Bath). PLoS Biol 2:e303. https://doi.org/10.1371/journal.pbio.0020303

86. Welander PV, Summons RE (2012) Discovery, taxonomic distribution, and phenotypic characterization of a gene required for 3-methylhopanoid production. Proc Natl Acad Sci U S A 109:12905–12910

87. del Cerro C, García JM, Rojas A, Tortajada M, Ramón D, Galán B, Prieto MA, García JL (2003) Genome sequence of the methanotrophic poly-β-hydroxybutyrate producer *Methylocystis parvus* OBBP. J Bacteriol 194:5709–5710. https://doi.org/10.1128/JB.01346-12

88. Pieja AJ, Sundstrom CCS (2011) Poly-3-hydroxybutyrate metabolism in the type II methanotroph *Methylocystis parvus* OBBP. Appl Environ Microbiol 77:6012–6019. https://doi.org/10.1128/AEM.00509-11

89. Stein LY, Yoon S, Semrau JD, Dispirito AA, Crombie A, Murrell JC, Vuilleumier S, Kalyuzhnaya MG, Op den Camp HJ, Bringel F, Bruce D, Cheng JF, Copeland A, Goodwin L, Han S, Hauser L, Jetten MS, Lajus A, Land ML, Lapidus A, Lucas S, Médique C, Pitluck S, Woyke T, Zeytun A, Klotz MG (2010) Genome sequence of the obligate methanotroph *Methylosinus trichosporium* strain OB3b. J Bacteriol 192:6497–6498. https://doi.org/10.1128/JB.01144-10

90. Drake H (1995) Acetogenesis, acetogenic bacteria, and the acetyl-CoA 'Wood/Ljungdahl' pathway: past and current perspectives. In: H. Drake (Ed) Acetogenesis. Chapman & Hall, New York, NY, S 3–60

91. Ljungdahl LG (1986) The autotrophic pathway of acetate synthesis in acetogenic bacteria. Annu Rev Microbiol 40:415–450. https://doi.org/10.1146/annurev.micro.40.1.415

92. Wood HG (1991) Life with CO or CO_2 and H_2 as a source of carbon and energy. FASEB J 5:156–163

93. Drake HL, Gößner AS, Daniel SL (2008) Old acetogens, new light. Ann N Y Acad Sci 1125:100–128. https://doi.org/10.1196/annals.1419.016

94. Ragsdale SW, Ljungdahl LG, DerVartanian DV (1983) Isolation of carbon monoxide dehydrogenase from *Acetobacterium woodii* and comparison of its properties with those of *Clostridium thermoaceticum* enzyme. J Bacteriol 155:1224–1237

95. Ragsdale SW, Pierce E (2008) Acetogenesis and the Wood-Ljungdahl pathway of CO_2 fixation. Biochim Biophys Acta Proteins Proteom 1784:1873–1898. https://doi.org/10.1016/j.bbapap.2008.08.012

96. Drake HL, Hu SI, Wood HG (1980) Purification of carbon monoxide dehydrogenase, a nickel enzyme from *Clostridium thermocaceticum*. J Biol Chem 255:7174–7180

97. Raybuck SA, Bastian NR, Orme-Johnson WH, Walsh CT (1988) Kinetic characterization of the carbon monoxide-acetyl-CoA (carbonyl group) exchange activity of the acetyl-CoA synthesizing carbon monoxide dehydrogenase from *Clostridium thermoaceticum*. Biochemistry 27:7698–7702. https://doi.org/10.1021/bi00420a019

98. Pezacka E, Wood HG (1984) Role of carbon monoxide dehydrogenase in the autotrophic pathway used by acetogenic bacteria. Proc Natl Acad Sci U S A 81:6261–6265. https://doi.org/10.1073/pnas.81.20.6261

99. Grahame DA (2003) Acetate C–C bond formation and decomposition in the anaerobic world: the structure of a central enzyme and its key active-site metal cluster. Trends Biochem Sci 28:221–224. https://doi.org/10.1016/s0968-0004(03)00063-x

100. Hess V, Schuchmann K, Müller V (2013) The ferredoxin:NAD^+ oxidoreductase (Rnf) from the acetogen *Acetobacterium woodii* requires Na^+ and is reversibly coupled to the membrane potential. J Biol Chem 288:31496–31502. https://doi.org/10.1074/jbc.m113.510255

101. Tremblay P, Zhang T, Dar SA, Leang C, Lovley DR (2012) The Rnf complex of *Clostridium ljungdahlii* is a protontranslocating ferredoxin:NAD^+ oxidoreductase essential for autotrophic growth. MBio 4:1. https://doi.org/10.1128/mbio.00406-12

102. Ljungdahl LG (1994) The acetyl-CoA pathway and the chemiosmotic generation of ATP during acetogenesis. In: Drake H (ed) Acetogenesis. Chapman & Hall, New York, NY, pp 63–87. https://doi.org/10.1007/978-1-4615-1777-1_2
103. Ciferri O (1983) *Spirulina*, the edible microorganism. Microbiol Rev 47:551–578
104. Dürre P, Eikmanns BJ (2015) C1-carbon sources for chemical and fuel production by microbial gas fermentation. Curr Opin Biotechnol 35:63–72

Microbial Processes for the Conversion of CO_2 und CO

8

Dirk Weuster-Botz and Ralf Takors

Abstract

CO_2 is gaining industrial interest as a carbon source for microbial production of multi-carbon organic chemicals. For gas fermentation with acetogenic microorganisms, bubble columns and gas-lift bioreactors are the preferred choice for large-scale applications. Knowledge of the details of growth kinetics is a prerequisite for successfully developing gas fermentation processes. Trickle-bed biofilm reactors also represent a promising alternative to conventional pneumatically stirred systems. In addition, this chapter also provides insight into synthetic co-cultures and the current challenges and activities in true syngas conversion.

Keywords

Microbial production · Gas fermentation · Acetogenic microorganisms · Bubble columns · Gas-lift bioreactors · Trickle-bed biofilm reactors

8.1 CO_2 as Microbial Carbon Source

Reducing the emissions of greenhouse gases like CO_2 is a major demand for mankind to embank the effect of climate change. In this context great interest has emerged in biological CO_2-fixing processes which are able to effectively convert CO_2 emissions into multi-carbon organic chemicals and may therefore open the door

D. Weuster-Botz (✉)
Technical University of Munich (TUM), Garching, Germany
e-mail: dirk.weuster-botz@tum.de

R. Takors
University of Stuttgart, Stuttgart, Germany
e-mail: ralf.takors@ibvt.uni-stuttgart.de

© The Author(s), under exclusive license to Springer Nature Switzerland AG 2023   131
M. Kircher, T. Schwarz (eds.), *CO2 and CO as Feedstock*, Circular Economy and
Sustainability, https://doi.org/10.1007/978-3-031-27811-2_8

for the establishment of a circular carbon economy [1]. Interestingly, microorganisms have a whole range of metabolic possibilities for providing electrons for CO_2 reduction and the biosynthesis of organic products:

Like plants, photoautotrophic microalgae (prokaryotic cyanobacteria and eukaryotic unicellular organisms) use energy generated by sunlight to reduce and fix CO_2. Photoautotrophic microalgae are already being used industrially to produce antioxidants, pigments, and proteins as intracellular products from CO_2 using large-scale, open photobioreactors [2]. Here, large-area photobioreactors are required due to the shallow penetration of sunlight into aqueous algal suspensions resulting in "two-dimensional bioprocesses".

Aerobic, carboxidotrophic microorganisms can provide electrons for CO_2 reduction under chemolithoautotrophic conditions [3] by metabolically reacting H_2 with O_2 to form water, and using some of the electrons made available to reduce CO_2. Although the famous carboxidotrophic bacterium *Cupriavidus necator* (formerly *Ralstonia eutropha*) has been known for many decades and studied as an efficient producer of polyhydroxyalkanoates (PHA), no industrial exploitation using CO_2 as a carbon source has been reported so far. The crucial, non-limiting supply of H_2 and O_2 to the microorganisms is still difficult to achieve in bioreactors from a safety point of view [4].

Strictly anaerobic, acetogenic microorganisms can use hydrogen gas (H_2) as an electron source under autotrophic conditions. These bacteria reduce CO_2 with H_2 via the so-called Wood-Ljungdahl pathway to the metabolic intermediate acetyl-CoA. Electrons of H_2 are transferred to the intracellular mediator ferredoxin Fd using charged Fd_{red} to reduce CO_2 to CO and to enable the formation of formate. Both intermediates fuel the two branches of the Wood-Ljungdahl pathway, i.e. the carbonyl and the methyl branch, leading to acetyl-CoA. Notably, the reaction network is flexible [5] also allowing microbial growth on gas compositions with dominating CO content. The latter is characteristic for steel mill off-gas. Since the natural product spectrum starting from acetyl-CoA is essentially limited to the C2 products acetic acid and ethanol with these bacteria, ethanol is produced on an industrial scale from steel mill exhaust gases using acetogenic bacteria since 2018, which was pioneered by the company LanzaTech [6, 7].

A possibility to use electrons directly for microbial CO_2 reduction was described first in 2010 [8]. Here, the provision of electrons takes place at a cathode covered with a biofilm of acetogenic bacteria. Alike industrial microalgae application, this can be considered as a "two-dimensional process" because direct transfer of electrons is only possible in thin biofilms and thus large cathode areas must be installed in a production reactor. Industrial applications are not known so far.

The electrochemical CO_2 reduction via electrolysis is an alternative to the direct microbial consumption. Different companies have already established demonstration reactors or pilot plants [9]. Generally, CO_2 electrolysis has two main commercially desirable scenarios: the production of syngas (CO/H_2O) for subsequent methanol production and the electrosynthesis of value-added organic chemicals such as acetate or alcohols [10]. Noteworthy, electrochemical CO_2 reduction may also be coupled with microbial conversion. An example is the integration of CO and

H_2 tolerant bacteria in an electrochemical setting which renders the technology stoichiometric independently. In a collaborative work, Siemens, Evonik, and Covestro reported that this approach is already at the current technological status relevant for industrial purposes shown on the example of 1-butanol and 1-hexanol bioelectrosynthesis from CO_2 and H_2O [11].

8.2 Chances and Challenges of Gas Fermentation with Acetogenic Microorganisms

Acetogenic bacteria are able to produce CO_2-based chemicals in aqueous media by autotrophic conversion of CO_2 and H_2 and/or CO with high energetic efficiency of up to 70–90% [12]. Natural products are mainly acetate/ethanol followed by butyrate/butanol or 2,3-butanediol with reduced yields and small amounts of hexanoate/hexanol. Recombinant acetogens have already been engineered among others to produce acetone, isopropanol, 3-hydroxypropionate, mevalonate, isoprene, farnesene, butanoic acid butyl ester, methylethylketone, and isobutanol [13]. Carbon-negative production of acetone and isopropanol has recently been shown by gas fermentation at industrial pilot scale with engineered *Clostridium autoethanogenum* at volumetric production rates of up to $3\,\mathrm{g\,L^{-1}\,h^{-1}}$ and 90% selectivity [14].

The autotrophic growth of acetogenic microorganisms with CO_2 or CO as carbon source is strongly energy-limited, since no ATP surplus can be obtained via the Wood-Ljungdahl pathway even with acetate as metabolic end product. Consequently, acetogens depend on gaining ATP chemiosmotically, i.e. by coupling the membrane bound ATP formation from ADP to the re-import of protons that had been translocated via a network of interacting hydrogenases, transhydrogenases, and an export system. ATP yields may vary widely under autotrophic conditions ranging from - 4.5 mol ATP (mol product)$^{-1}$ to +4.4 mol ATP (mol product)$^{-1}$ depending on the final product, metabolic pathway, and electron donor (H_2 or CO) [15]. Products with ATP demand (negative ATP yields) can therefore only be produced autotrophically if other products with ATP gain are simultaneously produced by the cells.

However, acetogens may provide the required amount of ATP by increasing metabolic activities which is reflected by rising cell-specific CO_2/H_2 or CO uptake and product formation rates. In such a case, high carbon throughput may be beneficial for providing sufficient ATP and reduction equivalents to build up microbial biomass. However, the achievement of high cell-specific conversion rates is often limited by the low solubility of the gaseous substrates H_2 and CO in water. For example, the water solubility of pure hydrogen (H_2) under standard conditions (1 bar, 20 °C) is 65% lower on molar basis compared to pure oxygen (O_2). For pure carbon monoxide gas (CO), the water solubility is reduced by 75% on molar basis compared to O_2. Compared to typical aerobic heterotrophic bioprocesses with air supply, the maximum microbial cell concentrations in the bioreactor without cell retention are therefore reduced by a factor of 10-100 in gas fermentations with acetogenic bacteria, resulting in about $0.5–2\,\mathrm{g\,L^{-1}}$ dry cell mass, which basically leads to comparatively low volumetric productivities (space-time yields).

$$n_{gas,i} = k_L a_i \cdot \left(c_{i,L}^* - c_{i,L}\right) \tag{8.1}$$

Equation (8.1) describes the mass transfer rates of the gas component *i* into the liquid phase ($n_{gas,\,i}$) considering the mass transfer coefficient $k_L a_i$, the equilibrium gas concentration in the liquid $c_{i,L}^*$, and the gas concentration in the bulk of the liquid $c_{i,\,L}$. Consequently, mass transfer rates may be increased either by improving $k_L a_i$ or by increasing the gas concentration difference of the bracket term. Very often, mass transfer rates are limited because of the poor water solubility of the gaseous substrates H_2 and CO. CO_2 limitations usually do not occur because CO_2 solubility in water is comparatively high due to the carbonic acid equilibrium, especially at a pH in the neutral range ($6.0 < pH < 8.0$), which many acetogenic microorganisms require for growth.

Given that $k_L a_i$ values are proportional to the volumetric power input and to the superficial gas velocity, H_2 and CO transfer from a dispersed gas phase into the water may be improved by increasing the volumetric power input and by increasing gas flows. Furthermore, Eq. (8.1) outlines that rising the partial pressures of gas i is equally supportive as the equilibrium concentration $c_{i,L}^*$ is elevated. However, increasing the volumetric power input, for example by increasing the stirrer speed in stirred-tank bioreactors, quickly reaches economic limits, especially when low value-added products such as short-chain organic acids or alcohols are produced.

Increasing the partial pressures of H_2 and CO in the dispersed gas phase is therefore the method of choice, which is also applied industrially, using bioreactors with a liquid height of 30 m and above. The hydrostatic pressure of the water column above the gas distributor is used to increase the partial pressures in the dispersed gas phase at the bottom of the bioreactor. Therefore, bubble column or gas-lift reactors are preferred industrially for gas fermentations with suspended acetogenic microorganisms. Bubble column reactors represent simple mass transfer and reaction apparatus in which a gas phase is brought into contact with a liquid phase. These reactors are characterized by their simple design and, in contrast to the stirred-tank reactor, by the complete absence of mechanically moving internals. On an industrial scale, the power input of bubble column reactors is achieved solely by the isothermal expansion of the gas phase dispersed at the bottom of the reactor. As a result, the power input increases linearly with increasing gas flow rate and nonlinearly with the height of the water column above the gas distributor. Increasing partial pressures and increasing power input by an increase of the water column thus provide, in principle, improved mass transfer between the gas and liquid phases (in the lower part) of bubble column or gas-lift reactors.

However, improving the gas-liquid mass transfer needs acetogens able to survive at elevated H_2 or CO partial pressures, which can reach several bar at the bottom of industrial bubble column reactors depending on the gas mixture supplied. Unfortunately, no or less quantitative data has been published so far on substrate inhibition kinetics of acetogenic microorganisms [16–19]. The kinetics of acetogenic microorganisms as a function of gaseous substrates can differ greatly (Fig. 8.1). For example, the acetogenic bacterium *Clostridium aceticum* shows strong CO

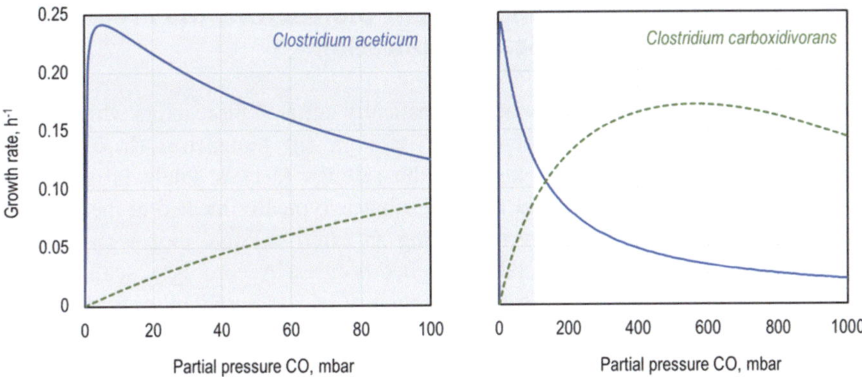

Fig. 8.1 Autotrophic growth kinetics of *Clostridium aceticum* (blue) und *Clostridium carboxidivorans* (green, broken line) as function of the limiting CO partial pressure [20, 21]

excess inhibition with a very low inhibition constant of $K_I = 87$ mbar CO and a growth optimum at 5.4 mbar CO [20], whereas *Clostridium carboxidivorans* has a much higher inhibition constant of $K_I = 1259$ mbar CO with a growth optimum at 565 mbar CO, which is two orders of magnitude higher than *Clostridium aceticum* [21]. The inhibition constant indicates when the half-maximal growth rate is reached with increasing substrate concentration.

The partial pressures of CO, H$_2$, and CO$_2$ of the gas bubbles rising in the bubble column reactor change not only due to the decreasing hydrostatic pressure of the remaining water column, but also due to the locally different consumption or formation of gas components by microbial metabolic processes. Besides, many acetogenic microorganisms are unable to simultaneously utilize H$_2$ and CO$_2$, especially at higher CO partial pressures. Only when the local CO partial pressure has become low enough, inhibition of CO$_2$ reduction with H$_2$ does no longer occur. Notably, the preferred use of CO compared to CO$_2$ + H$_2$ also mimics the increased thermodynamic driving force for CO finally leading to maximum electron catch in the reduced metabolic products and a maximum availability of the internal redox mediator reduced ferredoxin Fd_{red} [5].

Furthermore, some acetogenic bacteria show a typical two-phase process behavior: If the pH in the aqueous phase is high enough (usually $>$ pH 6), the anaerobic bacteria can grow and organic acids, preferably acetic acid and, with significantly lower concentrations, butyric acid or even hexanoic acid, are formed initially. In this "acetogenic phase", the pH decreases because of the formation of the acids. As a consequence, the growth of the acetogenic bacteria continuously decreases until a pH minimum is reached, usually below pH 5.0. At this pH, alcohol formation starts, whereby not only the formation of alcohols from the gaseous substrates takes place, but also, in particular, the previously formed organic acids are reduced to the corresponding alcohols. In this "solventogenic phase", growth no longer occurs and, due to the microbial consumption of the acids, the pH in the aqueous medium rises again.

8.3 Bubble Columns and Gas-Lift Bioreactors, the Preferred Choice for Large Scale Application

Figure 8.2 provides an overview of pneumatically agitated bioreactors which is the design category that bubble columns (a) and gas-lift bioreactors (b, c) belong to. Notably, volumetric power input of the settings (a-c) is solely provided by dispersion of compressed gas via a sparger that is typically located at the bottom. The jet loop setup (d) differs by integrating an external liquid loop with a pump injecting gas and liquid via specially designed nozzles. Accordingly, settings (a-c) share the common characteristic that the quality of gas/liquid mass transfer and mixing directly depends on power input via compressed gas. That differs from conventional stirred tanks where mixing quality is decoupled from aeration because of the additional mechanical power input. In bubble columns, a parabolic upward liquid velocity exists in the center of the column that turns to a downward liquid velocity close to the walls. There, no liquid velocity is found. The liquid velocity profile mimics the effects of rising bubbles displacing the surrounding liquid finally creating a large circulation loop in the bioreactor. Notably, rising bubbles also cause trailing eddies that diminish slowly. As a consequence, longitudinal dispersion is observed that depends on the gas flow regime. The latter may be categorized as homogeneous bubbly flow, heterogeneous churn-turbulent flow, and slug flow. The first occurs at low superficial gas velocity comprising (small) bubbles with narrow size and velocity distributions and is predominately observed in columns with large diameters. If superficial gas velocities exceed the flooding threshold, which is found at approximately 0.05 m s^{-1} for bubble sizes of 0.05–0.1 m, bubble coalescence is getting dominating. Bubble clusters with relatively high rising velocity occur. Accordingly, bubble size and velocity distribution show multimodal distributions. This churn-turbulent flow often happens in industrial applications. If column diameters are small, the flow regime may even lead to slug flow with bubbles filling large parts of the column diameter.

Said flow regimes are not only characteristic for bubble columns (Fig. 8.2a) but also for the riser section in gas-lift bioreactors (b, c). Typically, downcomers are operating in flow regime. At low power inputs, liquid may even circulate without any bubbles. However, preferred industrial applications apply high specific power inputs leading to re-circulating gas fractions. As outlined in Fig. 8.2, gas-lift bioreactors may consider internal draft tubes for hosting the riser or may comprise external downcomers. There, additional heat exchangers may be installed, if necessary. The jet loop setup 8.2 d considers mechanical power input via a pump, integrated in the external loop. Together with smart nozzle design, such concepts may improve gas/liquid mass transfer and may reduce mixing times.

Bubble columns and gas-lift bioreactor typically show lean geometries, i.e. height/diameter ratios >3. Such designs enable high solubility of the gas thereby improving the gas/liquid mass transfer according to Eq. (8.1). Notably, small diameters also cause relatively high superficial gas velocities which equally contribute for improving mass transfer and for reducing mixing times. On the downside, efforts for compressing gas rise together with foaming tendencies. Thanks to the

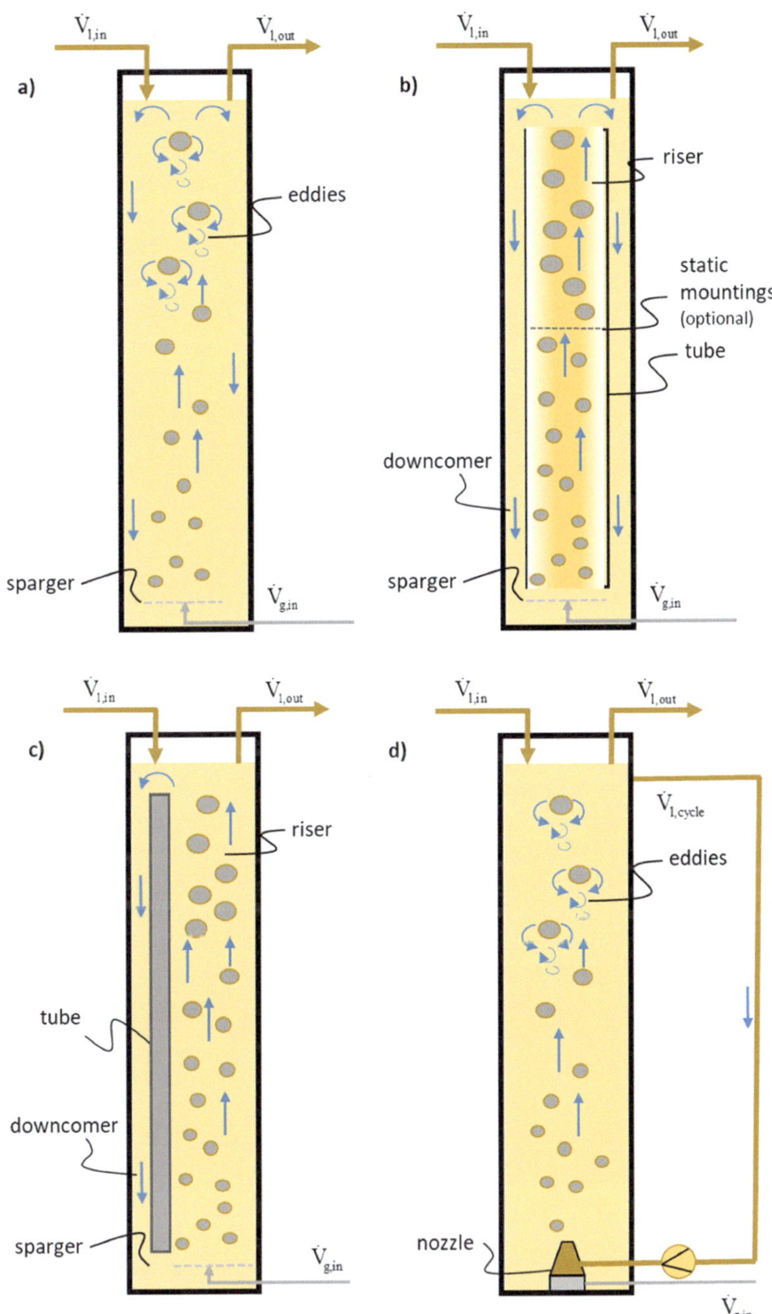

Fig. 8.2 Basic settings of pneumatically agitated bioreactors. (**a**) bubble column with sparger at the bottom, circulating flows and eddies, created by rising bubbles; (**b**) gas-lift reactor with internal loop based on the riser inside the draft tube and the downcomer; (**c**) gas-lift reactor with external downcomer; (**d**) jet loop reactor with external pump and nozzle for gas dispersion

relatively simple setup, bubble columns and gas-lift reactors require significantly lower investments and operational costs than conventional stirred tank reactors. However, their conceptual design is challenging. Obviously, large vertical gradients of dissolved gases exist that are superimposed by local gradients around moving bubbles. Given the complex bubble flow regimes local liquid velocity heterogeneities disturb the fairly simplifying picture of parabolic up- and downwards flows. Accordingly, the necessity to accurately describe bubble breakage and coalescence is a prerequisite for predicting reaction conditions in bubble columns and gas-lift bioreactors.

Basic concepts for designing bubble column bioreactors have been developed in the 80s and 90s predominately focusing on aerobic cultivations studying columns in an integral manner [22–24]. In the focus of investigations were—and still are—models to predict liquid and gas phase dispersion coefficients D_L and D_G, circulation and mixing times and k_La values. All are important parameters to design large-scale bubble columns and gas-lift bioreactors.

Irrespective of the merit of those approaches their transfer to different columns and operation conditions turned out to be rather problematic [25]. The advent of *computational fluid dynamics* (*cfd*) coinciding with boosting computational power revealed that 3-dimensional (3D) grid modelling is essential to describe fluid dynamics properly [26]. Improvement of previously applied 1D modelling could be achieved by implementing novel mass transfer approaches [27] and by iteratively feeding the results of fully unsteady hydrodynamics on the convective and dispersive mass transport back to the (homogeneously mixed) zones of the column [25]. Recently, this principle has been revisited by [28] outlining its applicability to probe the parameter space of column design and operation. First model-based investigations of large-scale bubble column performance for syngas fermentations have been published by [29, 30] who linked 1D hydrodynamic and mass transfer modelling with stoichiometric flux balancing considering *Clostridium ljungdahlii*. Those studies were extended by *cfd* simulations of [31] who applied the tool of *lifeline analysis* for *Clostridium ljungdahlii* recording and analyzing the movement of 120,000 cells in a 125 m³ bubble column. Alternatively, biothermodynamics and mass transfer can be combined to a hybrid large-scale bubble column model for simulating microbial syngas fermentations [32].

8.4 Getting a Grip on Growth Kinetic Details, a Prerequisite for Successful Bioprocess Development

Acetogenic microbes show a broad range of growth characteristics that need to be known quantitatively before starting any type of bioprocess developments. Stirred tank reactors that are ideally mixed are particularly suitable on a laboratory scale for kinetic studies of the metabolic performance of acetogenic microorganisms under autotrophic conditions. Controlling the gas supply with individually adjustable partial pressures and volume flows of H_2, CO, and CO_2 and on-line off-gas analysis to determine the consumption rates of the individual gas components play an important role in addition to the control of the standard reaction conditions like pH, temperature, and power input, respectively.

Gas fermentations at controlled conditions provide basic information on the metabolic performance of acetogenic microorganisms without gas-liquid mass transfer limitations, since high volumetric power inputs are possible with lab-scale stirred tank reactors (see for example [33–37]). Growth and product formation kinetics can thus be identified as a function of dissolved gas concentrations or gas partial pressures at low cell concentrations [20, 21]. For the identification of growth kinetics, steady-state investigations in a continuously operated stirred tank reactor are more suitable, in principle, even for gas fermentations, since arbitrary growth rates can be specified by setting different hydraulic residence times (e.g. [38]). However, the low growth rates of acetogenic microorganisms under autotrophic conditions represent a major obstacle, especially under unfavorable reaction conditions (pH, temperature, …), since, on the one hand, very long experimental times are required until steady-state conditions are reached, and, on the other hand, achieving stable operating states in continuously operated stirred tank reactors is only possible to a very limited extent in the case of substrate inhibition.

Continuous process modes are essential to achieve sufficiently high volumetric productivities (space-time yields) in gas fermentations, which may be crucial to ensure economically viability. This holds particularly true for the production of mass products with low added value such as organic acids and alcohols. In the case of suspended microorganisms, cell retention in the continuously operated bioreactor can be achieved by submerged microfiltration membranes, as already used on a large scale in biological wastewater treatment [39]. Using the example of gas fermentation (CO$_2$/H$_2$) with *Acetobacterium woodii*, it was indeed possible to achieve increased cell densities and thus also significantly increased volumetric acetate productivities of more than 147 g L^{-1} d^{-1} in the continuously operated stirred tank reactor with submerged microfiltration membranes compared with 19 g L^{-1} d^{-1} in the batch process or in the continuous process without cell retention. However, the product concentrations were significantly reduced from 59 g L^{-1} acctate in the batch process to 17.6–23.5 g L^{-1} acetate due to the continuous withdrawal of the products solved in the aqueous phase in the continuous process [40].

In two-phase gas fermentation processes, higher volumetric productivities than in batch modes may also be achieved by connecting two continuously operated bioreactors in series. For this purpose, growth and acid formation take place in the first reactor at a higher pH ("acetogenic phase"). Thereof, cultivation media containing acetogenic microorganisms is drained into the second reactor where organic acids are reduced to the corresponding alcohols at lower pH ("solventogenic phase"). Furthermore, they form additional alcohols from the gaseous substrates, as shown in the example of continuous ethanol production with *Clostridium ljungdahlii* [41]. In the continuous conversion of CO to produce alcohols (ethanol, 1-butanol, 1-hexanol) with *Clostridium carboxidivorans* in a stirred tank cascade operated at pH 6 in the first reactor and pH 5 in the second reactor (Fig. 8.3), higher space-time yields were obtained as expected compared with the batch process (increase by a

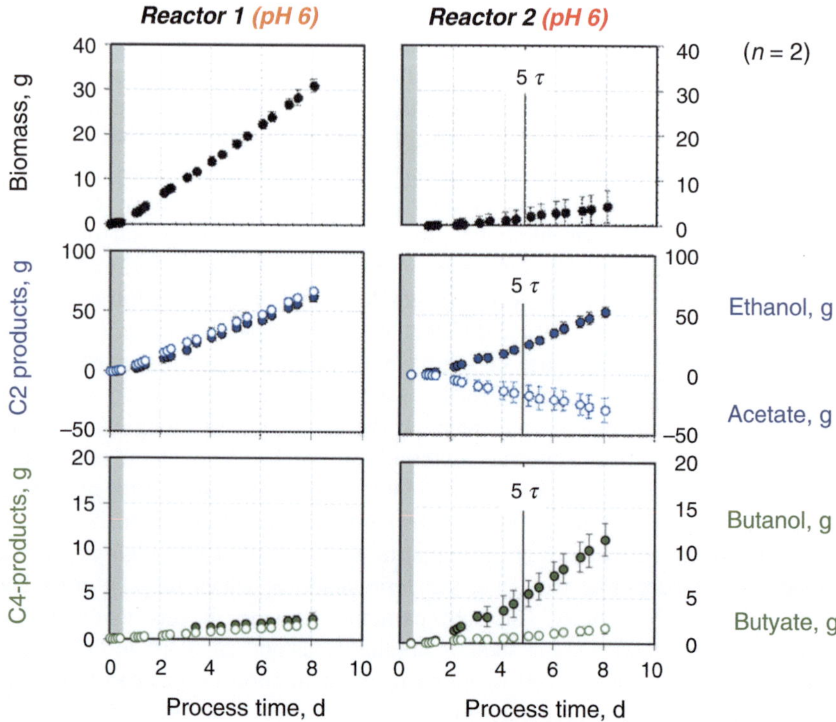

Fig. 8.3 Cumulative production of acids and alcohols from CO in a continuously operated stirred-tank reactor cascade with *C. carboxidivorans* with a dilution rate of 0.12 h^{-1} in the first reactor (V = 1 L) and 0.08 h^{-1} in the second reactor (V = 1.5 L), pH 6.0 in the first reactor and pH 5.0 in the second reactor, continuous gassing with CO = 600 mbar, CO_2 = 400 mbar, F_{Gas} = 0.083 vvm, T = 37 °C. Biomass and alcohols are shown in full circles, acids are shown in open circles. Standard deviations result from two independent experiments performed in the bioreactor cascade. The batch phase is shaded in grey. Data from Doll (2018) [21]

factor of 6 for the alcohols ethanol, 1-butanol, and 1-hexanol). Surprisingly, however, the product concentrations in the continuous cascade process were also increased by a factor of 3 on average [42]. This is due to the unusual property of *Clostridium carboxidivorans* to form equal amounts of alcohols (ethanol and 1-butanol) and organic acids (acetate and butyrate) already in the "acetogenic phase" in the first reactor during growth.

In 2018, the first industrial gas fermentation plant started operation for ethanol production from steel mill waste gases with a capacity of 46,000 metric tons per year in Caofeidian, China [7]. The gas fermentation process developed by LanzaTech, a biotech company headquartered in Chicago, Illinois/USA, uses *Clostridium autoethanogenum* in bubble column reactors. In the pilot plant in Shougang (China), LanzaTech was able to convert 70–75% of the CO contained in the steel mill off-gas to ethanol [43]. The first industrial gas fermentation plant in Europe is currently under construction at the world's largest steel producer ArcelorMittal in

Ghent (Belgium). The plant is designed to handle 50,000 Nm^3 h^{-1} of steel mill waste gas and is expected to have an annual capacity of 60,000 m^3 of ethanol [44]. Carbon-negative production of acetone and isopropanol has recently been shown by gas fermentation at industrial pilot scale with engineered *Clostridium autoethanogenum* at volumetric production rates of up to 3 g L^{-1} h^{-1} and 90% selectivity [14]. Aspects on future industrial applications of gas fermentations with acetogenic microorganisms considering existing industrial infrastructure and market requirements are summarized in [45].

8.5 Trickle-Bed Biofilm Reactors, a Promising Alternative for Conventional Pneumatically Agitated Settings

Trickle-bed biofilm reactors with immobilized acetogenic bacteria represent an interesting alternative for continuous gas fermentations. In the classical design, a carrier material is filled in a cylindrical vessel, the (inner) surface of which is covered with a biofilm. The liquid phase is evenly distributed on the surface of the filling and flows over the carrier material with the immobilized microorganisms as a thin trickling film. At the bottom of the trickle-bed biofilm reactor, the liquid phase is collected and discharged. In contrast to bubble column reactors, the gas phase is not dispersed in the liquid phase, but represents the continuous phase in trickle-bed biofilm reactors. The pressure drop of the gas phase transported either cocurrently or countercurrently through the trickle-bed biofilm reactor is thus negligible, and the power input of a trickle-bed biofilm reactor is determined solely by the transport of the liquid phase to the head of the reactor. Despite the low volumetric power input, the mass transfer rates of CO and H$_2$ from the gas phase to the liquid film and to the biofilm surface are usually sufficiently high due to the thin liquid film. Due to the low operating costs at high mass transfer rates, trickle-bed biofilm reactors and special designs, such as so-called submerged trickle bed reactors (horizontally rotating fixed-bed reactors that are partially "submerged" in the liquid phase), are widely used on a technical scale in biological wastewater treatment.

A limiting factor is the diffusive mass transport in the biofilm due to the low gas solubility in water, so that theoretically, depending on cell-specific gas uptake rates and gas partial pressures, only a small biofilm thickness of a few 10 µm can be supplied with the gaseous substrates. Very few studies have investigated the use of trickle-bed biofilm reactors (Fig. 8.4) in laboratory-scale gas fermentations [46–51]. Biofilm formation seems to be easily achievable with most acetogenic microorganisms, although no systematic studies on this exist to date. High CO conversions of more than 90% were observed when *Clostridium ragsdalei* was used in a tricke-bed biofilm reactor with simple, non-porous glass beads of 6 mm in diameter [48]. Hydrogen conversion of more than 80% was obtained when *Clostridium carboxidivorans* was used in a submerged trickle-bed biofilm reactor with non-porous HDPE support material with a volume-specific surface area of 500 m^2 m^{-3}. Compared to a gas fermentation in a continuously operated stirred

Fig. 8.4 In-situ sterilizable trickle-bed biofilm reactor for reaction engineering studies on continuous gas fermentations with immobilized acetogens at total pressures of up to 3.5 bar at the Chair of Biochemical Engineering, Technical University of Munich (Photo: Tobias Hase)

tank reactor under comparable operating conditions, the volumetric ethanol productivity was increased by a factor of 3.3 [50].

Axial gradients of pH and product concentrations in the aqueous trickling film are unavoidable as a function of trickle-bed biofilm reactor height on an industrial scale (10–15 m). Axial gradients could be reduced by recirculating the aqueous phase, but this would result in an increase in volumetric power input. Compared to bubble column or gas-lift reactors, the liquid volume (reaction or working volume) in trickle-bed biofilm reactors is reduced to about 20% of the total volume due to the volume of the inert support material and the gas volume, compared to about 80% of the total volume in bubble column reactors with a gas content of 20%. The reduced working volume of trickle-bed biofilm reactors is partially balanced by improved mass transfer at high gas conversion and by higher biocatalyst concentrations in the working volume due to immobilization. In addition to solving fundamental questions about scale-up, a technical application of trickle-bed biofilm reactors in gas fermentation will only become possible if stable operation over long process times is made possible, i.e., if a stable biofilm can be formed, especially if recombinant acetogenic microorganisms are applied for improved production of natural, but also non-natural products. To date, no studies have been published on this subject.

8.6   Synthetic Co-cultures and Challenges in Converting Real Syngas, a Glimpse on Current Activities

The value of the fermentation product increases with the number of carbon atoms, e.g. hexanoate has a two times higher market value than that of ethanol [52]. In addition to the higher product value, the recovery of middle chain fatty acids (>C6) is less costly due to their lower solubility in water in contrast to ethanol and short chain fatty acids [53, 54]. But the production rates and concentrations of butyrate/ butanol or hexanoate/hexanol are usually low in gas fermentations with acetogens [55]. The combined application of pathway optimization, strain optimization, carbon flux optimization and fermentation process optimization as shown recently by Liew et al. (2022) [14] on the example of the autotrophic acetone and isopropanol production with production rates of up to 3 g L^{-1} h^{-1} (72 g L^{-1} d^{-1}) is a promising way to overcome these limitations, especially if volatile products with low carbon numbers are produced. As an alternative approach, microbial formation of middle chain carboxylic acids can be achieved from acetate and ethanol by chain elongation via the reversed β-oxidase pathway and has attracted interest in the recent years [56–59].

The chain elongating *Clostridium kluyveri* is an anaerobic microorganism converting a mixture of acetate and ethanol to butyrate, hexanoate, hydrogen gas (H$_2$) and small amounts of octanoate [60]. Carbon dioxide is necessary for growth of *Clostridium kluyveri* [61]. C^{14}-labeling studies showed that more than 25% of the cellular carbon comes from carbon dioxide [62] and the availability of CO$_2$ was reported to be necessary for chain elongation and resulted in highest hexanoate concentration in a batch process applying a stirred-tank bioreactor with controlled pH [63, 64].

Chain elongation of a syngas fermentation effluent with pure cultures of *Clostridium kluyveri* was studied in continuously operated bioreactors. Carbon conversion efficiencies of 90% were possible and hexanoate was produced with a space-time yield of 4.6 g L^{-1} d^{-1}. Even octanoate was produced, but concentrations were below 0.3 g L^{-1} [58].

Instead of operating two bioreactors in series, co-cultivation of the acetogen *Clostridium ljungdahlii* with the chain elongating *Clostridium kluyveri* in a continuously operated bioreactor system gassed with 60% CO, 35% H$_2$, and 5% CO$_2$, respectively, resulted in a hexanoate production rate of 4.2 g L^{-1} d^{-1}. Two-third of the ethanol produced by the acetogen was converted by *Clostridium kluyveri* into hexanoate, which was reduced again by *Clostridium ljungdahlii* to 1-hexanol with a volumetric productivity of 3.2 g L^{-1} d^{-1} [65]. The correct pH was found to be an important parameter for this co-culture to function (pH 6). This can be expected, because the proportion of ethanol to acetate formed by *C. ljungdahlii* is pH dependent [66] and chain elongation as well as growth of *C. kluyveri* is not possible at pH 5.5 or less [67, 68].

The partial pressure of carbon monoxide was shown to be another critical factor for successful co-cultivation as the metabolism of a pure culture of *C. kluyveri* was already inhibited at 50 kPa CO in the gas phase [69], whereas in co-cultures with acetogens high CO partial pressures are necessary to avoid CO-limitations and reduced acetate/ethanol production rates.

Synthetic co-cultures of autotrophic *C. carboxidivorans* and chain elongating *C. kluyveri* were studied in batch operated stirred-tank bioreactors with continuous CO/CO_2-gassing and monitoring of the cell counts of both clostridia by flow cytometry after fluorescence *in situ* hybridization (FISH-FC). Two specific 23S rRNA oligonucleotide probes were designed for the monitoring of *C. kluyveri* and *C. carboxidivorans* in co culture [70]. At 800 mbar CO, chain elongation activity was observed at pH 6.0, although growth of *C. kluyveri* was restricted. Organic acids produced by *C. kluyveri* were reduced by *C. carboxidivorans* to the corresponding alcohols butanol and hexanol. This resulted in a three-fold increase in final butanol concentration compared with a mono-culture of *C. carboxidivorans* and enabled hexanol production. [71].

Recently, a synthetic co-culture of an *Acetobacterium woodii* mutant converting CO_2 and H_2 to lactate combined with the chain-elongating *Clostridium drakei* for the production of hexanoate from lactate was shown to function for the first time [72]. For autotrophic lactate production, a recombinant *A. woodii* strain was used with a deleted lactate dehydrogenase complex and an inserted D-lactate dehydrogenase (LdhD) originating from *Leuconostoc mesenteroides*.

Although co-cultivation processes with acetogens and chain elongating bacteria seem to be very promising approaches for the microbial production of middle chain carboxylic acids/alcohols from C1-gases, the small operating windows (e.g. pH, CO partial pressure and others) for optimal interaction of a synthetic co-culture needs further research for better understanding and simulation of the mutual microbial interactions in bioreactors on an industrial scale.

Industrial residual gas streams containing CO_2 and CO, but also synthesis gases from the gasification of residual biomass or non-recyclable plastic waste, contain a whole range of gaseous impurities. In addition to the non-critical inert gas N_2, the gasification of residual biomass can produce, for example, methane, various gaseous hydrocarbons, such as ethene and ethane, and various gaseous sulfur and nitrogen compounds (such as H_2S, COS, SO_2 or HCN, NH_3, NO_X) in addition to CO_2, CO and H_2 [73].

In general, it is found that the requirements for syngas purification in gas fermentation are significantly lower than in the classical Fischer-Tropsch process for the catalytic production of hydrocarbons (see, for example, [44, 74]). However, quantitative studies on inhibitory effects of individual impurity components of synthesis gases are few. In *Clostridium ragsdalei*, the effect of NH_3 on metabolism was investigated and, as a result, it was postulated that medium costs could be saved if the nitrogen supply of acetogenic microorganisms could be at least partially covered by syngas contamination NH_3 [75].

Recently it was shown that sulfide supply enhanced autotrophic production of alcohols with *Clostridium ragsdalei* in stirred-tank bioreactors with continuous gassing [76]. Furthermore, it has been shown that increased NO concentrations in the gas phase inhibit the hydrogenase of *Clostridium carboxidivorans*, thereby inhibiting growth and eventually affecting the product spectrum. Complete inhibition of the hydrogenase occurs at 160 ppm (0.288 μM) NO content in the gas phase [77]. The syngas contaminants NH_3 and H_2S had a positive effect on both autotrophic growth and alcohol formation (ethanol, 1-butanol, 1-hexanol) in batch processes with *C. carboxidivorans*, but NO_x contaminants in syngas should be selectively removed [78]. The effects of defined additions of the syngas contaminants were studied in autotrophic batch processes with *C. autoethanogenum*, *C. ljungdahlii*, and *C. ragsdalei*. *C. ljungdahlii* showed the highest tolerance to ammonium and nitrate, whereas *C. ragsdalei* was positively influenced by the presence of H_2S. Quantitative goals for the purification of syngas were identified for each of the acetogens studied. Syngas purification should in particular focus on the NO_X impurities that caused the highest inhibiting effect and maintain the concentrations of NH_3 and H_2S within an acceptable range (e.g., $NH_3 < 4560$ ppm and $H_2S < 108$ ppm). [79].

Even "low" HCN concentrations in syngas from biomass gasification were the reason of the failed start-up of a syngas fermentation plant for ethanol production at the former Ineos Bio Indian River Biorefinery in Vero Beach, Florida, USA [80]. In contrast, steel mill waste gases, such as those used at LanzaTech, are characterized by very low impurities of less than 1 ppm HCN [44]. Syngas purification prior to gas fermentation seems necessary, but details are unknown due to the lack of quantitative data on inhibitory effects of individual impurity components of synthesis gases on acetogenic microorganisms.

References

1. Mohan SV, Modestra JA, Amulya K, Butti SK, Velvizhi G (2016) A circular bioeconomy with biobased products from CO$_2$ sequestration. Trends Biotechnol 34:506–518
2. Khan MI, Shin JH, Kim JD (2018) The promising future of microalgae: current status, challenges, and optimization of a sustainable and renewable industry for biofuels, feed, and other products. Microb Cell Factories 17:36
3. Yu J (2018) Fixation of carbon dioxide by a hydrogen-oxidizing bacterium for value added products. World J Microbiol Biotechnol 34:89
4. Schlegel HG, Gottschalk G, von Bartha R (1961) Formation and utilization of poly-[beta]-hydroxybutyric acid by Knallgas bacteria (*Hydrogenomonas*). Nature 191:463–465
5. Hermann M, Teleki A, Weitz S, Niess A, Freund A, Bengelsdorf F, Takors R (2020) Electron availability in CO$_2$, CO, and H$_2$ mixtures constrains flux distribution, energy management and product formation in *Clostridium ljungdahlii*. Microb Biotechnol 13:1831–1846
6. LanzaTech (2018) LanzaTech press release. https://www.lanzatech.com/2018/06/08/worlds-first-commercial-waste-gas-ethanol-plant-starts/. Accessed 18.10.2019
7. Gehrmann S, Tenhumberg N (2020) Production and use of sustainable C2-C4 alcohols – an industrial perspective. Chem Ing Tech 92:1444–1458
8. Nevin KP, Woodard TL, Franks AE, Summers ZM, Lovley DR (2010) Microbial electrosynthesis: feeding microbes electricity to convert carbon dioxide and water to multicarbon extracellular organic compounds. MBio 1(2):e00103–e00110
9. Sánchez OG, Birdja YY, Bulut M, Vaes J, Breugelmans T, Pant D (2019) Recent advances in industrial CO$_2$ electroreduction. Curr Opin Green Sustain Chem 16:47–56
10. Wu J, Sharifi T, Gao Y, Zhang T, Ajayan PM (2018) Emerging carbon-based heterogeneous catalysts for electrochemical reduction of carbon dioxide into value-added chemicals. Adv Mater 31:1804257

11. Haas T, Krause R, Weber R, Demler M, Schmid G (2018) Technical photosynthesis involving CO_2 electrolysis and fermentation. Nat Cat 1:32–39
12. Claassens NJ, Cotton CAR, Kopljar D, Bar-Even A (2019) Making quantitative sense of electromicrobial production. Nat Cat 2:437–447
13. Bengelsdorf FR, Beck MH, Erz C, Hoffmeister S, Karl MH, Riegler P, Wirth S, Poehlein A, Weuster-Botz D, Dürre P (2018) Bacterial anaerobic synthesis gas (syngas) and $CO_2 + H_2$ fermentation. Adv Appl Microbiol 103:143–221
14. Liew FE, Nogle R, Abdalla T, Rasor BJ, Canter C, Jensen RO, Wang L, Strutz J, Chirania P, De Tissera S, Mueller AP, Ruan Z, Gao A, Tran L, Engle NL, Bromley JC, Daniell J, Conrado R, Tschaplinski TJ, Giannone RJ, Hettich RL, Karim AS, Simpson SD, Brown SD, Leang C, Jewett MC, Köpke M (2022) Carbon-negative production of acetone and isopropanol by gas fermentation at industrial pilot scale. Nat Biotechnol 40:335–344
15. Bertsch J, Müller V (2015) Bioenergetic constraints for conversion of syngas to biofuels in acetogenic bacteria. Biotechnol Biofuels 8:210
16. Vega JL, Holmberg VL, Clausen EC, Gaddy JL (1988) Fermentation parameters of *Peptostreptococcus productus* on gaseous substrates (CO, H_2/CO_2). Arch Microbiol 151:65–70
17. Chang I-S, Kim D-H, Kim B-H, Shin P-K, Sung H-C, Lovitt RW (1998) CO fermentation of *Eubacterium limosum* KIST612. J Microbiol Biotechnol 8:134–140
18. Skidmore BE, Baker RA, Banjade DR, Bray JM, Tree DR, Lewis RS (2013) Syngas fermentation to biofuels: effects of hydrogen partial pressure on hydrogenase efficiency. Biomass Bioenergy 55:165–162
19. Mohammadi M, Mohamed AR, Najafpour GD, Younesi H, Uzir MH (2014) Kinetic studies on fermentative production of biofuel from synthesis gas using *Clostridium ljungdahlii*. Sci World J 910590
20. Mayer A, Schädler T, Trunz S, Stelzer T, Weuster-Botz D (2018) Carbon monoxide conversion with *Clostridium aceticum*. Biotechnol Bioeng 115:2740–2750
21. Doll (2018) Reaction engineering analysis of *Clostridium carboxidivorans* for autotrophic production of alcohols. Doctoral thesis, Technical University of Munich
22. Schügerl K, Lübbert A (1994) Pneumatically agitated bioreactors. In: Merchuk JC, Asenjo JA (eds) Bioreactor system design. CRC Press, Boca Raton, FL. https://doi.org/10.1201/9781482277470
23. Decker WD (1985) Bubble column bioreactors In: Fundamentals of biochemical engineering (editor: Brauer H) Vol 2 of Biotechnology (Rehm HJ and Reed G) VCH Weinheim, p 445
24. Merchuk JC (1991) Tower reactor models. In: Measuring, modelling and control (editor: Schügerl K) Vol 4 of Biotechnology (Rehm HJ and Reed G) VCH Weinheim, p 349
25. Bauer M, Eigenberger G (1999) A concept for multi-scale modeling of bubble columns and loop reactors. Chem Eng Sci 54:5109–5117
26. Ekambara K, Mahesh T, Dhotre T, Joshi JB (2005) CFD simulations of bubble column reactors: 1D, 2D and 3D approach. Chem Eng Sci 60:6733–6746
27. Haut B, Cartage T (2005) Mathematical modeling of gas-liquid mass transfer rate in bubble columns operated in the heterogeneous regime. Chem Eng Sci 60:5937–5944
28. Siebler F, Lapin A, Takors R (2020) Synergistically applying 1-D modeling and CFD for designing industrial scale bubble column syngas bioreactors. Eng Life Sci 20:239–251
29. Chen J, Gomez JA, Höffner K, Barton PI, Henson MA (2015) Metabolic modeling of synthesis gas fermentation in bubble column reactors. Biotechnol Biofuels 8:89
30. Chen J, Henson MA (2016) *In silico* metabolic engineering of *Clostridium ljungdahlii* for synthesis gas fermentation. Metab Eng 38:389–400
31. Siebler F, Lapin A, Hermann M, Takors R (2019) The impact of CO gradients on *C. ljungdahlii* in a 125 m^3 bubble column: mass transfer, circulation time and lifeline analysis. Chem Eng Sci 207:410–423
32. Benalcazar EA, Noorman H, Maciel R, Posada JA (2020) Modeling ethanol production through gas fermentation: a biothermodynamics and mass transfer-based hybrid model for microbial growth in a large-scale bubble column bioreactor. Biotechnol Biofuels 13:59

33. Demler M, Weuster-Botz D (2011) Reaction engineering analysis of hydrogenotrophic production of acetic acid by *Acetobacterium woodii*. Biotechnol Bioeng 108:470–474

34. Groher A, Weuster-Botz D (2016) Comparative reaction engineering analysis of different acetogenic bacteria for gas fermentation. J Biotechnol 228:82–94

35. Groher A, Weuster-Botz D (2016) General medium for the autotrophic cultivation of acetogens. Bioprocess Biosyst Eng 39:1645–1650

36. Kantzow C, Weuster-Botz D (2016) Effects of hydrogen partial pressure on autotrophic growth and product formation of *Acetobacterium woodii*. Bioprocess Biosyst Eng 39:1325–1330

37. Mayer A, Weuster-Botz D (2017) Reaction engineering analysis of the autotrophic energy metabolism of *Clostridium aceticum*. FEMS Microbiol Lett 364:fnx219

38. Mohammadi M, Younesi H, Najafpour G, Mohamed AR (2012) Sustainable ethanol fermentation from synthesis gas by *Clostridium ljungdahlii* in a continuous stirred tank bioreactor. J Chem Technol Biotechnol 87:837–843

39. Xiao K, Liang S, Wang XM, Chen CS, Huang X (2019) Current state and challenges of full-scale membrane bioreactor applications. Bioresour Technol 271:473–481

40. Kantzow C, Mayer A, Weuster-Botz D (2015) Continuous gas fermentation by *Acetobacterium woodii* in a submerged membrane reactor with full cell retention. J Biotechnol 212:11–18

41. Richter H, Martin ME, Angenent LT (2013) A two-stage continuous fermentation system for conversion of syngas into ethanol. Energies 6:3987–4000

42. Doll K, Rückel A, Kämpf P, Weuster-Botz D (2018) Two stirred-tank bioreactors in series enable continuous production of alcohols from carbon monoxide with *Clostridium carboxidivorans*. Bioprocess Biosyst Eng 41:1403–1416

43. Heijstra BD, Leang C, Juminaga A (2017) Gas fermentation: cellular engineering possibilities and scale-up. Microb Cell Factories 16:60

44. Molitor B, Richter H, Martin ME, Jensen RO, Juminaga A, Mihalcea C, Angenent LT (2016) Carbon recovery by fermentation of CO-rich off gases – turning steel mills into biorefineries. Bioresour Technol 215:386–396

45. Takors R, Kopf M, Mampel J, Bluemke W, Blombach B, Eikmanns B, Bengelsdorf F, Weuster-Botz D, Dürre P (2018) Using gas mixtures of CO, CO$_2$, and H$_2$ as microbial substrates: The dos and don'ts of successful technology transfer from lab to production scale. Microb Biotechnol 11:606–625

46. Bredwell MD, Srivastava P, Worden RM (1999) Reactor design issues for synthesis gas fermentations. Biotechnol Prog 15:834–844

47. Yasin M, Jeong Y, Park SJ, Jeong J, Lee EY, Lowitt RW, Kim BH, Lee J, Chang IS (2015) Microbial synthesis gas utilization and ways to resolve kinetic and mass-transfer limitations. Bioresour Technol 177:361–374

48. Devarapalli M, Atiyeh HK, Phillips JR, Lewis RS, Huhnke RL (2016) Ethanol production during semi-continuous syngas fermentation in a trickle bed reactor using *C. ragsdalei*. Bioresour Technol 209:56–65

49. Schulte MJ, Wiltgen J, Ritter J, Mooney CB, Flickinger MC (2016) A high gas fraction, reduced power, syngas bioprocessing method demonstrated with a *Clostridium ljungdahlii* OTA1 paper biocomposite. Biotechnol Bioeng 113:1913–1923

50. Shen Y, Brauwn RC, Wen Z (2017) Syngas fermentation by *Clostridium carboxidivorans* P7 in a horizontal rotating packed bed biofilm reactor with enhanced ethanol production. Appl Energy 187:585–594

51. Riegler P, Bieringer E, Chrusciel T, Stärz M, Löwe H, Weuster Botz D (2019) Continuous conversion of CO$_2$/H$_2$ with *Clostridium aceticum* in biofilm reactors. Bioresour Technol 291: 121760

52. Reddy MV, Mohan SV, Chang Y-C (2018) Medium-chain fatty acids (MCFA) production through anaerobic fermentation using *Clostridium kluyveri*: effect of ethanol and acetate. Appl Biochem Biotechnol 185:594–605

53. Steinbusch KJJ, Hamelers HVM, Plugge CM, Buisman CJN (2011) Biological formation of caproate and caprylate from acetate: fuel and chemical production from low grade biomass. Energy Environ Sci 4:216–224

54. Spirito CM, Richter H, Rabaey K, Stams AJM, Angenent LT (2014) Chain elongation in anaerobic reactor microbiomes to recover resources from waste. Curr Opin Biotechnol 27:115–122

55. Li F, Hinderberger J, Seedorf H, Zhang J, Buckel W, Thauer RK (2008) Coupled ferredoxin and crotonyl coenzyme A (CoA) reduction with NADH catalyzed by the butyryl-CoA dehydrogenase/Etf complex from *Clostridium kluyveri*. J Bacteriol 190:843–850

56. Cavalcante WD, Leitao RC, Gehring TA, Angenant LT, Santaella ST (2017) Anaerobic fermentation for n-caproic acid production: a review. Process Biochem 54:106–119

57. Gildemyn S, Molitor B, Usack JG, Nguyen M, Rabaey K, Angenent LT (2017) Upgrading syngas fermentation effluent using *Clostridium kluyveri* in a continuous fermentation. Biotechnol Biofuels 10:83

58. Kucek LA, Spirito CM, Angenent LT (2016) High n-caprylate productivities and specificities from dilute ethanol and acetate: chain elongation with microbiomes to upgrade products from syngas fermentation. Energy Environ Sci 9:3482–3494

59. Yin YN, Zhang YF, Karakashev DB, Wang JL, Angelidaki I (2017) Biological caproate production by *Clostridium kluyveri* from ethanol and acetate as carbon sources. Bioresour Technol 241:638–644

60. Seedorf H, Fricke F, Veith B, Bruggemann H, Liesegang H, Strittimatter A, Miethke M, Buckel W, Hinderberger J, Li FL, Hagemeier C, Thauer RK, Gottschalk G (2008) The genome of *Clostridium kluyveri*, a strict anaerobe with unique metabolic features. PNAS 105:2128–2133

61. Tomlinson N, Barker HA (1954) Carbon dioxide and acetate utilization by *Clostridium kluyveri*: I. Influence of nutritional conditions on utilization patterns. J Biol Chem 209:585–595

62. Grootscholten TIM, Steinbusch KJJ, Hamelers HVM, Buisman CJN (2013) Chain elongation of acetate and ethanol in an upflow anaerobic filter for high rate MCFA production. Bioresour Technol 135:440–445

63. González-Cabaleiro R, Lema JM, Rodríguez J, Kleerebezem R (2013) Linking thermodynamics and kinetics to assess pathway reversibility in anaerobic bioprocesses. Energy Environ Sci 6:3780

64. Roghair M, Hoogstad T, Strik DPBTB, Plugge CM, Timmers PHA, Weusthuis RA, Bruins ME, Buisman CJN (2018) Controlling ethanol use in chain elongation by CO_2 loading rate. Environ Sci Technol 52:1496–1505

65. Richter H, Molitor B, Diender M, Souza DZ, Angenent LT (2016) A narrow pH range supports butanol, hexanol, and octanol production from syngas in a continuous co-culture of *Clostridium ljungdahlii* and *Clostridium kluyveri* with in-line product extraction. Front Microbiol 7:1773

66. Abubackar HN, Veiga MC, Kennes C (2015) Carbon monoxide fermentation to ethanol by *Clostridium autoethanogenum* in a bioreactor with no accumulation of acetic acid. Bioresour Technol 186:122–127

67. Kenealy WR, Cao Y, Weimer PJ (1995) Production of caproic acid by cocultures of ruminal cellulolytic bacteria and *Clostridium kluyveri* grown on cellulose and ethanol. Appl Microbiol Biotechnol 44:507–513

68. Lonkar S, Fu Z, Holtzapple M (2016) Optimum alcohol concentration for chain elongation in mixed-culture fermentation of cellulosic substrate. Biotechnol Bioeng 113:2597–2604

69. Diender M, Olm IP, Gelderloos M, Koehorst JJ, Schaap PJ, Stams AJM, Souza DZ (2019) Metabolic shift induced by synthetic co-cultivation promotes high yield of chain elongated acids from syngas. Sci Rep 9:18081

70. Schneider M, Bäumler M, Lee NM, Weuster-Botz D, Ehrenreich A, Liebl W (2021) Monitoring co-cultures of *Clostridium carboxidivorans* and *Clostridium kluyveri* by fluorescence in situ hybridization with specific 23S rRNA oligonucleotide probes. Syst Appl Microbiol 44:126271

71. Bäumler M, Schneider M, Ehrenreich A, Liebl W, Weuster-Botz D (2022) Synthetic co-culture of autotrophic *Clostridium carboxidivorans* with chain elongating *Clostridium kluyveri* monitored by flow cytometry. Microb Biotechnol 15:1471–1485

72. Herzog J, Mook A, Guhl L, Bäumler M, Beck MH, Weuster-Botz D, Bengelsdorf FR, Zeng A-P (2022) Novel synthetic co-culture of *Acetobacterium woodii* and *Clostridium drakei* using CO$_2$ and *in situ* generated H$_2$ for the production of caproic acid via lactic acid. Eng Life Sci. https://doi.org/10.1002/elsc.202100169

73. Xu D, Tree DR, Lewis RS (2011) The effects of syngas impurities on syngas fermentation to liquid fuels. Biomass Bioenergy 35:2690–2696

74. Liew FM (2016) Gas fermentation - a flexible platform for commercial scale production of low-carbon-fuels and chemicals from waste and renewable feedstocks. Front Microbiol 7:694

75. Phillips JR, Atiyeh HK, Huhnke RL (2014) Method for design of production medium for fermentation of synthesis gas to ethanol by acetogenic bacteria. Biol Eng Trans 7(3):113–128

76. Oliveira L, Röhrenbach S, Holzmüller V, Weuster-Botz D (2022) Continuous sulfide supply enhanced autotrophic production of alcohols with *Clostridium ragsdalei*. Biores Bioproc 9:15

77. Ahmed A, Lewis R (2007) Fermentation of biomass-generated synthesis gas: effects of nitric oxide. Biotechnol Bioeng 97:1080–1086

78. Rückel A, Hannemann J, Maierhofer C, Fuchs A, Weuster-Botz D (2021) Studies on syngas fermentation with *Clostridium carboxidivorans* in stirred-tank reactors with defined gas impurities. Front Microbiol 12:655390

79. Oliveira L, Rückel A, Nordgauer L, Schlumprecht P, Hutter E, Weuster-Botz D (2022) Comparison of syngas-fermenting *Clostridia* in stirred-tank bioreactors and the effects of varying syngas impurities. Microorganisms 10:681

80. Voegele E (2018) Former Ineos Bio site purchased for conversion into eco-district. Biomass magazine:15089

Microbial Processes: Current Developments in Gas Fermentation

9

Frank Kensy

Abstract

Microorganisms that can convert C1 gases into industrially relevant products (e.g. ethanol or butanol) are already being used commercially today. For example, LanzaTech is currently the market leader among companies that produce ethanol biotechnologically from C1 gases. Acetogens naturally produce organic acids and alcohols (often $\leq$ C4) from C1 substrates. By establishing new molecular biology methods, it has been possible to recombinantly modify acetogens and expand the product spectrum of gas fermentation to very interesting compounds such as isobutene or isoprene. Further optimization of these strains and the corresponding fermentation approaches will determine in the future whether gas fermentation can become competitive with chemical synthesis in the production of required platform chemicals, but also specialty chemicals.

Keywords

Microorganisms · Acetogens · Organic acid · Alcohol · Isobutene · Isoprene

9.1 Introduction

Gas fermentations have become attractive as climate change has gained importance and CO_2 emissions are one of the reasons for the intensification of climate change. Industry is a major emitter of various carbonaceous gases. The steel, cement, and power industries are particularly strong emitters (Chaps. 13, 14, 15). Gas fermentation can be a consumer of these so-called waste gases and convert them into valuable products. Furthermore, these waste gases represent a cheap source of carbon. Nature

F. Kensy (✉)
b.fab GmbH, Cologne, Germany
e-mail: kensy@bfab.bio

M. Kircher, T. Schwarz (eds.), *CO2 and CO as Feedstock*, Circular Economy and Sustainability, https://doi.org/10.1007/978-3-031-27811-2_9

provides a variety of microorganisms that metabolize gases: via (1) hydrogenesis, (2) methanogenesis, or (3) acetogenesis [1]. This section will focus on acetogenesis and acetogens as they provide the broadest range of applications for the chemical industry.

Acetogens are obligate anaerobic microorganisms capable of growing on CO or CO_2 and H_2 alone (autotrophy), usually forming acetate and/or ethanol. Many acetogens also grow heterotrophically on sugars, which broadens the range of applications. Acetogens were already used industrially in the last century. The first beginnings of commercialization started in World War I with the Weizman process (also called the ABE process), but still with heterotrophic substrates such as starch or glucose. The process produces acetone, butanol, and ethanol in an anaerobic fermentation with Clostridium acetobutylicum [2]. Production was discontinued in the 1970s as the process became increasingly uneconomical and the three solvents could be produced more cheaply from oil. The foundation for today's knowledge of acetogens was strongly built in the 1990s [3]. It was not until the 2000s that more thought was given to commercialization, and companies such as LanzaTech, Coskata, and Ineos Bio emerged [4]. It was these companies that developed and built the first commercial plants. The Coskata and Ineos Bio plants are now out of operation. Only LanzaTech is currently massively advancing the commercialization of its ethanol technology with partners. The first commercial production plant, with an annual capacity of 46,000 tons, was just commissioned in May 2018 with joint venture partner Shougang Group [5]. A second plant with AcelorMittal and an annual capacity of 60,000 tons is under construction in Ghent, Belgium, and is expected to come on stream in mid-2020 [6]. Further plants are planned in South Africa, Japan and India [7]. This makes LanzaTech the leading company worldwide in commercializing gas fermentation. Unfortunately, few other companies are following its footsteps. The example of LanzaTech also shows that the products introduced industrially today are mostly still native products such as ethanol and no recombinant microorganisms have yet made it into industrial practice in this field. This chapter gives an overview of other products that can be produced with anaerobic gas fermentation and which molecular biological tools have now been established.

9.2   State of the Art

Acetogens are known to convert the substrates CO, CO_2 and H_2 in a very energy efficient manner. The Wood-Ljungdahl pathway (WLP) is most prevalent in these organisms and enables the utilization of the gaseous substrates. The metabolism branches into a methyl and a carbonyl arm, with CO being utilized as a substrate in the first branch and CO_2 being reduced to CO via ferredoxin in the second branch. The main intermediate of metabolism is acetyl-CoA, from where product syntheses are derived. If acetate is formed by the acetogens, one ATP is gained per mole of acetate. Since the WLP also requires one ATP for the synthesis of methyl-THF from formate, no energy in the form of ATP is left over for other metabolic activities, so further energy conservation mechanisms, such as Na^+ or proton gradients, are required. Only in this way the WLP can form further reduction equivalents. Acetogens in natural form also form interesting C4 basic building blocks for chemical applications, in addition to formate, acetate and ethanol. These include 2,3-butanediol, butyrate, and butanol [3]. Table 9.1 provides an overview of the products expressed natively or recombinantly in acetogens to date.

To produce new products in acetogens, genetic tools are needed to modify the organisms to generate products that are unusual for them. In recent years, many scientific research groups have devoted themselves to the subject. The first genetic manipulation of acetogens was described by Köpke et al. [15]). Nowadays, techniques such as ClosTron have become widely accepted as gene disruption tools [32] and are continuously developed [33]. The establishment of the CRISPR/ Cas9 method in acetogens is very promising, as it allows for much faster generation of mutants and more precise genome intervention [34, 35]. Furthermore, the construction and use of genome-wide metabolic models (GEMs), as first applied to Clostridium ljungdahlii [36], are very promising and allow prior in silico simulation of a metabolic pathway before cloning.

These genetic tools were used to generate the first recombinant strains and to establish new product routes. Acetone is one of the first products. In this context, a recombinant Acetobacterium woodii strain was able to achieve 0.87 g L^{-1} acetone in batch preparation and 26 mg L^{-1} h^{-1} in continuous culture [20]. Acetone can be further converted to isopropanol by a native alcohol dehydrogenase in C. autoethanogenum [20]. Another product is lactate. Lactate can be produced in A. woodii by the carboxylation of acetyl-CoA to pyruvate and the subsequent reduction of pyruvate by lactate dehydrogenase. It is suggested that primarily the substrate CO is used, since only on this substrate the ATP yield is positive and thus the biosynthesis proceeds preferentially [6]. Furthermore, the biosynthesis of butyrate and butanol was described in recombinant strains of C. autoethanogenum [25] and C. ljungdahli [15], and product titers of 2 mM butanol and 16 mM butyrate, respectively, were obtained. A completely new synthetic pathway starting from acetone was described for isobutene [28]. Acetone is first converted to 3-hydroxyisovaleric acid by acetylation and then decarboxylated to isobutene by an ATP-dependent decarboxylase. Another interesting product is isoprene. Isoprene is synthesized most efficiently via the mevalonate pathway (Chap. 30). However, energetic calculations by Bertsch and Müller [6] showed that the synthesis of isoprene in anaerobes from H_2 and CO_2 (-4.5 mol ATP per mol isoprene) and from CO (-0.3 mol ATP per mol isoprene) is ATP negative and therefore not beneficial.

The products mentioned in Table 9.1 are often not produced as the only product by the acetogens, but mixtures of different products occur (e.g., acetate together with ethanol). To produce pure products, the metabolic pathway or the cultivation condition must be optimized. Furthermore, Table 9.1 provides only a general overview of possible products of acetogens. A more detailed overview is provided in Takors et al. [4], Humphreys and Minton [37], and Schiel-Bengelsdorf and Dürre [3].

Table 9.1 Overview of native and recombinant products in acetogens

Microorganism	Substrate	Products	Product origin	Reference
Acetobacterium woodii, Clostridium formicoaceticum, Clostridium methoxybenzovornas, Eubacterium aggregans	CO, $H_2 + CO_2$ CO $H_2 + CO_2$ $H_2 + CO_2$	Formate	Native	[8–11]
Acetobacterium woodii	CO, $H_2 + CO_2$	Acetate	Native + recombinant	[12–14]
Clostridium authoethanogenum, Clostridium ljungdahlii	CO, $H_2 + CO_2$ CO, $H_2 + CO_2$	Ethanol	Native + recombinant	[15–17]
Acetobacterium woodii	CO, $H_2 + CO_2$	Lactate	Recombinant	[6, 18]
Clostridium autoethanogenum, Clostridium ljungdahlii, Clostridium ragsdalei	CO, $H_2 + CO_2$ CO, $H_2 + CO_2$ CO, $H_2 + CO_2$	2,3-butanediol	Native + recombinant	[15, 19]
Clostridium aceticum, Acetobacterium woodii	CO, $H_2 + CO_2$ CO, $H_2 + CO_2$	Acetone	Recombinant	[3, 20]
Clostridium autoethanogenum, Clostridium ljungdahlii	CO, $H_2 + CO_2$ CO, $H_2 + CO_2$	Isopropanol	Recombinant	[21, 22]
Acetonema lnogum, Butyribacterium methylotrophicum, Clostridium carboxyvorans	$H_2 + CO_2$ (CO), $H_2 + CO_2$ CO, $H_2 + CO_2$	Butyrate	Native + recombinant	[4, 23, 24]
Butyribacterium methylotrophicum, Clostridium autoethanogenum, Clostridium carboxyvorans	CO, $H_2 + CO_2$ CO, $H_2 + CO_2$ CO, $H_2 + CO_2$	Butanol	Native + recombinant	[25–27]
Clostridium ljungdahlii, Acetobacterium woodii	CO, $(H_2 + CO_2)$ CO, $(H_2 + CO_2)$	Isobutene	Recombinant	[6, 28, 29]
Clostridium ljungdahlii, Acetobacterium woodii	CO, $(H_2 + CO_2)$ CO, $(H_2 + CO_2)$	Isoprene	Recombinant	[6, 30, 31]

The product examples show that to date products with a maximum chain length of C5 could be synthesized in anaerobes. The achieved product concentrations have so far been in the lower mM or mg/L range, which currently makes commercialization of the processes a distant prospect. A fundamental problem is the energy budget of acetogens. The organisms are designed for minimal energy consumption, which makes it difficult for them to synthesize more energetically expensive products.

9.3   Solutions to Limitations of Gas Fermentation

Alternative ATP-Generating Metabolic Pathways

A first approach to solve the energy limitation of acetogens was described by Valgepea et al. [38] by using arginine as a nitrogen source via the arginine deiminase pathway, providing additional ATP for energy-demanding products. Modeling of the metabolic pathway in the GEM of C. autoethanogenum was very beneficial in this regard.

Aerobic Gas Fermentation

The addition of oxygen to syngas components (CO, H_2, CO_2) as substrates of gas fermentation enables aerobic CO oxidation and promises energetic advantages for the synthesis of ATP-intensive products [1]. The gain in cellular energy comes at the cost of higher safety precautions (explosion protection) when O_2 and H_2 are used simultaneously in the bioreactor. Furthermore, carboxydotrophic microorganisms that are O_2-tolerant and grow on syngas mixtures must be used. These microorganisms have been poorly studied and hardly sequenced. Candidate species are: Oligotropha, Bradyrhizobium, Mesorhizobium, Hydrogen-ophaga, Burkholderia, and some species of Mycobacterium, Pseudomonas, Alcaligenes, and Acinetobacter [4].

Chain Elongation by Reverse β-Oxidation

Reverse β-oxidation metabolism can elongate the products of gas fermentation, e.g., the standard products acetate and ethanol can be prolonged to butanol and hexanol. Often these fermentations run in two stages: 1. syngas fermentation with C. ljungdahli or C. autoethanogenum to acetate and ethanol, 2. fermentation with C. kluyveri to butanol and hexanol. Open cultures or pure cultures of the organisms are used. So far, chain extensions up to n-caprylic acid (n-octanoic acid, C8) have been demonstrated [39]. The pH optimum for alcohol production and chain extension are in different ranges and need to be further investigated for the overall productivity of the process [40].

9.4 Conclusion

Acetogens represent an interesting biotechnological species for the utilization of synthesis gas-like, off-gas streams (mixture of CO, H_2 and CO_2). In anaerobic gas fermentation, the microorganisms produce useful basic chemicals such as acetate and ethanol. So far, only LanzaTech's ethanol process has been established to production scale (in 2018) [41]. Other products show only low productivities so far, and the product range is still very limited. The recent establishment of efficient genetic tools (especially CRISPR/Cas9) to design modified and new metabolic pathways raises hope for a broader product range and higher productivities. The latent ATP limitation of acetogens for synthesis of higher value products ($>$ C4) currently limits its applications. However, possible solutions are already being intensively researched, such as aerobic gas fermentation, implementation of new ATP-generating metabolic pathways, and chain elongation with reverse β-oxidation. The increased adoption of molecular biology methods (synthetic biology) and the elucidation and expansion of genome-wide metabolic models of acetogens suggest a more rapid advancement of gas fermentation and its industrial application in the future. The high energetic and material efficiency of gas fermentation is a promising way to reduce climate gases (CO_2, CO) and to detach the chemical industry from fossil feedstocks.

References

1. Diender M, Stams AJM, Sousa DZ (2015) Pathways and bioenergetics of anaerobic carbon monoxide fermentation. Front Microbiol 6:1275. https://doi.org/10.3389/fmicb.2015.01275
2. Weizmann C (1919) Improvements in the bacterial fermentation of carbohydrates and in bacterial cultures for the same. Patent, GB191504845
3. Schiel-Bengelsdorf B, Dürre P (2012) Pathway engineering and synthetic biology using acetogens. FEBS Lett 586(15):2191–2198
4. Takors R, Kopf M, Mampel J, Bluemke W, Blombach B, Eikmanns B, Bengelsdorf F, Weuster-Botz D, Dürre P (2018) Using gas mixtures of CO, CO_2, and H_2 as microbial substrates: The do's and don'ts of successful technology transfer from lab to production scale. Microb Biotechnol 11:606–625
5. Bioenergy International (2018) Press release. https://bioenergyinternational.com/lanzatech-commission-worlds-first-commercial-waste-gas-ethanol-plant-china/. Accessed July 2022
6. Steelanol (2018) Press release. http://www.steelanol.eu/en/news/arcelormittal-and-lanzatech-break-ground-on-150million-project-to-revolutionise-blast-furnace-carbon-emissions-capture. Accessed July 2022
7. Lanzatech (2018) Press release. News section under https://lanzatech.com/investor-relations/. Accessed July 2022
8. Schuchmann K, Müller V (2013) Direct and reversible hydrogenation of CO_2 to formate by a bacterial carbon dioxide reductase. Science 342(6164):1382–1385
9. Andreesen JR, Gottschalk G, Schlegel HG (1970) Clostridium formicoaceticum nov. spec. Isolation, description and distinction from C. aceticum and C. thermoaceticum. Arch Microbiol 72:154–174
10. Mechichi T, Labat M, Patel BKC, Woo THS, Thomas P, Garcia J-L (1999) Clostridium methoxybenzovorans sp. nov., a new aromatic o-demethylating homoacetogen from an olive mill wastewater treatment digester. Int J Syst Bacteriol 49:1201–1209

11. Mechichi T, Labat M, Woo THS, Thomas P, Garcia J-L, Patel BKC (1998) Eubacterium aggregans sp. nov., a new homoacetogenic bacterium from olive mill wastewater treatment digestor. Anaerobe 4:283–291

12. Balch WE, Schoberth S, Tanner RS, Wolfe RS (1977) Acetobacterium, a new genus of hydrogen-oxidizing, carbon dioxide-reducing anaerobic bacteria. Int J Syst Bacteriol 27:355–361

13. Straub M, Demler M, Weuster-Botz D, Dürre P (2014) Selective enhancement of autotrophic acetate production with genetically modified Acetobacterium woodii. J Biotechnol 178:67–72

14. Kantzow C, Mayer A, Weuster-Botz D (2015) Continuous gas fermentation by Acetobacterium woodii in a submerged membrane reactor with full cell retention. J Biotechnol 212:11–18

15. Köpke M, Held C, Hujer S, Liesegang H, Wiezer A, Wollherr A, Ehrenreich A, Liebl W, Gottschalk G, Dürre P (2010) Clostridium ljungdahlii represents a microbial production platform based on syngas. Proc Natl Acad Sci U S A 107:13087–13092

16. Abrini J, Naveau H, Nyns EJ (1994) Clostridium autoethanogenum, sp. nov., an anaerobic bacterium that produces ethanol from carbon monoxide. Arch Microbiol 161:345–351

17. Tanner RS, Miller LM, Yang D (1993) Clostridium ljungdahlii sp. nov., an acetogenic species in clostridial rRNA homology group I. Int J Syst Bacteriol 43:232–236

18. Weghoff MC, Bertsch J, Müller V (2015) A novel mode of lactate metabolism in strictly anaerobic bacteria. Environ Microbiol 17(3):670–677

19. Huhnke RL, Lewis RS, Tanner RS (2008) Isolation and characterization of novel clostridial species, Patent, WO2008028055 A2, Applicant: The Board of Regents for Oklahoma State University

20. Hoffmeister S, Gerdom M, Bengelsdorf FR, Linder S, Flüchter S, Öztürk H, Blümke W, May A, Fischer RJ, Bahl H (2016) Acetone production with metabolically engineered strains of Acetobacterium woodii. Metab Eng 36:37–47

21. Köpke M, Simpson S, Liew F (2012) Fermentation process for producing isopropanol using a recombinant microorganism. Patent, US20120252083 A1, Applicant: LanzaTech New Zealand Ltd.

22. Bengelsdorf FR, Poehlein A, Linder S, Erz C, Hummel T, Hoffmeister S, Daniel R, Dürre P (2016) Industrial acetogenic biocatalysts: a comparative metabolic and genomic analysis. Front Microbiol 7:1036

23. Ueki T, Nevin KP, Woodard TL, Lovley DR (2014) Converting carbon dioxide to butyrate with an engineered strain of clostridium ljungdahlii. MBio 5(5):e01636–e01614

24. Lynd L, Kerby R, Zeikus JG (1982) Carbon monoxide metabolism of the methylotrophic acidogen Butyribacterium methylotrophicum. J Bacteriol 149.255–263

25. Köpke M, Liew F (2012) Production of butanol from carbon monoxide by a recombinant microorganism. Patent, WO2012053905 A1, Applicant: LanzaTech New Zealand Ltd.

26. Liou JS, Blakwill DL, Drake GR, Tanner RS (2005) Clostridium carboxidivorans sp. nov., a solvent producing clostridium isolated from an agricultural settling lagoon, and reclassification of the acetogen Clostridium scatologenes strain SL1 as Clostridium drakei sp. nov. Int J Syst Evol Microbiol 55:2085–2091

27. Grethlein AJ, Worden RM, Jain MK, Datta R (1991) Evidence for production of n-butanol from carbon monoxide by Butyribacterium methylotrophicum. J Ferment Bioeng 72:58–60

28. Van Leeuwen BN, van der Wulp AM, Duijnstee I, van Maris AJ, Straathof AJ (2012) Fermentative production of isobutene. Appl Microbiol Biotechnol 93:1377–1387

29. Güntner B (2016) Recombinant microorganism producing alkenes from acetyl-CoA. Patent, WO2016034691A1, Applicant: Syngip Bv

30. Furutani M, Uenishi A, Iwasa K, Jennewein S, Fischer R (2013) Recombinant cell and production method for isoprene. Patent, EP2913392A1, Applicant: Fraunhofer Gesellschaft zur Forderung der Angewandten Forschung, Sekisui Chemical Co Ltd.

31. Beck Z, Cervin M, Chotani G, Diner B, Fan J, Peres C, Sanford K, Scotcher M, Wells D, Whited G (2014) Recombinant anaerobic acetogenic bacteria for production of isoprene and/or

industrial bio-products using synthesis gas. US20140234926A1, Applicant: Goodyear Tire and Rubber Co., Danisco US Inc.

32. Heap JT, Pennington OJ, Cartman ST, Carter GP, Minton NP (2007) The ClosTron: a universal gene knock-out system for the genus Clostridium. J Microbiol Methods 70:452–464

33. Liew F, Henstra AM, Köpke M, Winzer K, Simpson SD, Minton NP (2017) Metabolic engineering of Clostridium autoethanogenum for selective alcohol production. Metab Eng 40:104–114

34. Huang H, Chai C, Li N, Rowe P, Minton NP, Yang S, Jiang W, Gu Y (2016) CRISPR/Cas9-based efficient genome editing in Clostridium ljungdahlii, an autotrophic gas-fermenting bacterium. ACS Synth Biol 5(12):1355–1361

35. Nagaraju S, Davies NK, Walker DJF, Köpke M, Simpson SD (2016) Genome editing of Clostridium autoethanogenum using CRISPR/Cas9. Biotechnol Biofuels 9:219

36. Nagarajan H, Sahin M, Nogales J, Latif H, Lovley DR, Ebrahim A, Zengler K (2013) Characterizing acetogenic metabolism using a genome-scale metabolic reconstruction of Clostridium ljungdahlii. Microb Cell Factories 12:118

37. Humphreys CM, Minton NP (2018) Advances in metabolic engineering in the microbial production of fuels and chemicals from C1 gas. Curr Opin Biotechnol 50:174–181

38. Valgepea K, Loi KQ, Behrendorff JB, Lemgruber RSP, Plan M, Hodson MP, Köpke M, Nielsen LK, Marcellin E (2017) Arginine deiminase pathway provides ATP and boosts growth of the gas-fermenting acetogen Clostridium autoethanogenum. Metab Eng 41:202–211

39. Molitor B, Marcellin E, Angenent LT (2017) Overcoming the energetic limitations of syngas fermentation. Curr Opin Chem Biol 41:84–92

40. Ganigué R, Sánchez-Paredes P, Baneras L, Colprim J (2016) Low fermentation pH is a trigger to alcohol production, but a killer to chain elongation. Front Microbiol 7:702

41. Kane MD, Breznak JA (1991) Acetonema longum gen. Nov. sp. nov., an H_2/CO_2 acetogenic bacterium from the termite Pterotermes occidentis. Arch Microbiol 156:91–98

Microbial Processes: Production of Polyhydroxyalkanoates from CO_2

10

Heleen De Wever and Linsey Garcia-Gonzalez

Abstract

The bacterium *Cupriavidus necator* is capable of using CO_2 as a carbon source for the production of polyhydroxyalkanoate (PHA) in a gas fermentation process. Heterotrophic-autotrophic PHA production is characterized as more sustainable and safer compared to exclusively heterotrophic or autotrophic fermentations for the production of biopolymers from CO_2. The bacteria are propagated heterotrophically under feeding with an organic carbon source. This is followed by the autotrophic production phase of PHA under gassing with CO_2 and H_2. The process passed the practical test using industrial waste gas streams containing CO_2. The same performance data were achieved as under gassing with pure laboratory gases.

Keywords

Cupriavidus necator · Polyhydroxyalkanoate (PHA) · Gas fermentation

10.1 Biosynthesis of Complex Molecules from CO_2

When comparing biological CO_2 conversion processes with chemical processes, it is noticeable that the former complement the chemical processes mainly in terms of the chemicals that can be produced. In particular, *de novo* syntheses of certain complex molecules, such as polymers, proteins, etc., can only be carried out by biological

H. De Wever (✉) · L. Garcia-Gonzalez
VITO - Vlaamse Instelling voor Technologisch Onderzoek, Mol, Belgium
e-mail: heleen.dewever@vito.be; linsey.garcia-gonzalez@vito.be

routes and are only possible because of the wide range of biocatalysts available in a single organism [1]. In this context, H_2-oxidizing bacteria represent a very interesting group. These organisms are able to fix CO_2 and convert nutrients into valuable biomass rich in proteins, biopolymers or microbial oil. As part of energy conservation, they require H_2 and O_2 as electron donor and acceptor, respectively. Since both gases can be generated by electrolysis from water and with the help of renewable energy, the process can also be included in the broader context of power-to-gas technologies, which address H_2 production with green energy. H_2-producing organisms have also gained much interest as producers of single cell proteins (SCPs) and polyhydroxybutyrate (PHB) [2].

Numerous companies are currently exploring and scaling up fermentation-based platforms for flexible production of complex molecules. As an example, Kiverdi focuses on the conversion of CO_2 into high-value oils, renewable feedstocks, and proteins, which in turn are used in numerous applications, such as consumer products, industrial products, and fish feed. Similarly, Oakbio has developed a proprietary biotechnology platform for CO_2 conversion into chemical feedstocks, biofuels, bioplastics and animal feed. Furthermore, the Belgian company Avecom is developing power-to-protein concepts to upgrade ammonium into microbial proteins using mixed bacterial cultures. The concept is coupled to wastewater treatment processes where the required ammonium, CO_2, and energy are available [3].

Interest in Polyhydroxyalkanoates (PHA)

Polyhydroxyalkanoates (PHAs) are synthesized by a variety of organisms as intracellular energy and carbon storage compounds. Particularly interesting properties of these macromolecular polyesters are biodegradability, biocompatibility, and a diverse molecular structure and/or composition. Therefore, the compounds are promising materials for a wide variety of applications in the packaging and food industries, as well as for medical applications, as bioplastics, etc. [4]. The homopolymer PHB is the best studied PHA.

VITO's focus in numerous funded projects has been mainly, but not exclusively, on PHA applications in the textile industry, as prebiotics in aquaculture [5] and as a component of 3D printer filaments, as well as the depolymerization of PHA to the high-value biomonomer hydroxybutyric acid [6].

10.2 A Two-Step Process for Heterotrophic-Autotrophic PHA Production

The most commonly used cultivation method for PHA production is based on two distinct stages. First, a sufficiently high concentration of catalytically active biomass is generated in a balanced medium. In this phase, PHA accumulation is negligible. Then, stress is induced by nutrient deficiency by limiting at least one essential growth component. Under these conditions, PHA is finally generated as intracellular storage material.

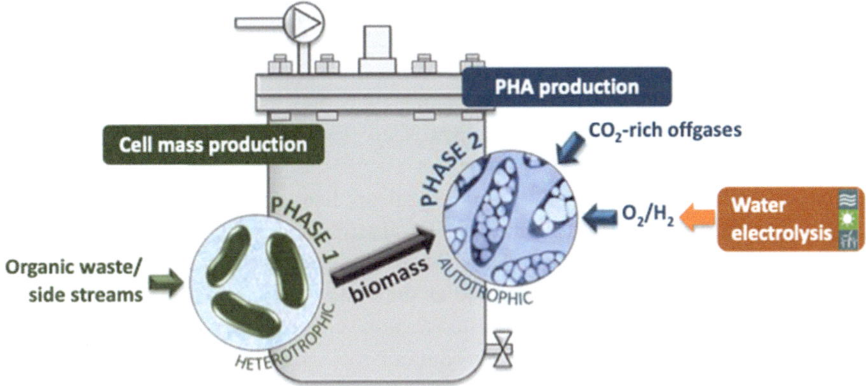

Fig. 10.1 Two-stage heterotrophic-autotrophic approach investigated by VITO using organic and inorganic waste streams or side streams as substrates

While most of the methods used organic substrates as carbon sources, some isolated experiments were also performed in which CO_2 was converted to PHA using H_2 as energy source and O_2 as electron acceptor [7, 8]. In principle, there are two cultivation methods to use inorganic substrates for PHA production:

1. Fully autotrophic approach: a gas mixture of CO_2, H_2 and O_2 forms the substrate for both cell growth and PHA accumulation. To ensure sufficient cell growth, the gas has a composition of H_2: O_2: CO_2 of 7: 1: 1. This composition has a large explosion potential and is therefore too dangerous to work with [7]. A possible solution to this problem is the heterotroph-autotrophic approach [9].
2. Heterotrophic-autotrophic approach: biomass formation occurs under heterotrophic conditions with organic carbon sources, followed by PHA accumulation under autotrophic conditions with a gas mixture of CO_2, H_2 and O_2. The advantage of this cultivation method is that high biomass concentrations and associated high productivities can be maintained during the growth phase, since O_2 can be added under non-limiting conditions. In contrast, in the second, autotrophic phase, where PHA is synthesized, the O_2 concentration is lowered below the lower explosive limit (LEL) of O_2 in H_2 (estimated to be between 4 and 6.9%).

VITO has developed an innovative fermentation alternative based on the heterotrophic-autotrophic approach. This focuses on the use of organic and inorganic waste streams as feedstocks for biomass production and PHA production (Fig. 10.1). This allowed two goals to be pursued at once: reducing feedstock/substrate costs, which now still amount to about 50% of total production costs, and improving the sustainability profile of PHA production.

10.3 Production of the Homopolymer PHB from Pure CO_2 and CO_2-Rich Waste Gas Streams

The technical feasibility of the presented two-step process was successfully demonstrated using the microorganism *Cupriavidus necator* DSM 545 and described in detail by Garcia-Gonzalez et al. (2015; [10]). In order to maintain conditions relevant for subsequent industrial application, a safety margin below the lower explosive limit (LEL) of O_2 was always maintained during autotrophic PHB synthesis. First, different organic waste streams were tested for the biomass growth phase, while pure CO_2 was used as substrate for the generation of PHB. The best results were obtained with waste streams containing glycerol. Growth on glycerol resulted in the highest PHB production rates reported so far in the literature that could be achieved with CO_2 under industrially relevant conditions. Moreover, the properties of the resulting PHB were similar to those exhibited by PHB from organic substrates.

Subsequently, the influence of real CO_2-rich gases on PHB production was investigated, which has never been described before in the literature. Two different industrial point sources—from an ethanol production plant and a biogas plant, respectively—were sampled, and the collected gas from the plants was used in the fermentation process, analogous to the pure CO_2 gas before. No effect of these gases compared to pure CO_2 on the metabolic performance of the bacteria was observed. The biopolymer content and productivity reached values of 73% and 0.227 g L^{-1} h^{-1}, respectively, which were similar to the values obtained with pure CO_2. Surprisingly, the properties of the biopolymer, such as the molecular weight, thermal properties, degree of crystallinity, and polydispersity index, were not only similar among the different CO_2 sources tested, but also comparable to PHA generated starting from organic carbon sources [11].

By combining previously established models describing fully heterotrophic or fully autotrophic biomass growth and PHB production, a heterotrophic-autotrophic model was developed and validated that can accurately predict the dynamic behavior during heterotrophic biomass growth and autotrophic PHA production. A scenario analysis allowed the determination of the optimal oxygen and ammonium concentration ranges for maximum PHB production [12].

Production of PHA Copolymers from CO_2

To ensure significant market entry and profitable business, PHA applications should be developed in diverse areas, including those applications with particularly high added value. PHA is an interesting material because its composition and properties can be tailored through the selection of appropriate growth conditions and environmental factors, and thus adapted to the desired product specifications of the end user. Therefore, additional market potential can be expected from the production of tailor-made copolymers [4].

To generate PHA copolymers, an organic cosubstrate in combination with CO_2 can be added to the PHA production phase. As expected, such a mixotrophic approach is a complex interplay of different processes, which include the interaction between the substrates, monomer synthesis from the substrates, intracellular accumulation of PHA and its consumption. The main challenge here is to obtain a specific distribution and fraction of monomers in the copolymers. To date, there are very few publications on the co-metabolism of organic substrates and CO_2 [13, 14].

10.4 Conclusion

Initially, VITO aimed at the synthesis of the copolymer poly(3-hydroxybutyrate-co-2-hydroxyvalerate) (PHBV). The compound can be produced from 3-hydroxybutyrate and 3-hydroxyvalerate monomers, which are synthesized from CO_2 and valeric acid, respectively. Mathematical modeling and a deep understanding of the dynamic process behavior and underlying mechanisms were used to optimize PHBV production. An operating diagram was created and validated, allowing to shift the fermentation towards PHBV copolymers and predict PHBV production with desired properties. It was also shown that random monomer distribution could be achieved by limiting the valeric acid concentration below a critical concentration. This methodology is also applicable to other PHA copolymers.

Acknowledgements The authors gratefully acknowledge Environmental and Energy Technology Innovation Platform (MIP) for funding most of the experimental work described here.

References

1. Jajesniak P, Omar Ali HE, Wong TS (2014) Carbon dioxide capture and utilization using biological systems: opportunities and challenges. J Bioprocess Biotech 4:155. https://doi.org/10.4172/2155-9821.1000155
2. Matassa S, Boon N, Verstraete W (2015) Resource recovery from used water: the manufacturing abilities of hydrogen oxidizing bacteria. Water Res 68:467–478. https://doi.org/10.1016/j.watres.2014.10.028
3. Avecom (2017) https://www.powertoprotein.eu/#info. Accessed Aug 2017
4. Kaur G, Roy I (2015) Strategies for large-scale production of polyhydroxyalkanoates. Chem Biochem Eng Q 29(2):157–172. https://doi.org/10.15255/CABEQ.2014.2255
5. Franke A, Clemmesen C, De Schryver P, Garcia-Gonzalez L, Miest JJ, Roth O (2017) Immunostimulatory effects of dietary poly-β-hydroxybutyrate in European sea bass postlarvae. Aquac Res 48(12):5707–5717
6. MIP (2017) http://www.mipvlaanderen.be/nl/webpage/150/biopol.aspx and http://www.i-cleantechvlaanderen.be/co2mpass. Accessed Aug 2017
7. Ishizaki A, Tanaka K, Taga N (2001) Microbial production of poly-D-3-hydroxybutyrate from CO_2. Appl Microbiol Biotechnol 57:6–12. https://doi.org/10.1007/s002530100775
8. Akaraonye E, Keshavarz T, Roy I (2010) Production of polyhydroxyalkanoates: the future green materials of choice. J Chem Technol Biotechnol 85:732–743. https://doi.org/10.1002/jctb.2392
9. Tanaka K, Ishizaki A (1994) Production of poly-D-3-hydroxybutyric acid from carbon dioxide by a two-stage culture method employing Alcaligenes eutrophus ATCC 17697T. J Ferment Bioeng 77:425–427. https://doi.org/10.1016/0922-338X(94)90017-5

10. Garcia-Gonzalez L, Mozumder MSI, Dubreuil M, Volcke EIP, De Wever H (2015) Sustainable autotrophic production of polyhydroxybutyrate (PHB) from CO_2 using a two-stage cultivation system. Catal Today 257:237–245. https://doi.org/10.1016/j.cattod.2014.05.025

11. Garcia-Gonzalez L, De Wever H (2017) Valorization of CO_2-rich off-gases to biopolymers through biotechnological process. FEMS Microbiol Lett 364(20):fnx196. https://doi.org/10.1093/femsle/fnx196

12. Mozumder MSI, Garcia-Gonzalez L, De Wever H, Volcke EIP (2016) Model-based process analysis of heterotrophic-autotrophic poly(3-hydroxybutyrate) (PHB) production. Biochem Eng J 114:202–208. https://doi.org/10.1016/j.bej.2016.07.007

13. Volova TG, Kiselev EG, Shishatskaya EI, Zhila NO, Boyandin AN, Syrvacheva DA, Vinogradova ON, Kalacheva GS, Vasiliev AD, Peterson IV (2013) Cell growth and accumulation of polyhydroxyalkanoates from CO_2 and H_2 of a hydrogen-oxidizing bacterium, Cupriavidus eutrophus B-10646. Bioresour Technol 146:215–222. https://doi.org/10.1016/j.biortech.2013.07.070

14. Park I, Jho EH, Nam K (2014) Optimization of carbon dioxide and valeric acid utilization for polyhydroxyalkanoates synthesis by Cupriavidus necator. J Polym Environ 22(2):244–251. https://doi.org/10.1007/s10924-013-0627-6

Microbial Processes: Photosynthetic Microalgae   11

Stefan Verseck

Abstract

Algal biotechnology is enjoying increasing attention as a methodology for biomass production. Commercially interesting valuable substances from algae include proteins, pigments as well as lipids. The applications of the ingredients range from nutritional and fish feed additives to cosmetic ingredients, medical products, and fertilizers. Although a large number of microalgae have been identified as feedstocks for commercial applications, commercial production is limited to β-carotene and astaxanthin and biomass production, respectively. Microalgae producers or production processes for other specialty chemicals require further significant improvements in terms of efficiency and production cost. Starting with improved reactor concepts via optimized cultivation of the organisms to more effective processing methods, much development work still needs to be done.

Keywords

Algae · Proteins · Pigments · Lipids · Feed additives · Cosmetic ingredients · Medical products · Fertilizer · β-carotene · Astaxanthin

11.1 Introduction

Since the last two decades, algal biotechnology has gained increasing attention as a methodology for biomass production. This development is driven by the identification of CO_2 as a major contributor to climate change. Microalgae are unicellular eukaryotic organisms capable of building biomass through photosynthesis with CO_2,

S. Verseck (✉)
BluCon Biotech GmbH, Cologne, Germany
e-mail: stefan.verseck@blucon-biotech.com

© The Author(s), under exclusive license to Springer Nature Switzerland AG 2023   165
M. Kircher, T. Schwarz (eds.), *CO2 and CO as Feedstock*, Circular Economy and
Sustainability, https://doi.org/10.1007/978-3-031-27811-2_11

nitrogen, inorganic nutrients, and light energy. The ubiquitous and cheap availability of CO_2 as a carbon source from the air as well as from power plants and other industrial side streams also make algae interesting as production organisms for valuable substances for specialty chemicals.

The advantages of unicellular microalgae can be summarized as follows [1, 2]:

- For the same cultivation area, microalgae achieve higher biomass productivity compared to conventional plants.
- They use readily available basic materials, such as inorganic precursors and CO_2 as a carbon source from the air or industrial emitters, to build biomass.
- No arable land is required to grow the algae, thus eliminating competition with cultivable land and its nutrients.

The commercially interesting valuable substances from algae (Sect. 11.3) include proteins, pigments (natural β-carotene or astaxanthin) and lipids (in particular poly-unsaturated omega-3 fatty acids). The applications of the ingredients range from nutritional and fish feed additives to cosmetic ingredients, medical products, and fertilizers [3, 4]. Despite the wide range of applications, the production of substances of industrial interest is at an early stage of development, and only few reliable data on biomass or product productivities and the costs of different production systems are public domain.

11.2 Current Status of the Commercial Use of Microalgae

To understand the advantages and limitations of producing valuable materials using microalgae, one first considers the existing cultivation systems. A basic distinction is made between "open systems" and "closed reactor systems" [5–7].

"Open systems" are large-scale cultivation ponds, or so-called raceway ponds being widely used by established algal biomass producers (e.g., Cyanotech). Open systems are characterized by low complexity and therefore low maintenance efforts as well as investment costs. Productivity is relatively low in these systems, due to an unfavorable surface-to-volume ratio and limited mixing. In open systems, mixing of the growth medium is achieved, for example, by injecting CO_2 or by moving the fluid by paddles. Open systems are also susceptible to unwanted contamination by foreign organisms or other contaminants. Applications of this form of cultivation include the production of *Chlorella* or *Spirulina* biomass as well as robust production processes such as that of β-carotene using *Dunaliella salina* [5–7]. Depending on the organism cultured and the system used, the productivity of algal biomass ranges from 0.02 to 0.2 g L^{-1} d^{-1} [8].

Closed reactor systems, mainly represented by tubular tube systems or so-called "flat-panel" reactors, represent a different concept. In tubular reactors, the cultivation medium is pumped through a transparent piping system, e.g., lying flat on the floor or arranged in stacks. In "flat panel reactors", the liquid containing the algae is passed between transparent plates. In most cases, the cultivation medium is mixed by gassing with a CO_2 mixture (airlift principle). Due to their design, both reactor types have a much better surface-to-volume ratio. Together with improved mixing and the ability to control the cultivation process, biomass production reaches serveral times higher titers compared to the open cultivation systems [5–7]. Again, productivity is highly dependent on the cultivated organism and reaches algal biomass yields of 0.12–2.4 g L^{-1} d^{-1} depending on the reactor type [8]. The advantages of the closed systems are bought by higher investment and operating costs, e.g., for the transparent material and the more complex technology. Since the closed reactor systems are less susceptible to contamination, this type allows the cultivation of sensitive algal species such as *Haematococcus* species, the producer of astaxanthin.

In the following paragraph selected examples are listed in which microalgae are used for the commercial production of industrial valuable substances.

Algatechnologies was founded in 1998 and is based in the Arava Desert in Israel. The company primarily produces astaxanthin of high purity by using *Haematococcus pluvialis* in closed cultivation systems. The tubular reactors cover an area of approximately 6 ha of non-arable desert soil. The transparent cultivation tubes adding up to a length of more than 500 km. The cultivation process is fully automated and centrally controlled. Natural astaxanthin finds application in various dosage forms in cosmetics, human nutrition, healthcare or as nutraceuticals [9].

Commercial production of *Dunaliella* salina as a source of natural β-carotene was established in 1986 by Western Biotechnology and Betatene in Australia. The production facilities are now owned by BASF, making it the world's largest producer of natural β-carotene using *Dunaliella salina*. The open cultivation systems are located in Hutt Lagoon (Western Australia) and Whyalla (South Australia) and cover approximately 400 ha each. Cultivation takes place in an assembly of artificial basins ranging from 40 to 125 ha. Since these cultivation basins are simply excavated, the investment costs are very low. Mixing is done by the wind, and the saline water comes from the adjacent sea. Strong solar radiation and the increasing salinity in the basins due to evaporation induce the formation of β-carotene in the microalgae. Natural β-carotene finds application as a food colorant, as an antioxidant to protect against oxidative stress (in cardiovascular disease, skin aging, etc.), as a precursor for vitamin A, and as an additive in animal feed [10].

Cellana was founded in 2008 as a joint venture between HR BioPetroleum and Shell Venture Funding. The company maintains a demonstration plant in Kailua-Kona on the Big Island (Hawaii) and cultivates microalgae for the production of omega-3 oils (eicosapentaenoic acid and docosahexaenoic acid), animal feed additives (mainly proteinogenic algal meal), and bio-based fuels. The patented ALDUOTM hybrid system is used, in which microalgae are first grown in a closed tubular system and then transferred to raceway ponds to form biomass on a commercial scale. The company sees this process as an opportunity to combine the advantages of both systems (open system: low cost; tubular reactors: contamination-free biomass as inoculum for raceway ponds). The increased biomass production should thus allow to reduce the overall cost of the products. The current production facility covers approximately 2.4 ha with a capacity of 700,000 liters of cultivation volume [11]. Cellana is seeking investors to produce the above three product classes in facilities worldwide.

Cyanotech was founded in 1983 in Kailua-Kona, Hawaii (Big Island). Cyanotech's main product is spirulina biomass for the human health and dietary supplement sectors. On a site of approximately 36 ha, 60 large open raceway ponds are located for the cultivation of *Spirulina*. A second product of the company is natural astaxanthin from *Haematococcus pluvialis*. For the cultivation of this microalgae, Cyanotech uses a hybrid system, where the first step of cultivation takes place in a closed reactor and the algae are transferred to an open system for final biomass production [12].

The difficulty of the financial environment and cost pressures in the field of commercial algal biotechnology can be seen in the closure of Aurora Algae [13] and Sapphire Energy [14] in recent years. Both companies were planning large production facilities that would use "economy of scale" costs for biomass production to lead to areas that would make microalgae attractive even for low-cost products (e.g., biofuels). A general limitation of production using microalgae is the same for all systems presented: all value products are formed as part of the algal biomass. This means that the primary products of interest must be separated from the rest of the biomass and the remaining residues disposed or other uses identified. Together with the relatively low yields, e.g., compared to classical heterotrophic fermentation, this results in production costs that are currently only viable for specialty or higher priced products such as natural pigments, omega-3 oils, etc. [2, 15]. One solution to the challenges described above lies in the consistent and complete utilization of algal biomass, e.g., in specially designed biorefinery concepts [16]. Another approach to reduce production costs is to use genetically optimized organisms whose product formation is increased by targeted "pathway engineering" (e.g., lipids), or to allow the organisms to excrete the desired product into the medium [16–18]. The latter approach allows the decoupling of direct product formation from algal biomass. An example is provided by Algenol (Fort Myers, Florida), which grows optimized cyanobacteria in foil reactors and extracts secreted ethanol from the medium [19].

One of the largest international research activities investigating and analyzing the holistic optimization of microalgae and cyanobacteria production is the AlgaeParc consortium at Wageningen University in the Netherlands. The consortium is a collaboration between the Wageningen University Research Center (WUR) and 19 industrial partners, supported by the Dutch Ministry of Economic Affairs. The AlgaeParc research program aims to bridge the gap between basic research and production-scale manufacturing of algae products. The focus is on providing and developing knowledge, technologies and process strategies to scale up production plants of microalgae and their products under industrial conditions and to optimize productivity. In doing so, production costs and energy consumption are to be reduced [20, 21]. The following research areas are being addressed:

- The performance of different reactor types (open and closed systems) are investigated on a pilot scale by operating them in parallel under the same conditions and additionally considering climatic influences.

- Influences on the productivity of microalgae are investigated using the example of lipid formation. This includes a screening program for suitable algae strains and the optimization of processes using process engineering methods.
- In order to be able to create commercial scenarios of economic models, all data concerning the different reactor types will be analyzed and corresponding life cycle analyses will be carried out.

With the aim of developing an optimized biorefinery concept, options for the holistic use of algal biomass are being explored by investigating different processing techniques for different organisms and products. The results of this consortium should make it possible to understand bottlenecks and areas for optimizing commercial algae production and to derive appropriate measures or provide technical solutions.

11.3   Specialty Chemicals from Algae and Their Market Volumes

The earliest human use of microalgae, such as *Nostoc* and *Spirulina* (*Arthrospira*), dating back nearly 2000 years, is as food. However, the first commercial large-scale cultivation began only in the early 1960s with the cultivation of Chlorella biomass in Japan [22, 23].

The total worldwide sales of all valuable materials and products produced by microalgae are currently estimated at US$1.2 billion per year [24].

11.3.1 Microalgae Biomass

The biomass of the green alga Chlorella finds application as a food supplement, in aquaculture, and in cosmetics. In this regard, the largest producer is Taiwan Chlorella Manufacturing and Co. (Taiwan) with a capacity of 400 t/a [25, 26]. Other producers, specialized in growing the cyanobacteria *Arthrospira platensis* and *A. maxima* for the food sector, are e.g., Hainan DIC (China) or Earthrise Nutraceuticals (USA). The algal biomass of these genera is mainly used for the production as food supplements (health food) due to its high content of protein, vitamin A and carotenoids. However, it is also used in the form of extracts, for example, as an additive in pasta, beverages or cosmetics [26, 27].

Cultivation of microalgae takes place in open systems such as raceway ponds or small lakes. Productivity reaches 550 t/a at Earthrise, the largest producer of *Spirulina* biomass, and 350 t/a at Hainan DIC [28]. Earthrise has published production costs of about US$5 per kilogram of biomass [24].The market for Spirulina biomass is estimated to be about 3000 t/a dry matter and that for *Chlorella* about 2000 t/a. Together, these products generate annual sales of about US$80 million worldwide [3, 4, 29, 30].

11.3.2 Carotenoids

Microalgae are able to produce different classes of carotenoids such as astaxanthin, β-carotene, lutein or zeaxanthin. These products find application in the food sector, in animal nutrition or in cosmetics as additives with antioxidant effects or as colorants [25, 27, 31]. The global market for carotenoids in general (of natural origin or synthetically produced) was estimated at approximately US$ 1.2 billion in 2010 [27]. Since there is a specific demand for carotenoids of natural origin, these target markets are very attractive for products derived from microalgae.

Commercial production of *Dunaliella salina* as a source of natural β-carotene was established in 1986 by Western Biotechnology and Betatene in Australia. Both companies are now part of BASF. This was followed by additional production facilities in Israel (Nature Beta Technologies), India (Parry Nutraceuticals) and the USA (Cyanotech). Natural β-carotene is mainly used as a food additive as a vitamin A precursor or as an antioxidant but is also used in animal nutrition [25–27]. The total annual production of *Dunaliella* biomass is estimated to be 1200 tons per year [25]. The price of β-carotene ranges from US$300 to US$3000 per kilogram depending on the application, concentration and formulation. The total market for β-carotene was estimated to be approximately US$270 million (2010), with a 20–30% share of natural β-carotene [25–27, 30].

The production of the carotenoid astaxanthin from *Haematococcus pluvialis* is established in companies such as Algatechnologies (Israel), Parry Nutraceuticals (India) and Cyanotech (USA). Astaxanthin is mainly used in aquaculture as a source of pigmentation in fish, for example in salmon and trout farming. Other applications of astaxanthin are in cosmetics, as a food colorant or nutraceuticals with antioxidant activity [24, 27, 32]. The total market for astaxanthin is estimated to be US$200–250 million with main application in aquacultures. In this regard, the price of astaxanthin varies from US$2000 to US$2500 per kilogram [24, 27, 29, 31, 32]. The annual production of *Haematococcus* biomass is reported to be 300 tons [26]. It should be emphasized that more than 90–95% of this demand is met via synthetically produced astaxanthin, as it can be produced at significantly lower cost compared to the carotenoid from algae [26, 29].

Increasing customer demand for astaxanthin of natural origin, and thus for larger and less expensive production facilities, could eventually help that the *Haematococcus* technology achieve a breakthrough.

Other carotenoids from microalgae, such as canthaxanthin (*Chlorella sp.*), zeaxanthin (*Chlorella ellopsoidea*), or fucoxanthin (*Phaeodactylum tricornutum*), are being intensively investigated for their commercialization as colorants or antioxidants. Lutein, for example, has an estimated market potential of over US$230 million [27, 31, 34]. Natural lutein is mainly derived from the *Tagetes* plant, but microalgae such as *Murellopsis spp, Scenedesmus almeriensis*, and *Chlorella spp.* have also been identified as alternative producers of lutein. However, no commercially significant production levels of the above carotenoids from algae are currently known [27, 31, 33, 34].

11.3.3 Products for Special Applications

Niche markets or special applications of commercialized algal products include phycobiliproteins such as phycoerthrin and phycocyanin. Phycobiliproteins are used as natural dyes in the food industry and in diagnostics as fluorescent markers or markers for antibodies and receptors [26]. Especially phycocyanin is used as a natural dye in food or cosmetics because of its blue color. Phycocyanin is produced using *Spirulina*, e.g., by the company Cyanotech (USA). In this process, the proportion of the dye in the biosolids can be as high as 46% [35]. The world market for phycocyanin is estimated at about US$ 10–60 million. The price of the pigment depends mainly on the purity and ranges from US$500 to US$100,000 per kilogram [25–27]. In the application field of labeled antibodies and fluorescent markers, derivatives of phycobiliproteins can fetch up to 1500 US$ per milligram [26].

11.3.4 Future Challenges

As described above, a large number of microalgae have already been identified as starting materials for a wide range of commercial applications. In addition, an intensive search for new applications continues. Nevertheless, the production of commercially established specialty chemicals is limited to β-carotene from *Dunaliella salina* and astaxanthin from *Haematococcus pluvialis*. Microalgae producers or production processes for other carotenoids, pigments, or polyunsaturated fatty acids (e.g., omega-3 fatty acids) require further significant improvements in terms of production costs and efficiency to compete with existing products for applications in the food sector, cosmetics, aquaculture, and agribusiness. The vast majority of these specialty chemicals are currently produced synthetically or extracted from natural sources [24, 27, 32].

One of the main challenges to establish algae as industrially useful production organisms lies in the development of cheap and efficient processes. Starting with improved reactor concepts, optimized cultivation of the organisms, and more effective downstream processing methods, much development work remains to be done.

At the same time, new production organisms capable of forming higher concentrations of the valuable substances described above has to be developed. The search for new products or applications in the fields of nutrition, health and cosmetics must also be intensified in order to establish algae technology on a broader basis. Furthermore, processes should be established that make it possible to utilize several valuable substances of the algae at the same time in order to make better use of the biomass. High productivity (yield per time) of individual valuable substances is essential in the long term to be able to produce profitably, also in relation to already existing products.

References

1. Chisti Y (2007) Biodiesel from microalgae. Biotechnol Adv 25:294–306
2. Leu S, Boussiba S (2014) Advances in the production of high-value product by microalgae. Ind Biotechnol 10(3):169–183
3. Spolaore P, Joannis-Cassan C, Duran E, Isambert A (2006) Commercial applications of microalgae. J Biosci Bioeng 101(2):87–96
4. Pulz O, Gross W (2004) Valuable products from biotechnology of microalgae. Appl Microbiol Biotechnol 65:635–648
5. Carvalho AP, Meireles LA, Malcata FX (2006) Microalgal reactors: a review of enclosed system designs and perfomances. Biotechnol Progress 22:1490–1506
6. Posten C (2009) Design principles of photo-bioreactors for cultivation of microalgae. Eng Life Sci 9(3):165–177
7. Pulz O (2001) Photobioreactors: production systems of phototrophic microorganisms. Appl Microbiol Biotechnol 57:287–293
8. Acién Fernández FG, Fernández Sevilla JM, Molina Grima E (2013) Photobioreactors for the production of microalgae. Rev Environ Sci Biotechnol 12:131–151
9. Algatech (2017) http://algatech.com. Accessed 16 May 2022
10. BASF (2017) Pretty in pink. https://www.basf.com/au/en/media/blog/pretty-in-pink.html. Accessed 16 May 2022
11. Cellana (2017) http://cellana.com. Accessed 16 May 2022
12. Cyanotech (2017) http://www.cyanotech.com. Accessed 19 May 2022
13. Lane J (2015) RIP, Aurora Algae: Algae and the Never-Never. Biofuels Digest (Hrsg) http://www.biofuelsdigest.com/bdigest/2015/07/22/rip-aurora-algae-algae-and-the-never-never/
14. Wikipedia: Sapphire Energy (https://en.wikipedia.org/wiki/Sapphire_Energy). Accessed 19 May 2022
15. Norsker NH, Barbosa MJ, Vermue MH, Wijffels RH (2011) Microalgal production – a close look at the economics. Biotechnol Adv 29:24–27
16. Wijffels RH, Barbosa MJ, Eppink MHM (2010) Microalgae for the production of bulk chemicals and biofuels. Biofuels Bioprod Biorefin 4:287–295. https://doi.org/10.1002/bbb.215
17. Gimpel JA, Henriquez V, Mayfield SP (2015) In metabolic engineering of eukaryotic microalgae: potential and challenges come with great diversity. Front Microbiol 6:1376–1376
18. Lü J, Sheahanb C, Fu P (2011) Metabolic engineering of algae for fourth generation biofuels production. Energy Environ Sci 4:2451–2466
19. Algenol (2017) http://algenol.com/sustainable-products/. Accessed 17 May 2022
20. AlgaePARC (2017) http://www.algaeparc.com/project/1/algaeparc. Accessed 17 May 2022
21. Bosma R, de Vree JH, Slegers PM, Janssen M, Wijffels RH, Barbosa MJ (2014) Design and construction of the microalgal pilot facility AlgaePARC. Algal Res 6:160–169. https://doi.org/10.1016/j.algal.2014.10.006
22. Borowitzka MA (1999) Commercial production of microalgae: ponds, tanks, tubes and fermenters. J Biotechnol 70:313–321
23. Jensen GS, Ginsberg DI, Drapeau MS (2001) Blue-green algae as an immuno-enhancer and biomodulator. J Am Nutraceuticals Assoc 3(4):24–30
24. Leu S, Boussiba S (2014) Advances in the production of high-value products by microalgae. Ind Biotechnol 10(3):169–183. https://doi.org/10.1089/ind.2013.0039
25. Pulz O, Gross W (2004) Valuable products from biotechnology of microalgae. Appl Microbiol Biotechnol 65:635–648. https://doi.org/10.1007/s00253-004-1647-x
26. Spolaore P, Joannis-Cassan C, Duran E, Isambert A (2006) Commercial application of microalgae. J Biosci Bioeng 101(2):87–96. https://doi.org/10.1263/jbb.101.87
27. Borowitzka MA (2013) High-value products from microalgae – their development ans commercialisation. J Appl Phycol 25:743–756. https://doi.org/10.1007/s10811-013-9983-9
28. DIC Lifetec (2017). http://www.dlt-spl.co.jp/business/en/spirulina/ecology.html#company. Accessed 19 May 2022

29. Milledge JJ (2012) Microalgae: commercial potential for fuel, food and feed. Food Sci Technol 26(1):28–31
30. Vigani M, Parisi C, Rodríguez-Cerezo E, Barbosa MJ, Sijtsma L, Ploeg M, Enzing C (2015) Food and feed products from microalgae: market opportunities and challenges for the EU. Trends Food Sci Technol 42:81–92. https://doi.org/10.1016/j.tifs.2014.12.004
31. Del Campo JA, García-González M, Guerrero MG (2007) Outdoor cultivation of microalgae for carotenoid production: current state and perspectives. Appl Microbiol Biotechnol 74:1163–1174. https://doi.org/10.1007/s00253-007-0844-9
32. Boussiba S, Aflalo C (2005) An insight into the future of microalgal biotechnology. Innov Food Technol 28:37–39
33. Blanco AM, Moreno J, Del Campo JA, Rivas J, Guerrero MG (2007) Outdoor cultivation of lutein-rich cells of Muriellopsis sp. in open ponds. Appl Microbiol Biotechnol 73:1259–1266. https://doi.org/10.1007/s00253-006-0598-9
34. Lin JH, Duu-Jong Lee DJ, Chang JS (2007) Lutein production from biomass: marigold flowers versus microalgae. Bioresour Technol 184:421–428. https://doi.org/10.1016/j.biortech.2014.09.099
35. Sekar S, Chandramohan M (2008) Phycobiliproteins as a commodity: trends in applied research, patents and commercialization. J Appl Phycol 20:113–136. https://doi.org/10.1007/s10811-007-9188-1

Challenges in Down-Streaming from Chemical and Biotechnological Processes

12

Bettina Sayder, Kerstin Schwarze-Benning, Hans-Jürgen Körner, and Ute Merrettig-Bruns

Abstract

In the processing of chemical and biotechnological products, two groups of separation processes in particular are industrially established: thermal and mechanical. In order to optimize investment and operating costs, different separation processes are combined where possible by interconnecting separate separation operations in terms of equipment. Methanol is presented as an example of the processing of a basic chemical. Its production from synthesis gas leads to a proportion of by-products of up to 5%. Biotechnological processes are more demanding in this respect, because the products usually occur in dilute aqueous solution, contain by-products as well as complex media components and also require cell disruption in some cases.

Keywords

Separation processes · Methanol · Synthesis gas · Biotechnological processes

12.1 From Chemical Processes

Production in the chemical industry is characterized by a large number of different manufacturing processes based on different reaction pathways. Chemical processes do not exclusively lead to a product, but depending on the reaction system, by-products are formed due to the stoichiometry of the main reaction. Furthermore, by-products are formed by parallel and subsequent reactions, the formation of which

B. Sayder (✉) · K. Schwarze-Benning · H.-J. Körner · U. Merrettig-Bruns
Fraunhofer Institute for Environmental, Safety and Energy Technology UMSICHT, Oberhausen, Germany
e-mail: bettina.sayder@umsicht.fraunhofer.de; kerstin.schwarze-benning@umsicht.fraunhofer.de; hans-juergen.koerner@umsicht.fraunhofer.de

M. Kircher, T. Schwarz (eds.), *CO2 and CO as Feedstock*, Circular Economy and Sustainability, https://doi.org/10.1007/978-3-031-27811-2_12

can be influenced by the reaction conditions but not completely suppressed. In addition, starting materials also remain in the reaction product due to incomplete conversion in the synthesis. To isolate the desired product, various separation processes are used for product separation and purification following the reaction.

Basic and bulk chemicals, such as inorganic base stocks, petrochemicals and their derivatives, are produced in large, continuously producing unit operations with high annual tonnages. The manufacturing processes are based on unit operations that are mature and optimized. The individual plant complexes are closely interlinked to make optimum use of energy and material flows. Product separation is usually achieved by distillation and rectification. Only when separation is difficult, e.g. in the case of azeotropic and narrow-boiling mixtures, alternative or hybrid separation processes are used. In contrast, fine and specialty chemicals are produced in smaller tonnages and in discontinuous plants. In addition, many fine and specialty chemical and pharmaceutical products have high boiling and melting points and are thermally sensitive. Here, processes that can be carried out with low energy input, such as liquid-liquid extraction, are superior to rectification [1].

Separation processes are of great economic importance as they are responsible for 40–70% of the capital expended and operating costs of industrial production. In addition, separation processes consume on average 43% of the process energy generated. Efficient design of product reprocessing processes is enormously important; their design should be simple and require low energy input. The coupling of different separation processes to solve a separation task (hybrid separation processes) as well as the integration of reaction and mass separation in one apparatus have been research topics in the chemical industry for years and find application in non-ideal organic-aqueous multicomponent mixtures or equilibrium-limited reactions [2].

12.1.1 Mechanical and Thermal Separation Processes

Separation processes exploit the different physical and chemical properties of substances mixed with each other in order to separate them from each other (Table 12.1). On the basis of the different separation techniques, the known separation processes can be assigned to particle and phase separation technology on the one hand, and to fluid process technology on the other. Particle and phase separation technology includes separation techniques such as filtration, centrifugation or crystallization, which enable several phases to be separated from one another or product properties to be adjusted. Fluid process engineering includes separation operations such as distillation, adsorption, absorption, evaporation, rectification, extraction, etc., which lead to a change in the concentration of one or more components in a phase.

The processes used on a large scale and in industrial applications include, in particular, those described below.

Table 12.1 Separation methods

Category	Separation method	Phases involved	Separation principle
Thermal separation processes	Rectification, distillation	Liquid/gas	Different volatilities
	Adsorption	Liquid/gas	Different accumulation on solid surfaces
	Crystallization	Liquid/gas	Different solubilities
	Extraction	Liquid/ liquid Solid/ liquid	Different solubilities
	Pervaporation	Liquid/gas	Different molecule sizes
	Adsorption, gas scrubbing	Liquid/gas	Different solubilities
Mechanical separation processes	Flotation	Solid/ liquid	Different surface wettability
	Sedimentation, centrifugation	Solid/gas Solid/ liquid	Different density
	Filtration, membrane separation process, reverse osmosis	Solid/ liquid Solid/gas	Different particle sizes
	Centrifugal separator	Solid/gas	Particle inertia
	Electrostatic dust collectors	Solid/gas	Electrical mobility

12.1.1.1 Adsorption

Adsorption is a physical process in which components of a gas or liquid adhere to the surface of another substance and accumulate on its surface. In principle, adsorption also includes the reverse process, desorption, because a system always strives for a balance between adsorbing and desorbing a substance. The effect of adsorption is based on the interaction forces between the solid surface and the adsorbed molecule and should not be confused with chemical adsorption (chemisorption) in heterogeneous catalysis. With the development of molecular sieves as adsorbents and improved regeneration techniques, adsorption has been more widely used as a separation process. It has advantages over rectification when the separation factor is close to 1 and can be influenced only slightly by selective additives. Furthermore, the process has advantages when very high or very low temperatures have to be realized during rectification or only very small amounts of a substance have to be separated.

Industrial applications of adsorption include the separation of solvent vapors from air, the drying of gases and solvents, natural gas processing, the separation of hydrocarbon mixtures, and the purification of wastewater and exhaust gas streams. In the large-scale production of basic chemicals, adsorptive separation (pressure swing adsorption, PSA) of the components obtained after the steam reformer and CO conversion produces almost pure hydrogen for hydrogenation processes and ammonia synthesis.

12.1.1.2 Absorption

Absorption as a separation process was developed more than 140 years ago. The aim is to selectively remove components from a gas stream using a suitable absorbent. Absorption is the underlying process in gas scrubbing. Since the absorbent is to be recirculated, absorption is followed by regeneration (desorption), which is as simple as possible. The gas to be purified enters an absorption column from below, from where it flows upward and comes into contact with the absorbent in countercurrent. In the process, the components are selectively washed out. The purified gas is obtained at the top of the column and the loaded absorbent is obtained at the bottom of the column, which is regenerated in a second column by changing the conditions. To achieve high enrichment during absorption, pressure and temperature are varied to increase the solubility of the component to be separated in the absorbent.

In addition to physical absorption, in which only the gas solubility is used, chemical absorption binds the gas both physically and chemically. Chemical absorption is used in the separation of carbon dioxide from process streams, known as potash scrubbing, in the separation of acid gases (CO_2, H_2S) in natural gas processing, and in various flue gas scrubbing operations. Physical absorption is used, among other things, for the separation of solvent vapors from air and in various washing processes with methanol (rectisol process).

12.1.1.3 Distillation and Rectification

Distillation is the most important thermal separation process in the chemical industry. It is used to recover vaporizable liquids or to separate solvents from substances that are difficult to vaporize and then collect them by condensation. Compared to other separation processes, distillation has the advantage that usually no further substances, such as adsorbents or solvents, have to be added.

In simple distillation, a suitable heat source is used to heat the sump to the desired temperature at which the target component begins to boil. Once this is reached, the substance rises in its gaseous state and is liquefied again in a condenser by cooling. The liquid condensate is then collected. Distillation does not lead to complete separation of the liquid mixture, but to separation into two mixtures, each with different contents of light and heavy volatile components. The separation effect is based on the different compositions of the boiling liquid and the vapor.

Rectification is an extension of distillation and can be understood as a series of many individual distillation steps. The main advantages of rectification are a continuous operation of the plant and a significantly higher separation effect compared to distillation, since the vapor is in countercurrent contact with the liquid several times in succession. The column is more energy-efficient, technically less complex and takes up less space than a series of single distillation units. A rectification column consists of several trays with special internals (bubble trays, column packings) to provide the contact area between vapor and liquid phases.

Typical applications of distillation and rectification are the distillation of alcohol and the rectification of crude oil in refineries or the production of distilled water. Multistage distillation in columns is a commonly used and well-controlled thermal separation process in the chemical industry. Other processes are used when the substance properties do not permit multistage distillation.

12.1.1.4 Extraction

Extraction is the term used for all separation processes in which one or more components are separated from a mixture of substances with the aid of a solid, liquid or gaseous extraction agent. The prerequisite for separation is that the two phases are not soluble in each other and the component to be separated has different equilibrium concentrations in the phases. Extraction also uses the countercurrent principle. After extraction, further separation processes usually have to be used to recover the extractant and return the extractant regenerated to the cycle. Extraction processes are divided into liquid-liquid extraction, solid-liquid extraction and high-pressure extraction.

Extraction is always used when distillation or rectification are out of the question for technical reasons, such as an unfavorable separation factor, "missing" vapor pressure (metal salts) or thermal instability, or when the extraction process is more cost-effective, e.g. when separating small quantities of a low-volatile component. Extraction is of great importance in the separation of metal salts from aqueous solutions and in the recovery of temperature-sensitive active ingredients in pharmaceuticals. Another large-scale application is the recovery of BTX aromatics, where extraction is used to separate aliphatic/aromatic mixtures.

12.1.1.5 Filtration

Filtration is a process for separating or purifying substances, usually a suspension or an aerosol. Filtration is one of the mechanical separation processes based solely on physics.

The mixture to be separated passes through a filter made, for example, of paper, textile fabric or metal, or through a container containing a filling of a filter material. All filter materials present resistance to all particles of the mixture to be separated. During filtration, not only particles larger than the pore size of the filter are retained, but through mechanisms such as particle inertia, diffusion effects, electrostatics or barrier effect, particles far smaller than the pore size of the filter are also separated in principle. Particularly in the field of gas filtration, filters have a particle size range called the filter gap, in which particles are only insufficiently separated—significantly larger and also significantly smaller particles are completely retained by inertia and barrier effect, or by diffusion.

After a certain time, depending on the filtering method used, either a layer (the so-called filter cake) is formed from the retained particles, or the pores of the filter mass are reduced by the deposition of the retained substances. After a sufficiently thick filter cake has been built up, complete separation of the particles is usually achieved, but the flow resistance of the filter also increases significantly. Depending on the design of the filter, the filter cake or the absorbed solids must be removed from time to time (e.g. by shaking, backwashing or a pressure pulse against the direction of flow), or the filter must be replaced. In liquid filtration, these periodic interruptions of operation (filtration) and cleaning (backwashing) are referred to as dead-end filtration. This procedure is necessary for many filter systems.

12.1.1.6 Centrifugal Separator and Centrifuge

Centrifugal separators (cyclones) and centrifuges serve as mass force separators in technical plants. Cyclones are suitable for separating solid or liquid particles from gases, e.g. for exhaust gas cleaning. A centrifuge can separate the components of suspensions, emulsions and gas mixtures. The centrifugal forces are created by the generation of a vortex flow. In a centrifuge, the necessary kinetic energy is transferred to the medium to be separated by the rotational movement of the vessel whereas in a cyclone, the gases as carriers of the particles to be separated are set in a rotational movement by their own flow velocity and appropriate design of the separator. The separation of substances takes place in the centrifuge by differences in density, in the cyclone by the particle mass.

12.1.1.7 Membrane Process

Membranes are filter media which have the property of selectively retaining components and allowing others to pass through. The selective properties are based on the one hand on the pore structure of the membrane, and on the other hand on the different solubility and diffusion coefficients of the components to be separated. Membranes enable the separation of substances in the particle range from 0.1 to 10 µm. Membrane materials consist of both organic materials, which can be hydrophilic or hydrophobic, or inorganic materials such as metal, glass or ceramics. The membrane used should have high selectivity, satisfactory permeability and high mechanical, chemical and thermal stability.

Membrane separation plants can be constructed in a modular way so that the plant can be adapted stepwise to the scope of a separation problem. However, since the total cost increases proportionally to the size of the plant, there is a critical size for most membrane processes beyond which the classical separation processes become economically more favorable.

The particular advantage of membrane separation processes, in addition to the simple modular design and low space requirements as well as the easy integration into individual production plants, is the gentle, purely physical separation principle. Separation takes place without chemical or thermal alteration of the components to be separated, thus enabling the use of both fractions (permeate and retentate). This is why the process has become established in food technology, biotechnology and pharmaceuticals in particular. Furthermore, membranes can be used to achieve separations that are not possible with thermal processes, e.g. for azeotropes that cannot be separated by classical distillation. A disadvantage can be the limited thermal, chemical and mechanical stability of some membrane materials, which are also sensitive to fouling.

Important technical applications include the clarification and concentration of beverages by microfiltration, the recovery of drinking water by reverse osmosis, and the recovery of valuable organic substances from fermentation broths by pervaporation. The recovery of organic vapors, for example gasoline vapor recovery, as well as the separation of gas mixtures, are carried out by means of gas permeation. Another field of application is wastewater treatment. With the aid of ultrafiltration and microfiltration, it is possible to remove particles, colloids and macromolecules so that wastewater can be disinfected in this way.

12.1.1.8 Stripping

In stripping, substances are removed from the liquid phase by passing gases (e.g. air or water vapor) through them and transferring the substances to the gas phase. For this purpose, the liquid phase is brought into contact with the gas using the countercurrent principle. Technically, stripping is usually carried out in packed columns. A nozzle finely distributes the liquid at the top of the column so that it trickles over the packing in the column into the sump. In countercurrent, the stripping gas (e.g. air) is conveyed through the column. The packing serves to finely distribute the liquid and thus maximize the phase interface. Purification efficiency is also determined by the gas/liquid ratio, with the transfer of contaminant from liquid to gas described by the Henry constant of each substance. Stripping is usually followed by purification of the off-gas, e.g. by adsorption, cooling or decomposition.

The process can be adapted to the requirements within wide limits so that the desired outlet concentration is obtained for a given inlet concentration of impurity in the liquid phase. The process is used, for example, in wastewater treatment for the removal of ammonia, hydrogen sulfide, phenols, organic halogen compounds and hydrocarbons. In addition, stripping is often used in petroleum processing to remove lighter components from the product drawn off the side of a column.

12.1.2 Combined Separation Processes

The conversion of the raw material base to renewable raw materials and the manufacture of diverse chemical products based on fewer chemical feedstocks, which make effective recovery possible, are just some of the challenges the chemical and pharmaceutical industries are facing in the coming years. Under the buzzword process intensification, methods have been developed in the last decade to make production plants more flexible and efficient.

The interconnection of at least two different apparatus-separated separation operations, both of which contribute to the solution of a defined separation task, is referred to as a hybrid separation process. One of the advantages of hybrid separation processes is that the separation takes place in the optimum working range of the respective separation operation. The specific separation costs of an optimal hybrid process should be lower than those of the individual basic operations. Azeotropic or narrow-boiling mixtures cannot be easily separated by means of rectification. Therefore, the hybrid process of rectification/crystallization allows separation without an additional auxiliary. In addition to rectification/crystallization, the combination of rectification and membrane processes is also used, e.g. for the separation of ETBE, ethanol and C4 hydrocarbons [3].

Extractive distillation is already used in petrochemistry, where a heavy-boiling auxiliary is added to the mixture to be separated by distillation to selectively influence the phase equilibrium [1]. New approaches include membrane distillation for solvent dehydration or extractive crystallization for the separation of mixtures of isomers [4].

The reactive separation process is another concept to overcome equilibrium limitations in synthesis and to avoid undesired subsequent reactions. The best known reactive separation processes are the membrane reactor and reactive distillation or reactive rectification. The membrane reactor is the integral coupling of chemical reaction and membrane process. It is used for selective product removal or to retain catalysts [5].

Reactive rectification is the most commonly used integral process. It is suitable for turnover enhancement of equilibrium-limited reactions or for breaking occurring azeotropes. Compared to conventional process steps, there are advantages in investment and operating costs and in reaction conversion. In addition, mass transfer and selectivity can be improved. The introduction of reactive rectification resulted in the saving of one reactor and nine distillation columns in methyl acetate synthesis compared to the conventional process [1].

Coupling heterogeneous catalysis with adsorptive mass separation leads to the adsorptive reactor. Adsorption allows the product to be removed simultaneously or a reactant to be presented as an adsorbate [4].

Reactive extraction is widely used in biotechnology and is suitable for the separation of heavy metals, organic acids, or pharmaceuticals from low-concentration aqueous solutions. Its advantages are low energy consumption, continuous operation and high selectivity [4].

To produce products with targeted properties, such as particle size, morphology and purity, reactive crystallization is suitable. Its applications range from diastereomer crystallization to wastewater treatment [4].

Chromatographic reactors combine chemical or biochemical reactions with chromatographic separation. Their field of application is in the area of fine chemistry, pharmacy and biotechnology. The principle of batch column and chromatographic reactor in countercurrent is particularly suitable for the production of enantiomerically pure products. The best-known reactor type is the "simulated-moving bed" reactor [4, 5].

12.1.3 Reprocessing of Products from the Use of C1 Raw Materials

Organic basic chemicals are among the most significant basic products in the chemical industry, as they are processed into a wide range of products, such as plastics, synthetic fibers, paints and coatings, polymers and pharmaceuticals. The production of these substances is based primarily on the processing of petrochemical primary compounds (C2, C3, C4 alkenes, aromatics) and synthesis gas, which are obtained from crude oil, natural gas or coal. Synthesis gas (CO, CO_2, H_2, N_2; Chap. 5) provides C1 building blocks for the synthesis of important basic chemicals.

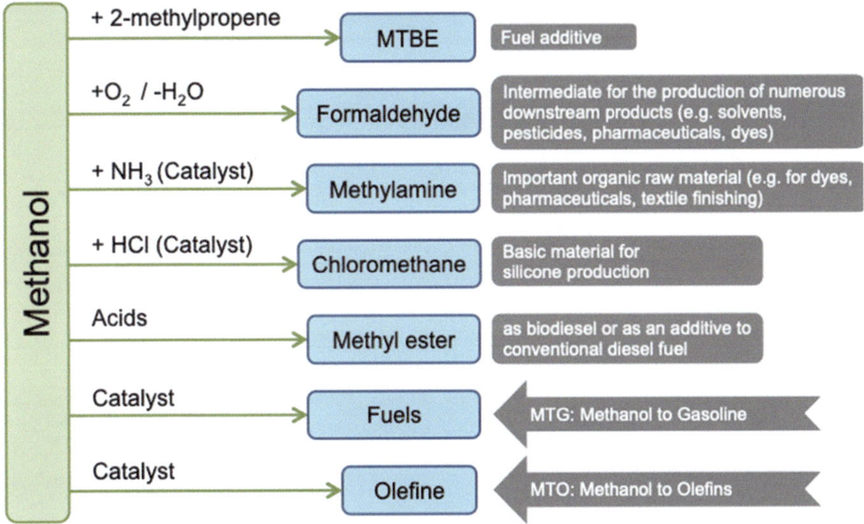

Fig. 12.1 Downstream syntheses of methanol (adapted from [1, 7]). © Wiley-VCH

The main consumers of syngas are ammonia and methanol synthesis, from which a variety of other products can be made. In addition, synthesis gas is used for the hydroformulation of olefins and for the synthesis of hydrocarbons according to the Fischer-Tropsch synthesis.

Methanol is the most important C1 platform chemical in the chemical industry. Global production volumes are expected to exceed 100 million metric tons by 2023. Methanol is produced via the synthesis gas route from coal and naphtha, but mainly from natural gas. A variety of other platform chemicals, such as dimethyl ether (DME), acetic acid, propylene, olefins, and aromatics, but also fuels, such as diesel and gasoline, and their additives can be produced from methanol (Fig. 12.1).

After synthesis, the crude methanol contains water, dissolved gases, and a variety of undesirable but unavoidable byproducts that have lower or higher boiling points than methanol. The nature and amount of the impurities depend primarily on the synthesis conditions, particularly the pressure and the hydrogen/carbon monoxide ratio. The water content depends mainly on the carbon dioxide content in the synthesis gas and can be up to 18 wt%, depending on the process. The total byproducts in high-pressure methanol is about 3–5 wt%, while low-pressure methanol contains less than 1 wt% byproducts. The requirements for pure methanol vary depending on the further application. When used for the synthesis of downstream products, it is important to ensure that the methanol does not contain catalyst-damaging components [6].

The crude methanol is worked up by distillation in three process steps with up to three distillation columns (Fig. 12.2). First, the raw methanol is expanded into a pressure vessel. The dissolved gases (CO, CO₂, H₂, CH₄) escape and are removed. The more volatile components (ethers, formates, aldehydes, ketones) are removed at

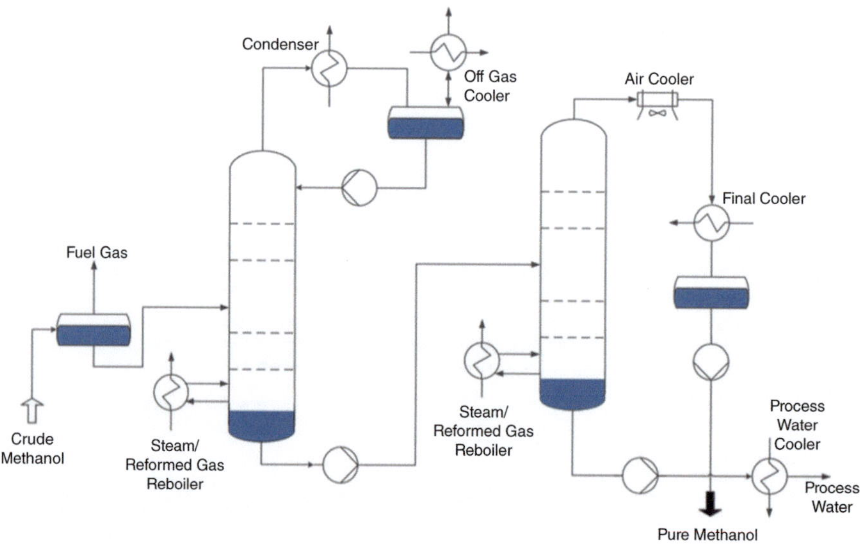

Fig. 12.2 Distillation of crude methanol (adapted from [7]). © John Wiley and Sons

Table 12.2 Products from biotechnological productions

Product	Example
Gas	Methane
Fermentation medium	Sauerkraut, milk products
Biomass	Yeast
Intracellular products	Some vitamins, polyhydroxyalkanoate
Extracellular products in fermentation media	
– High molecular	– Enzymes, polysaccharides
– Low molecular	– Ethanol, organic acids, amino acids

the head of the pre-column. Subsequently, the methanol is separated from the heavier boiling components (ethanol, higher alcohols) and the water in the methanol column. Different process variants are used in industry, which are usually embedded in a larger plant network. This can result in energy integration advantages, which can lead to cost savings in operating costs [7].

12.2 From Biotechnological Processes

A range of different products results from biotechnological production processes (Table 12.2). The product can be the biomass itself, it can be contained extracellularly in the fermentation medium or intracellularly in the microorganisms, or it can be present in gaseous form. Depending on the occurrence of the product, specific downstream processing of the fermentation broths and/or biomasses is required.

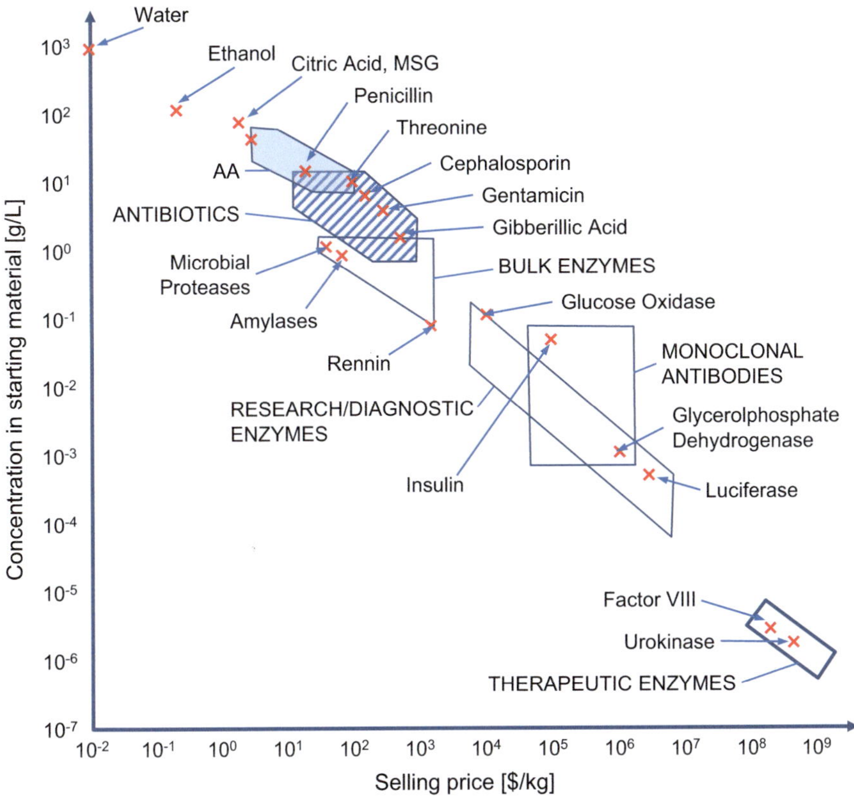

Fig. 12.3 Dependence of selling price on product concentration, acc. to Dwyer [10]. © Springer Nature

Product downstream processing includes all process steps from cell separation to the purified end product. The exact design of product downstream processing is tailor-made for each product. The form in which the product is present, the purity aimed for in the final product, and the level of impurities or the proportion of possible byproducts determine the further procedure.

While the cost of processing the biomass product is rather low, cell disruption and complex product isolation are necessary to obtain intracellularly formed products. Depending on the type of products, the share of product processing for fermentatively produced products in the total production costs can be up to 70% [8, 9]. In general, it can be stated that the processing costs and thus also the selling price of a product increase with decreasing product concentration (Fig. 12.3).

The recovery and processing of products from biotechnological processes are subject to special challenges:

Most products occur diluted in aqueous solution, so large volumes often need to be processed.

In addition to the desired product, fermentation media often contain organic impurities from complex medium components such as molasses, peptones, yeast extract and inorganic components such as salts.

Some products are formed intracellularly and require cell disruption.

12.2.1 Individual Steps in Product Processing

In general, product processing, as shown in Fig. 12.4, is divided into the steps of cell separation, cell disruption if necessary, product separation, concentration, purification, fine purification, and packaging.

12.2.1.1 Separation of the Cells/Biomass

In the production of biomass, the separation of cells from the fermentation medium is the most important processing step. For the production of other biotechnological products, the separation of cells is often a prerequisite for further processing. Intracellular products are extracted from the concentrated and subsequently digested biomass. Extracellular products are usually obtained from the cell-free medium, since the presence of solids impairs the separation performance of some work-up methods, such as extraction and adsorption. The processes used for cell separation are filtration (surface filtration, depth filtration), sedimentation, centrifugation, flocculation and flotation. Basically, here the separation of cells is done by the difference in cell size or by the difference in density of the cells in relation to the medium to be separated. For viruses and small bacteria, the expected particle sizes range from <0.1 to 1.0 µm, for average bacteria about 1 µm, for microalgae 5–10 µm, and for large bacteria, yeasts, and fungal mycelia 1–100 µm [11, 12].

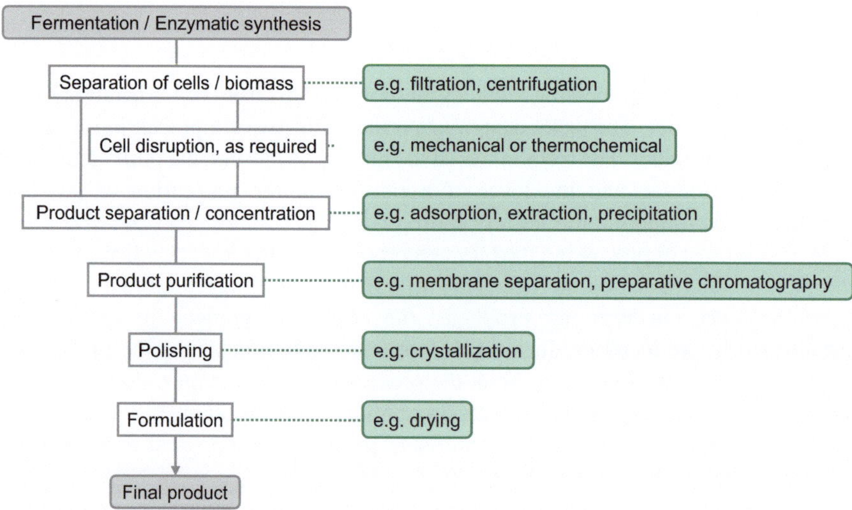

Fig. 12.4 Overview of the different steps in product reprocessing

12.2.1.2 Cell Disruption

For the recovery of intracellular products, cell disruption is required after cell separation from the fermentation medium. The effort of disruption depends on the type of cell. Animal cells are only enclosed by a cell membrane. Here, cell disruption is unproblematic. Plant cells and most bacteria have a cell wall in addition to the cell membrane. This lies outside the membrane and envelops the cell (Fig. 12.5). Digestion of Gram-positive bacteria is considerably more problematic, since their cell wall, in contrast to Gram-negative bacteria, consists of up to 40 layers of murein [13, 14].

Cell disruption processes are divided into physical (mechanical and non-mechanical), chemical and biological processes. An overview of the different methods is given in Table 12.3 [9, 13]. Cell disruption on a production scale (e.g., for yeasts) is mainly performed with agitated ball mills and high-pressure homogenizers. All other methods are mainly used on a laboratory scale [9].

12.2.1.3 Product Separation and Concentration

After release of the product and separation of the cell debris (in the case of intracellular products) or after separation of the cells (in the case of extracellular products), further processing steps follow for product separation and concentration from the cell-free aqueous fermentation broth. During precipitation, the product molecules are converted into particles, aggregates or flocs, which then sink to the bottom of the fermentation broth ("sedimentation"). For proteins, this can be done, for example, by the addition of salts, organic solvents or water-soluble polymers, and by changing the pH or temperature. The greater the difference in density between the particles and the liquid phase, the higher the rate of sinking. Precipitation represents an initial

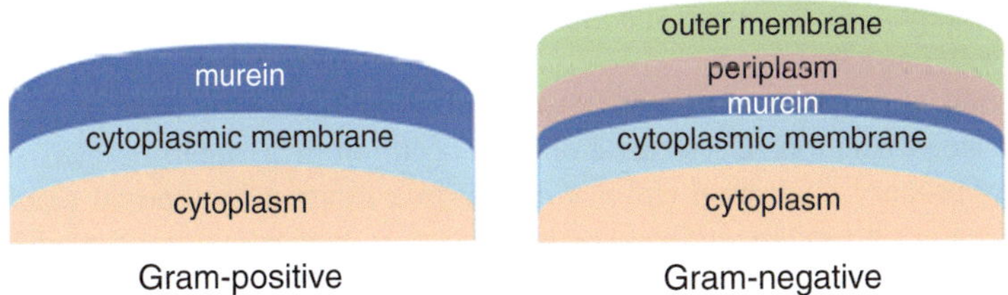

Fig. 12.5 Simplified schematic representation of the structure of a gram-positive and a gram-negative cell wall

Table 12.3 Subdivision of cell disruption methods

Mechanical methods			
Mechanical	Non-mechanical	Chemical methods	Biological methods
Ultrasonic	Osmotic shock	Acids/bases	Enzymes
Agitator ball mill	Freeze/defrost	Solvents	Phages
High-pressure homogenizer	Freeze-drying	Detergents	Viruses
	Decompression	Antibiotics	

coarse cleaning step and is used on an industrial scale for pre-cleaning and concentration [9].

Extraction represents another possibility for the coarse purification of the products. Here, it is important that the solubility of the product to be cleaned and the solubility of the impurities differ significantly. The different types of extraction are liquid-liquid extraction, solid-liquid extraction and high-pressure extraction. In liquid-liquid extraction, separation occurs due to different polarities and solubilities into two liquid phases that are immiscible with each other. The separation performance of solid-liquid extraction is also based on different solubilities of the components. Here, one or more components are extracted from a solid phase. High-pressure extraction is carried out with supercritical fluids, mainly supercritical carbon dioxide.

The separation of substances via the accumulation of the product on a solid surface is known as adsorption. A distinction is made here between physisorption (physical binding forces, such as Van der Waals forces, electrostatic interactions) and chemisorption (chemical "covalent" bonding). Examples of possible adsorbents include activated carbon, synthetic resins, silica gel, and zeolites. Possible designs for adsorption are either fixed or fluidized bed adsorbers or batch or stirred reactors (CSTR reactors) [15].

12.2.1.4 Fine Cleaning/Product Cleaning

After successful product separation and concentration, impurities remain which have to be separated in the course of product or fine purification. For product purification, chromatography, crystallization, membrane separation processes (e.g. ultra/ nanofiltration, electrodialysis) and electrophoresis are used.

In chromatography, there is a stationary phase and a mobile phase. The phase states solid, liquid or gas can be combined for the two phases. In adsorption chromatography, for example, the mixture to be separated passes through a column loaded with particles.

Crystallization refers to the process of crystal formation from a previously homogeneous solution. Crystallization can occur from a supersaturated solution either by primary nucleation (from a crystal-free solution by thermal agitation or by nucleation on foreign surfaces) or by secondary nucleation (via the addition of so-called seed crystals or via the formation of new crystal nuclei by abrasion). The different crystallization processes are divided into evaporation crystallization, cooling crystallization and precipitation crystallization.

Membrane processes, in particular ultrafiltration, can be used to separate large molecules and small particles via a membrane. The membrane retains molecules larger than the pore size (<0.1 μm). In contrast to ultrafiltration, electrophoresis takes advantage of the different charge of the molecules to be separated. Here, electrically charged particles migrate to the oppositely charged electrode when an electric field is applied.

12.2.1.5 Product Configuration

Product configuration is used to produce a stable final form of the product. In this process, any remaining solvent is removed by drying or freeze-drying and the end product is stabilized. Configuration can also include tabletting, packaging and labeling of the end product.

12.2.2 Integrated Product Separation

Innovative developments in product processing include integrated product separation already during fermentation—in situ product removal/recovery (ISPR). Integrated product recovery is particularly useful for products that have an inhibitory effect on microorganisms above a specific concentration or are chemically unstable. In principle, most of the processes already described above are applicable. A method for selecting suitable processes is shown Fig. 12.6.

The decision tree shows that ISPR is not worthwhile for non-inhibiting, stable products. Integrated product processing becomes interesting for inhibitory products and/or unstable products that are already subject to decomposition during fermentation. The focus for the development of ISPR strategies is on the production of alcohols (ethanol, butanol, acetone), organic acids, and fragrances and flavors [16].

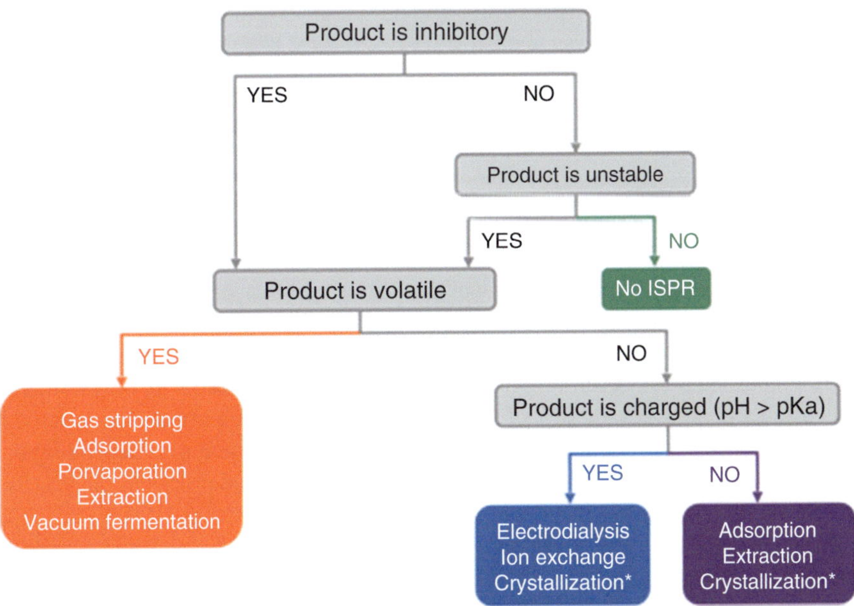

Fig. 12.6 Decision tree for the selection of suitable ISPR (in-situ-product removal/recovery) strategies, acc. to van Hecke et al. [16]. Urbanus et al. [17], provide a decision tree for different crystallization approaches. © Elsevier

For the appropriate selection of an ISPR strategy, the substance properties of the product, such as boiling temperature or charge, must first be determined. For products with low vapor pressure, such as ethanol, established processes such as gas stripping and pervaporation are suitable. For products with electrical charge, separation can be achieved by electrodialysis or ion exchange. Different configurations are possible for ISPR. The simplest option is direct product recovery during fermentation in the fermenter or bypass. If the presence of cells interferes with the work-up process, work-up is performed in cell-free medium.

12.2.3 Down-Stream Processing of Products from the Use of C1 Raw Materials

Methane, succinic acid and biomass (algae or single cell protein) are listed below as examples of products that can be produced biotechnologically using CO_2.

12.2.3.1 Methane

Methane is a combustible C1 gas that can be obtained, among other things, via biological methanation. In this process, CO_2 from biogas, biomethane, or wastewater treatment plants is biologically converted with hydrogen from the electrolysis of electricity from renewable sources into storable methane. The conversion of CO_2 and H_2 to methane is carried out by methanogenic microorganisms (so-called archaea) under anaerobic conditions. The challenge here is the solubility of the substrate hydrogen. Methane can be obtained in high purity in the gas phase via biological methanation. According to DVGW guideline G 262, a methane content of at least 95% is required for feeding into the natural gas grid [17]. If this value is reached or exceeded via biological methanation, no further treatment of the methane is necessary.

12.2.3.2 Succinic Acid

Succinic acid is a C4 dicarboxylic acid produced on the basis of petroleum from chemical processes or biotechnologically with the addition of carbon dioxide as a cosubstrate. It is chemically produced by catalytic hydrogenation of maleic acid, maleic anhydride or fumaric acid. According to a MarketsandMarkets market study, total global sales of succinic acid were 46,200 metric tons in 2014. Biobased succinic acid accounted for 28% of this total [18]. Large-scale fermentation is carried out either with yeasts at rather low pH or with bacteria at neutral pH. Depending on the process, the succinic acid is processed by direct crystallization, electrodialysis, or precipitation (Table 12.4) [19].

12.2.3.3 Biomass

Since the 1970s, biotechnologically produced biomass—single cell protein (SCP)—has been used as a protein- and vitamin-rich food or animal feed. SCP is derived from the protein of microorganisms, such as yeasts, fungi, algae, or bacteria, grown on various carbon sources [20]. CO_2 as a substrate is used together with light (via

Table 12.4 Process for the fermentative production of succinic acid [18]

Process	Technology	Down-streaming
BioSA-DC	Fermentation with yeasts at low pH to succinic acid	Direct crystallization
BioSA-ED	Fermentation with bacteria at neutral pH to the salt of succinic acid	Electrodialysis
BioSA-AS	Fermentation with bacteria at neutral pH to the salt of succinic acid	Precipitation with ammonium sulfate

photosynthesis) to grow algae. Algae can be broadly divided into microalgae and macroalgae (Chap. 7). Microalgae are important components of many ecosystems; they account for more than half of the primary production at the base of the food chain worldwide [21] and are cultivated in open reactors (open ponds), plate or tube reactors [22] (Chap. 11). Algal protein is of high quality and comparable to conventional vegetable protein. The main part of microalgae is used for the production of algae tablets or food supplements. Therefore, downstream processing focuses on separation of cells from the medium and drying [23].

References

1. Baerns M, Behr A, Brehm A, Gmehling J, Hofmann H, Onken U, Renken A (2006) Technische Chemie. Wiley-VCH, Weinheim
2. Eissen M, Metzger JO, Schmidt E, Schneidewind U (2002) 10 Jahre nach "Rio" – Konzepte zum Beitrag der Chemie zu einer nachhaltigen Entwicklung. Angew Chem 114(3):402–425
3. Franke M, Górak A, Strube J (2004) Auslegung und Optimierung von hybriden Trennverfahren. Chem Ing Tech 76:199–210. https://doi.org/10.1002/cite.200406150
4. Bazzanella A (2008) Prozessintensivierung - Eine Standortbestimmung. Positionspapier der ProcessNet Fachsektion Prozessintensivierung, DECHEMA e. V. (Hrsg). http://dechema.de/dechema_media/Roadmap+PI+FS+Stand+MV+2008-view_image-1-called_by dechema-original_site-dechema_eV-original_page-125076.pdf. Zugegriffen: 18 Jan 2018
5. Keller T, Roth T, Mackowiak JF, Kreis P, Górak A, Stankiewicz A (2011) Prozessintensivierung in der Fluidverfahrenstechnik. Chem Ing Tech 83:935–951. https://doi.org/10.1002/cite.201100010
6. Falbe J (1977) Chemierohstoffe aus Kohle. Georg Thieme Verlag, Stuttgart
7. Bertau M, Offermanns H, Plass L, Schmidt F, Wernicke H-J (2014) Methanol: the basic chemical and energy feedstock of the future. Asinger's vision today. Springer, Heidelberg
8. Zeikus JG, Jain MK, Elankovan P (1999) Biotechnology of succinic acid production and markets for derived industrial products. Appl Microbiol Biotechnol 51:545–552. https://doi.org/10.1007/s002530051431
9. Chmiel H (2006) Bioprozesstechnik, 2nd edn. Spektrum, Heidelberg
10. Dwyer JL (1984) Scaling-up bio-product separation with high performance liquid chromatography. Biotechnology 2:957–964
11. Starr MP, Stolp H, Trüper HG, Balows A, Schlegel HG (1981) The prokaryotes. A handbook on habitats, isolation and identification of bacteria. Springer, New York
12. Steinbusch SR (2015) Einfluss von Licht und Temperatur auf die Kultivierung von Mikroalgen – Auslegung und Betrieb von Freiland-Pilotanlagen zur Bestimmung prozessrelevanter Kinetiken. Dissertation, Karlsruher Institut für Technologie (KIT)

13. Kampen I (2006) Einfluss der Zellaufschlussmethode auf die Expanded Bed Chromatographie. Dissertation, TU Wilhelmina zu Braunschweig
14. Fuchs G, Schlegel HG, Eitinger T (2007) Allgemeine Mikrobiologie, 8th edn. Thieme, Stuttgart
15. Storhas W (2003) Bioverfahrensentwicklung. Wiley-VCH, Weinheim
16. van Hecke W, Kaur G, de Wever H (2014) Advances in in-situ product recovery (ISPR) in whole cell biotechnology during the last decade. Biotechnol Adv 32:1245–1255. https://doi.org/10.1016/j.biotechadv.2014.07.003
17. Urbanus J, Roeland CPM, Verdoes D, ter Horst JH (2012) Intensified crystallization in complex media: heuristics for crystallization of platform chemicals. Chem Eng Sci 77:18–25
18. MarketsandMarkets (2015) Succinic Acid Market – by source (petroleum & bio-based) and application (Polyurethane, food & beverage, resins, couatings & pigments, plasticizers, pharmaceuticals, de-icer solutions, PBS/PBST, solvents & lubricants, personal care) – Trends & Forecasts 2019
19. Smidt M (2011) A sustainable supply of succinic acid. Euro|Biotech|News 10:70–71
20. Najafpour GD (2008) Single-cell protein. In: Najafpour GD (ed) Biochemical engineering and biotechnology, 1st edn. Elsevier, Amsterdam, pp 332–341
21. Ghasemi Y, Rasoul-Amini S, Morowvat MH (2011) Algae for the production of SCP. In: Liong M-T (ed) Bioprocess sciences and technology. Nova Science Publishers, New York, pp 163–184
22. DECHEMA (2016) Mikroalgen-Biotechnologie. Gegenwärtiger Stand, Herausforderungen, Ziele. Positionspapier des DECHEMA e. V., Frankfurt am Main. https://dechema.de/dechema_media/Downloads/Positionspapiere/PP_Algenbio_2016_ezl-p-20001550.pdf
23. Nasseri AT, Rasoul-Amini S, Morowvat MH, Ghasemi Y (2011) Single cell protein: production and process. Am J Food Technol 6(2):103–116

Utilization of C1 Gas Streams from Steelworks

13

Marten Sprecher and Michael Hensmann

Abstract

In Germany, the steel industry in particular is characterized by high carbon-rich waste gas flows, although the composition and quality of the gas mixtures resulting from various process steps can vary greatly. In addition to the use of resulting gases to generate heat and electricity, which in turn are required in steel processing operations, approaches are being pursued to specifically reduce CO2 emissions through Carbon Direct Avoidance (CDA), Process Integration (PI), Carbon Capture and Storage (CCS) or Carbon Capture and Utilization (CCU).

Keywords

Steel industry · Steel processing operations · Carbon Direct Avoidance (CDA) · Process Integration (PI) · Carbon Capture and Storage (CCS) · Carbon Capture and Utilization (CCU)

13.1 Process Routes of Crude Steel Production

Crude steel production in Germany was approximately 42.4 MT in 2018. Two-thirds of the crude steel is produced in integrated steel mills via the blast furnace converter route, in which carbon carriers such as coal and coke are used as energy sources and reducing agents, and one-third via the electric furnace route, in which carbon carriers

M. Sprecher (✉)
Düsseldorf, Germany

M. Hensmann
VDEh - Betriebsforschungsinsitut GmbH, Düsseldorf, Germany
e-mail: michael.hensmann@bfi.de

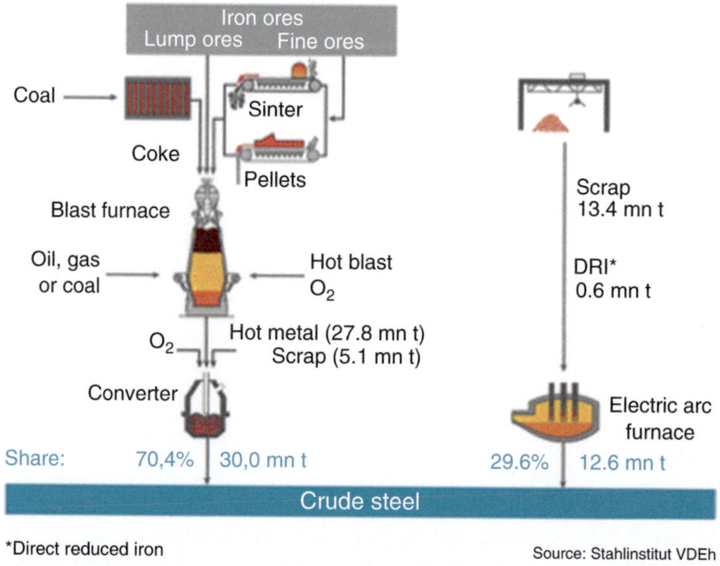

Fig. 13.1 Generation routes for steel production in Germany [1]

play only a minor role (Fig. 13.1). Accordingly, the co-product gases of the blast furnace converter route have process-related high concentrations of CO and CO_2.

The locations of the steel industry in Germany differ according to the production route and the regional location of the mills (Fig. 10.2). The six integrated steel mills, two of which are located in North Rhine-Westphalia, operate two to four blast furnaces at each of their sites. The pig iron from the blast furnaces has a carbon content of approx. 5% and is further processed into crude steel in the converters of the oxygen steel mills by decarburization with oxygen. The integrated steel mill usually includes a coke plant and a sinter plant. The co-produced gases from coke, pig iron and crude steel production are used to generate energy in the internal processes, e.g. for generating hot blast for the blast furnace, for underfiring the coke oven batteries and for firing the heating furnaces in the hot rolling mill. Surplus process gases are converted into electricity in the iron and steel mill power plants for self-supply. The locations of the integrated steel mills and electric steel mills in Germany can be seen in Fig. 13.2.

13.2   Co-Product Gases of the Integrated Steel Mills

The co-product gases are the process gases from the three process stages coke plant (Fig. 13.3), blast furnace and converter, which are produced in different quantities in relation to the ton of crude steel. The co-product gases contain CO, CO_2 and hydrogen, but in very different proportions (Table 13.1). Due to their composition,

Fig. 13.2   Steel production locations in Germany [2]

the co-product gases are used for different applications in heating and power generation.

> **Co-Product Gases**
>
> "By-products, gases or metallurgical gases resulting from the production of basic materials, including blast furnace gas, coke oven gas or converter gas, or a mixture of these gases", item 9 of Section 3 (2) Allocation Act 2012 of 07.08.2007, BGBI, I 07, 1788.

Fig. 13.3 Aerial view of a coking plant

The coke oven gas is produced during the conversion of coking coal into high-quality coke. In the coke ovens, the volatile components of the coking coal are released at temperatures of up to 1050 °C. The coke is then used as an energy source and reducing agent in the blast furnace process. This produces a porous, coarse-grained coke which is used in the blast furnace process as an energy source and reducing agent. After extensive treatment of the coke oven gas to recover tar, benzene and sulfur, the gas is used for underfiring the coke batteries and fed into the plant network in so-called district gas quality. Here it is used in particular as heating gas for the reheating furnaces.

Coke oven gas is produced during the dry distillation of hard coal in coke oven batteries. The coke produced serves as a reducing agent and energy carrier in the downstream blast furnace process. Blast furnace gas is generated during the blast furnace process. Blast furnace gas is captured at the top of the furnace.

Underlying reaction: $Fe_2O_3 + 2\,C \rightarrow 2\,Fe + CO + CO_2$

Blast furnace gas is produced during the smelting of iron ore in the blast furnace. The dust from the blast furnace gas is usually removed in a gravity separator with downstream wet fine dust removal in a cyclone or electrostatic precipitator. The purified blast furnace gas is mainly used for heating the hot blast stoves in the blast furnace area (Fig. 13.4) and, mixed with coke oven gas, for underfiring the coke ovens. Further surpluses are upgraded with other high-calorific gases and used as

Table 13.1 Composition and quantities of dome gases (blast furnace gas approx. 1500 m^3/t pig iron, converter approx. 75 m^3/t CS and coke plant 420 m^3/t coke)

		Blast furnace gas	Coke Oven Gas	Converter gas	Natural gas
Chemical composition					
CO_2	Vol.-%	22,0	2,3	17,8	0,08
N_2	Vol.-%	52,5	7,3	12,1	0,84
O_2	Vol.-%	–	0,9	0,1	< 0,01
CO	Vol.-%	21,4	5,8	62,5	–
H_2	Vol.-%	4,1	56,8	7,5	–
CH_4	Vol.-%	–	23,7	–	98,30
C_2H_6	Vol.-%	–	–	–	0,52
C_3H_8	Vol.-%	–	–	–	0,17
C_4H_{10}	Vol.-%	–	–	–	0,06
C_3H_6	Vol.-%	–	3,2	–	–
C_4H_8	Vol.-%	–	–	–	
Heating and calorific value					
Heating value $H_{o,n}$	kJ/m^3	3.226	20.413	8.852	39.809
Calorific value $H_{u,n}$	kJ/m^3	3.146	18.164	8.705	35.885
Combustion with air ($\lambda = 1,0$)					
Air requirement I_{min}	m_L^3/m_B^3	0,61	4,39	1,66	9,54
O_2-requirement $O_{2,min}$	m_L^3/m_B^3	0,128	0,922	0,349	2003
Exhaust gas (moist) for $\lambda = 1,0$					
Volume $v_{A,min,f}$	m_L^3/m_B^3	1,48	5,09	2,31	10,34
CO_2	Vol.-%	29,3	8,1	34,7	9,7
H_2O	Vol.-%	2,8	22,3	3,2	18,2
N_2	Vol.-%	67,9	69,6	62,1	72,1

mixed gas for heating the reheating furnaces. Surpluses are used in gas-fired power plants for electricity generation.

The converter gas is produced in the converters of the oxygen steel mills during the conversion of pig iron to crude steel (Fig. 13.5). In this process, the carbon contained in the pig iron is converted into CO with injected oxygen. The converter gas is dedusted in explosion-proof electrostatic precipitators or cloth filters and used to generate steam or, in the smelter's gas network, to heat the reheating furnaces or to generate electricity in the power plant.

Converter gas is produced during the "decarburization" of the liquide pig iron (approx. 4.7% carbon) in the converter of the steel mill. The target content in the steel is <0.1% carbon.

Underlying reaction: $2\,C + O_2 \rightarrow 2\,CO$ and partly: $2\,CO + O_2 \rightarrow 2\,CO_2$

Fig. 13.4 Blast furnace with three hot blast stoves

Fig. 13.5 Oxygen steel mill: Charging the oxygen steel converter with pig iron

As process gases, the co-product gases have different specific properties and differ in their chemical composition and consequently in their specific calorific value. Converter gas has the highest content of CO at approx. 60% and, like blast furnace gas, consists of approx. 20% CO_2 by volume. Coke oven gas, on the other

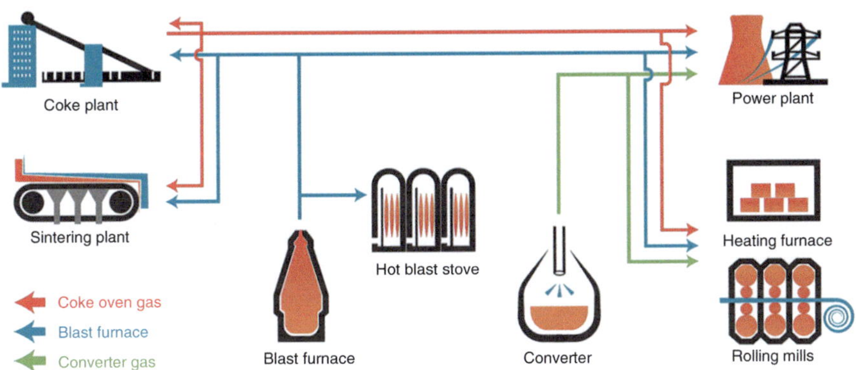

Fig. 13.6 Metallurgical gas network

hand, contains only small amounts of CO_2 and CO concentrations of up to approx. 10%. The main component of coke oven gas is hydrogen at over 60%, alongside methane at around 20–25%. All joint gases are made available for further use with low overpressure at temperatures of up to 50 °C.

Co-Generated Gases Are Used in the Energy Network (Fig. 13.6)

A special feature of the steel industry is the intensive use of co-generated energy, which includes co-generated gases, steam from waste heat and the company's own electricity. The gases produced in the coke plant, blast furnace and oxygen steel mill—coke oven gas, blast furnace gas and converter gas—are used in the steel mill's integrated energy system to generate heat and electricity. This minimizes the additional purchase of natural gas and electricity, saves fossil fuels and reduces CO_2 emissions. The integrated smelter's electricity, steam and heat requirements are thus fully covered.

Due to the local distribution of the co-product gases in the integrated steel mill to cover the electricity, steam and heat requirements in the respective main and auxiliary units, CO_2 emissions are largely generated on a decentralized basis. Figure 13.7 shows the location of the CO_2 emissions of the steel industry in Germany with regard to the entire steel production including the electric steel and blast furnace converter route.

After generation of the respective co-product gases in the corresponding plants, the gases are first transported via pipelines, briefly stored in gasometers if necessary and finally supplied to the respective intended use. The technical term for this is "dispatching". The scope of tasks for the dispatcher has increased considerably in recent years, especially due to the increased complexity of internal gas distribution and utilization. The reasons are the increased utilization possibilities due to improved plant technology, the variety of technical restrictions in the individual plants, energy efficiency changes due to fuel switching, the reduction of CO_2 emissions and other cost cross effects, e.g. the plant utilization factor.

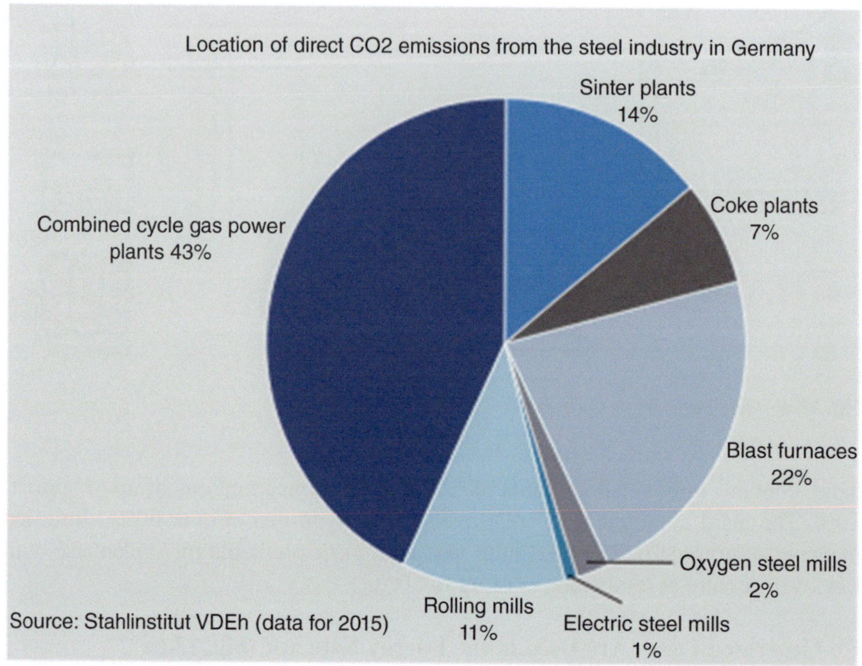

Fig. 13.7 Location of direct CO_2 emissions from the steel industry in Germany

13.3 CO_2 Reduction Measures in the Steel Industry

How much carbon has to be used in the steelmaking process via the integrated route and how much CO_2 is emitted in the production of pig iron and steel via the blast furnace/converter route can be calculated. There is a theoretical "ideal case" for this, but it cannot be achieved in reality. This is based on scientific principles and is a sound reference for assessing the level of CO_2 emissions in integrated steel mills.

CO_2 emissions are also generated during steel production by melting steel scrap in electric arc furnaces. In this case, the CO_2 emissions are essentially indirect because they correlate with the respective power generation mix.

The relationship between the two process routes is justified by the limited supply of steel scrap but is also related to global steel demand and material requirements such as the required purity of the steel. Ultimately, it reflects customer requirements and the production structure in the downstream value chains and thus the industrial structure in Germany.

Basically, four approaches can be identified for the future that could contribute to a significant reduction of CO_2 emissions in the steel production process:

- the direct reduction of CO_2 emissions (Carbon Direct Avoidance),
- the reduced use of fossil carbon through the application of new technologies (Process Integration)
- the storage of CO_2 (CO_2 Capture and Storage),
- the chemical conversion and utilization of CO_2 (Carbon Capture and Utilization).

13.3.1 Carbon Direct Avoidance (CDA)

Increasing the use of hydrogen in metallurgy is being discussed as a means of reducing or avoiding CO_2 emissions in steel production. Provided the hydrogen comes from CO_2-free production, hydrogen use could become one of the future key technologies for low-carbon steel production by reducing iron ores not with fossil carbon but with a significant proportion of hydrogen. If hydrogen is used for reduction in so-called direct reduction plants, the direct-reduced iron produced must be melted in a further process step.

13.3.2 Process Integration (PI)

There are already various approaches to saving fossil carbon in steel production and thus reducing CO_2 emissions by using new technologies in existing processes or in newly developed processes. Here, for example, electricity from renewable sources is fed in as an energy source or hydrogen is used as a reducing agent and substitute for pulverized coal in the blast furnace process. To further reduce the remaining CO_2 emissions, a combination with material CO_2 utilization (Sect. 13.3.4) or final storage of captured CO_2 (Sect. 13.3.3) is an option.

13.3.3 Carbon Capture and Storage (CCS)

Today, parts of the process gases from integrated steel mills are used in the power plants to generate electric power and steam in order to cover the power consumption of the integrated steel mill in a resource-saving way and to realize an operational mode that is as self-sufficient as possible from an economic point of view. When metallurgical gases are used in this way, all carbon carriers are emitted as CO_2. The CO_2 in the waste gas from these co-generation plants could be captured for final disposal, as has also been discussed for coal-fired power plants. As described earlier, this approach also lends itself to the final disposal of CO_2 from processes with already reduced carbon input.

13.3.4 Carbon Capture and Utilization (CCU)

Today, integrated steel mills are mainly optimized in terms of energy. Further potential for improving this process route in terms of CO_2 reductions is conceivable, for example, if sectors such as steel and chemicals work together in a network. For CO_2 reduction in the steel industry, the export gases of integrated steel mills could be used in such a cross-industrial network together with hydrogen from renewable generation (and to a small extent also from own generation) for the production of chemical products such as methanol, synthetic fuels or even urea.

References

1. Wirtschaftsvereinigung Stahl (2016) Fakten zur Stahlindustrie in Deutschland 2016. http://www.stahl-online.de/wp-content/uploads/2013/12/Fakten_Stahlindustrie_2016_V2.pdf
2. Wirtschaftsvereinigung Stahl (2017) Fakten zur Stahlindustrie in Deutschland 2017. https://www.stahl-online.de/wp-content/uploads/2017/12/Fakten_Stahlindustrie_2017_rz_web.pdf

Utilization of C1 Gas Streams from Cement Plants

14

Helmut Hoppe

Abstract

In order to be able to achieve the internationally agreed climate targets in the coming decades, significant CO_2 reductions are required in all relevant sectors. As a major CO_2 emitter, the cement industry is aware of this responsibility and is carrying out various projects in which the application of CO_2 capture processes is being investigated. However, it will be years before large-scale projects are realized and more widespread application is possible, since on the one hand there is still a considerable need for research and on the other hand the economic and political framework conditions for CCS (Carbon Capture and Storage) are not yet in place. In addition, the infrastructure for transporting and storing CO_2 is not yet in place.

In the medium and long term, however, the cement industry could be available as a starting point for new value chains for C1 gas streams, namely as a supplier of large mass flows of CO_2. The CO_2 sources at which CO_2 capture takes place are decentralized in the country, so that transport distances to the utilization plants in question can be relatively short. However, it is also clear that a significant portion of CO_2 from carbon capture projects will always need to be sent to geological storage, as markets for products from CCU projects are limited. It is not possible for plants in the cement industry to provide waste gas streams containing CO as a further C1 source.

Keywords

Cement industry · Carbon Capture and Storage (CCS) · Carbon Capture and Utilization (CCU)

H. Hoppe (✉)
VDZ gGmbH - Verein Deutscher Zementwerke, Düsseldorf, Germany
e-mail: helmut.hoppe@vdz-online.de

14.1 CO_2 Emissions of the Cement Industry

Cement is a building material that is essential for the construction and maintenance of modern infrastructure. To meet this demand, cement is produced in almost every country in the world. This gives rise to both process-related and fuel-related CO_2 emissions. More than 60% of CO_2 emissions are raw material-related, i.e. they occur during the calcination of limestone ($CaCO_3$), the most important raw material for cement production. Around 30% of the CO_2 emissions are caused by the combustion of the fuels used for the clinker burning process, and approx. 10% are indirect emissions due to the demand for electrical energy in the entire production process.

Clinker
Intermediate product of cement production formed in a high-temperature process in rotary kilns. When the clinker is ground with a small amount of gypsum and possibly other main constituents or additives, the final product, cement, is produced.

The cement industry thus makes a significant contribution to anthropogenic CO_2 emissions. According to current estimates, global CO_2 emissions from cement production are over 2 Gt/a [1], which corresponds to 7–9% of anthropogenic CO_2 emissions. In Germany, direct CO_2 emissions from the cement industry were just under 20 Mt in 2016. The emissions of the individual plants take a wider range— from just over 100,000 tons per year for small plants to 1.4 Mt per year for individual large cement plants. The location of the plants and thus of the CO_2 sources (plants with clinker production) is shown in Fig. 14.1a. It can be seen that the emitters are distributed from north to south - with certain focal points in Westphalia and southwestern Germany.

14.2   Production of Cement and Exhaust Gas of the Clinker Burning Process

Cement plants are large industrial facilities (Fig. 14.1b) in which raw materials extracted from quarries are processed and then burned in kilns to produce an intermediate product, clinker. This is followed by grinding to produce cement, which is stored in large silos until it is shipped.

Cement
Cement is a hydraulic binder which, after being mixed with water, hardens independently both in air and under water and remains permanently solid.

A flow diagram of the cement production process is shown in Fig. 14.2. The majority of CO_2 emissions occur during the energy-intensive production of the cement clinker, a high-temperature process in a rotary kiln plant, namely during the calcination of the limestone ($CaCO_3 \rightarrow CaO + CO_2$) and during the combustion of fossil and alternative fuels.

Fig. 14.1 (a) Location of cement plants in Germany; CO_2 sources: Cement plants with clinker production (source: VDZ). (b) View of a cement plant. © VDZ

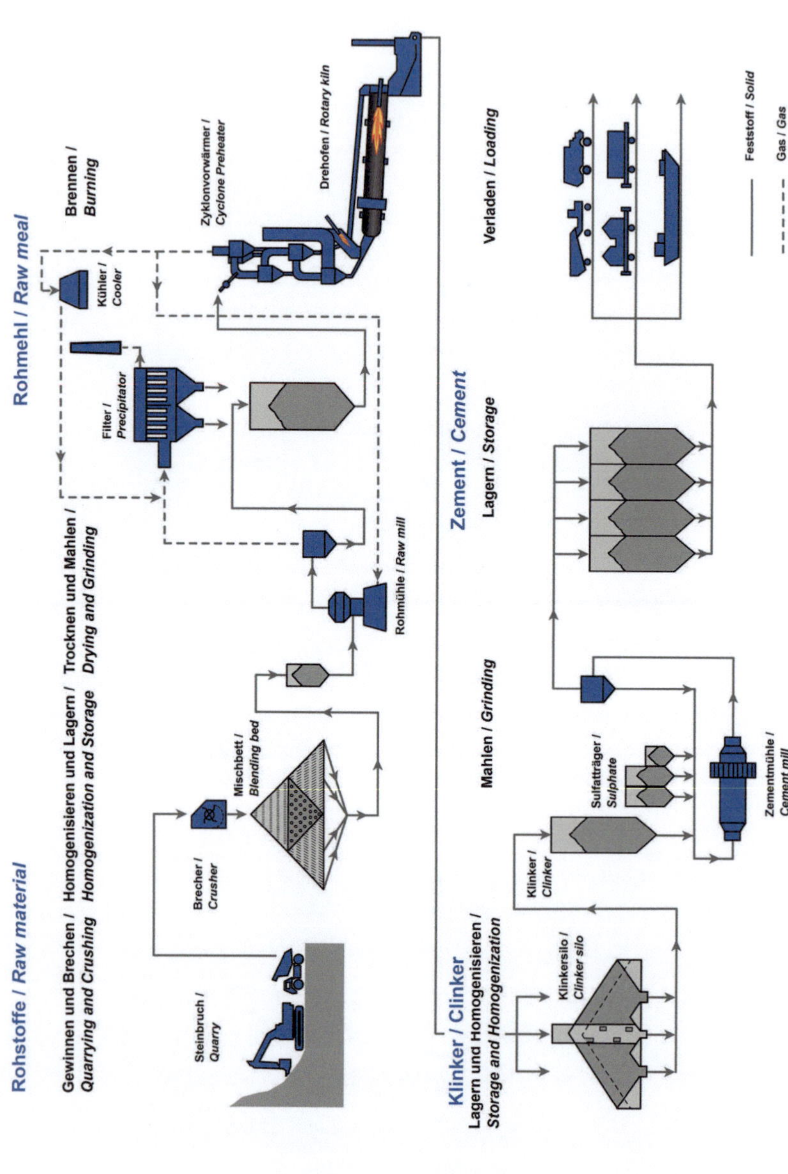

Fig. 14.2 Schematic representation of the cement manufacturing process. © VDZ

The waste gas from the clinker burning process is cleaned (after thermal utilization during grinding-drying of the raw material) and then emitted into the atmosphere via the plant's main stack. A typical exhaust gas volume flow, e.g. for a so-called BAT plant with a clinker capacity of 3000 tons per day, is in the range of 250,000–350,000 m^3/h (N., dry), depending on the mode of operation.

The clean gas temperature is in the range of 90–150 °C when the raw mill is operating, while temperatures in the range of 120–180 °C are reached when the raw mill is off (direct operation).

The emitted clean gas has the following typical composition in the compound operation (raw mill on) [2]:

CO_2: 14–22%
O_2: 5–14%
CO: < 0.1%
N_2: balance

Depending on the operating mode of the kiln plant (e.g. with or without raw mill operation), this composition of the waste gas stream changes to a certain extent, i.e. direct operation results in higher CO_2 and lower O_2 concentrations. Other components of the exhaust gas are various air pollutants (e.g. dust, NO_x, SO_2, total C, CO, trace elements), which are, however, only contained in the exhaust gas in very low concentrations and for which strict emission limits of the 17th BImSchV regulation apply. To achieve the required exhaust gas concentrations, the exhaust gas is dedusted and, as a rule, a NO_x reduction measure is also applied. In addition, further emission reduction measures are also applied if required, e.g. for SO_2 or for Hg. Further use of the emitted waste gas stream, e.g. for other industrial processes, is not yet possible.

14.3 Measures to Reduce CO_2 Emissions

For reasons of climate protection, however, considerable reductions in greenhouse gas emissions and in particular CO_2 emissions will be necessary in the future. In the cement industry, various ("conventional") measures have so far been applied to reduce CO_2 emissions, namely:

Reduction of the clinker content in cement,
Improvement of energy efficiency,
Use of alternative fuels with biogenic components.

In order to be able to achieve the climate targets for the cement sector, more far-reaching measures such as CCS (carbon capture and storage) are absolutely necessary, according to consistent information from various studies and "roadmaps". Since the use of CO_2 capture processes in the power sector is not absolutely necessary due to the ever increasing use of renewable energies, the application of these processes to reduce the practically unavoidable process-related CO_2 emissions may become increasingly important.

14.4   Carbon Capture Projects in the Cement Industry

To date, however, there is almost no operational experience on the application of CO_2 capture processes in cement plants. Therefore, further R&D projects and trials at pilot and demonstration plants are needed first. To enable later commercial application of these processes at many cement kiln plants in the world and to achieve the required global CO_2 reduction, the first demo projects must already be carried out in 2020 and more than 130 cement kiln plants must already be equipped with CO_2 capture processes in 2030, and almost 500 in 2050 [3].

So far, however, there are only a few projects in the cement industry investigating the application of CO_2 capture processes in the clinker burning process. One important research project is being carried out by the European Cement Research Academy (ECRA), under which the European cement industry is jointly conducting studies on CO_2 capture [4] and is now planning an oxyfuel demonstration project [5]. In another project by the HeidelbergCement cement group, various post-combustion processes were tested in pilot trials at a Norwegian cement plant [6], and plans are currently underway to build a demonstration plant for amine scrubbing. In addition, there have been research projects by individual cement companies, e.g., on the mineralization of CO_2, the cultivation of microalgae with waste gas streams containing CO_2, or even the calcium looping process.

Up to now, there are still considerable hurdles preventing such projects from being carried out. Sufficient funding is needed for the necessary R&D projects to install and operate appropriate CO_2 capture plants. Large-scale projects cannot be realized under current conditions, as the cost of cement production would increase significantly, threatening cement and clinker imports from more distant regions of the world. Only after the implementation of effective measures to prevent so-called "carbon leakage" would large-scale CO_2 capture projects in the cement industry be conceivable. According to current cost estimates, the specific costs for an application of CO_2 capture processes in cement plants (without CO_2 transport and storage) range from 40 to over 100 €/t CO_2 [7].

In addition, the fate of the captured CO_2 would be unclear, since no transport and storage infrastructure is yet available. At least in Germany and in some other European countries, there is no legal basis for storing large amounts of CO_2 in underground geological layers. A pipeline network for transporting the CO_2 is also not available.

14.5   Possibilities for the Utilization of Captured CO_2

As long as the framework conditions for storing large amounts of CO_2 in underground geological layers are not given, the utilization of captured CO_2 as a C1 source could gain some importance. In so-called CCU (carbon capture and utilization) projects, a CO_2 gas stream from the CO_2 capture of a cement plant could be transported via a pipeline network to another plant for utilization. In principle, plants for utilization of the CO_2 would also be possible in the immediate vicinity of the cement plant, i.e., in these cases one could dispense with a costly transport infrastructure. However, the potential of such utilization options is limited, since the amount of CO_2 from a cement plant could already cover a significant part of the market for various conceivable products.

Many CCU processes are based on the catalytic conversion of captured CO_2 with hydrogen, e.g. to products such as methanol, methane, fuels or formic acid. The hydrogen required for this would have to be produced by water electrolysis and the electricity needed would have to come from renewable sources. Power-to-gas projects, in which methane is produced with hydrogen from surplus electricity from renewable energies, could make an important contribution to energy storage in the future. However, the economic viability of such processes is not yet given, i.e., here too, the framework conditions must first be created (electricity prices, energy taxes, recognition as "green" products) to enable large-scale production [8, 9]. At present, there is only one project worldwide in which CO_2 is captured in a cement plant and utilized directly on site for the production of $NaHCO_3$ (sodium bicarbonate) [10].

14.6   Provision of CO Mass Flows Not Possible

While the clinker burning process is fundamentally suitable as a C1 source in the form of CO_2 mass flows, CO material flows can hardly be separated or even utilized. The goal of any plant operator is to keep the CO concentration in the off-gas as low as possible, since emission limits for this component must be complied with and elevated CO concentrations could cause significant problems in the process (e.g. CO shutdowns in electrostatic precipitators). Secondary abatement processes for CO are not used, however. In the cement industry, however, there are already initial projects in which secondary measures are being used to reduce emissions of organic substances (e.g., RTO, DeCONOx). During oxidative degradation of organic compounds, CO is also oxidized, further reducing emissions of this component. Against this background, it is not possible that CO gas streams will be separated from the clinker burning process in the future, which could serve as a C1 source for other processes.

References

1. IEA (2009) Cement Technology Roadmap: carbon emissions reductions up to 2050. IEA Technology Roadmaps, OECD Publishing, Paris. https://doi.org/10.1787/9789264088061-en
2. Hoenig V, Hoppe H, Fleiger K (2015) Zementindustrie. In: Fischedick M, Görner K, Thomeczek M (eds) CO_2: Abtrennung, Speicherung, Nutzung. Springer, Berlin/Heidelberg, pp 248–249
3. IEA (2011) Technology Roadmap – carbon capture and storage in industrial applications. IEA Technology Roadmaps, OECD Publishing, Paris. https://doi.org/10.1787/9789264130661-en
4. ECRA (2012) TR-ECRA-119/2012: ECRA CCS Project – Report on Phase III. European Cement Research Academy, Düsseldorf (siehe auch: www.ecra-online.org)
5. ECRA (2016) TR-ECRA-128/2016: ECRA CCS Project – Report on Phase IV.A. European Cement Research Academy, Düsseldorf (siehe auch: www.ecra-online.org)
6. Bjerge L-M, Brevik P (2014) CO_2 capture in the cement industry, Norcem CO_2 capture project (Norway). Energy Procedia 63:6455–6463
7. ECRA (2013) TR-0123/2013/E – Deployment of CCS in the Cement Industry. European Cement Research Academy, Düsseldorf (siehe auch: www.ecra-online.org)
8. Corcoran R (2013) ECRA Project – Report about CO_2 Reuse/MeOH and Methane Synthesis. European Cement Research Academy, Düsseldorf/University College Dublin
9. Spenner L (2015) Techno-ökonomische Bewertung einer Power-to-Gas Konfiguration in der Zementindustrie. Bachelorarbeit, Karlsruher Institut für Technologie (KIT)
10. Perilli D (2015) The Skyonic SkyMine: the future of cement plant carbon capture? Global Cement Magazine 2015:8–12

Jens Hannes

Abstract

The potential for utilization of CO_2 as a raw material is currently highest for coal-fired power plants, since CO_2 is present here in the flue gas as a point source with a relatively high concentration. The theoretically recoverable quantity is estimated at 220 million t/a in Germany. However, capture from the flue gas is associated with efficiency losses in power generation, and meaningful use of the CO_2 requires large quantities of hydrogen, which in turn must be produced using electricity. With the current electricity mix, this would not lead to a net reduction in CO_2 and would also not be able to compete commercially against conventional technology based on oil and gas.

Keywords

Power plants · Hydrogen · CO_2

15.1 Power Plant Processes and Their Specific Potential as CO_2 and CO Sources

Thermal power plants with combustion processes are essentially based on the principle of evaporating water under pressure and expanding it in a steam turbine, which then drives an electric generator. Today, the technology is only used in coal-fired and biomass power plants. The water is evaporated in a steam boiler with an integrated combustion chamber. Gas purification systems are connected to the combustion chamber so that the purified flue gas is ultimately discharged in the form of a point source through a stack or cooling tower (Fig. 15.1). The resulting

J. Hannes (✉)
RWE Power AG, Essen, Germany
e-mail: jens.hannes@rwe.com

M. Kircher, T. Schwarz (eds.), *CO2 and CO as Feedstock*, Circular Economy and Sustainability, https://doi.org/10.1007/978-3-031-27811-2_15

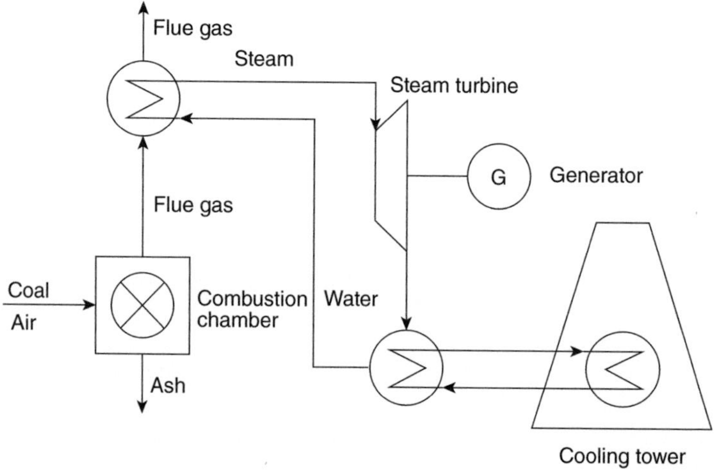

Fig. 15.1 Simplified block flow diagram of coal-fired power plant

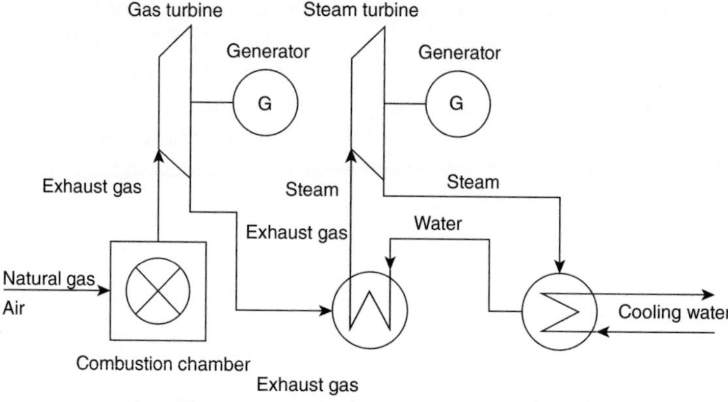

Fig. 15.2 Simplified block flow diagram of a combined cycle power plant

CO_2 can be separated from the flue gas there, preferably by chemical scrubbing. The CO_2 concentration in the flue gas is usually between 13 and 15% of the volume flow. CO does not play a role, since this should be avoided wherever possible.

Nowadays, gas-fired power plants are often designed as combined cycle power plants (CCPP), as higher efficiencies can be achieved here than with the pure steam process. The natural gas is burned in a gas turbine similar to a jet engine, which is coupled to a generator. The exhaust gas from the gas turbine, which is still sufficiently hot, passes through a heat exchanger which vaporizes water in a closed cycle. The water vapor is used to operate a steam turbine, which drives a separate generator or is coupled to the axis of the gas turbine (Fig. 15.2). The spot combustion of the gas in the combustion chamber of the gas turbine generates temperatures that would be too high for the materials used for the turbine blades. Therefore, the process is run

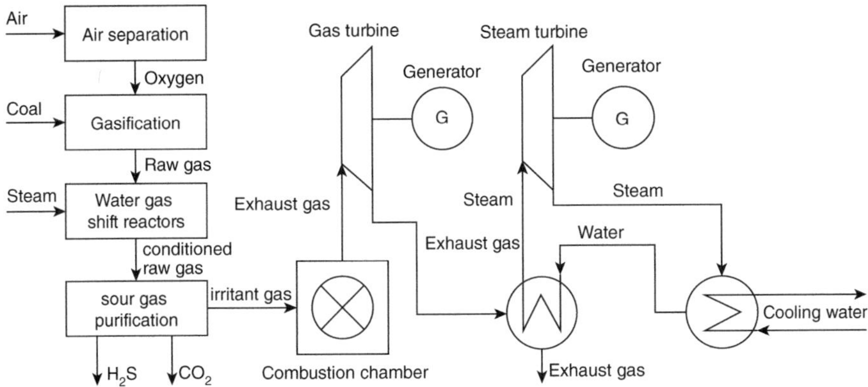

Fig. 15.3 Simplified block flow diagram IGCC power plant

with a high excess of air, which provides the necessary cooling. However, this leads to dilution of the exhaust gas, so that the CO_2 concentration in the stack of a combined cycle power plant is only 2–5%. The excess air produces virtually no CO.

The application of CCGT within gasification-based IGCC (Integrated Gasification Combined Cycle) power plants has also been tested for heterogeneous fuels (e.g. coal, petroleum coke), but is less frequently used because the process is technically complex and economic efficiency is more difficult to achieve. In this process, the synthesis gas produced and purified by gasification with pure oxygen—essentially consisting of CO, H_2, and CO_2—is fed to a combined cycle process (Fig. 15.3). This process enables CO_2 to be separated with less effort directly from the raw gas already between the gasifier and the gas turbine. By injecting steam into conversion reactors (water-gas shift reaction), the synthesis gas is converted into a mixture of mainly hydrogen and CO_2, and the CO_2 can then be scrubbed out even before the combustion process.

Since the raw gas contains large amounts of CO, this would also be directly usable for chemical processes. A combination of chemical and energy use, depending on the current electricity demand, can help to make electricity generation more flexible.

Another process is the so-called oxyfuel process, in which the starting fuel is burned with pure oxygen and CO_2 is partially circulated to cool the process. The resulting high CO_2 concentration in the exhaust gas has advantages in terms of CO_2 capture, but requires effort to separate the oxygen from air. The process has not yet reached commercial maturity.

15.2 Potential for CO_2 Extraction from Power Plant Flue Gases in Germany

Anthropogenic CO_2 emissions amount to about 35 billion tons of CO_2 annually, with global CO_2 emissions dominated by electricity generation [1–4].

In Germany, about 750 MT of CO_2 are emitted annually [5]. Of this, fossil-fired power plant processes account for about 41%, or 310 million t/a of CO_2 [5]. Of these, usable sources are divided mainly into lignite (51%) and hard coal (30%), as shown in Table 15.1. Of these, 220 million t/a could reasonably be recovered from lignite and hard coal-fired power plants for the time being, assuming a separation efficiency of about 90%. The somewhat higher CO_2 concentration in the flue gas of hard-coal-fired power plants, which is advantageous for capture, is put into perspective by the significantly lower number of operating hours compared with lignite.

An overview of the spatial distribution of power plants in Germany is given in Fig. 15.4. Since infrastructure must also be created for CO_2 capture and utilization, the greatest potentials are shown in the Ruhr region, the Rhineland, Lusatia, and some metropolitan areas.

15.3 Separation Process

Separation and utilization of CO_2 places high demands on the purity of the flue gas. In particular, sulfur components can trigger undesired reactions in the scrubbing solutions and must be separated beforehand. German coal-fired power plants have desulfurization systems that meet these requirements in some cases, but post-cleaning to the required purity is often still necessary. Usually, capture is followed by compression of the CO_2 to liquefy it and thus store and transport it in manageable volumes.

For CO_2 capture from flue gases, chemical scrubbing with aqueous amine solutions has the highest level of technical maturity [6]. Commercial solutions are available from several suppliers for this purpose [7–9]. Alternative washing processes in the development stage are based, for example, on aqueous ammonia solutions [10]. Membrane processes are also in the development stage and are not yet available for industrial applications.

Table 15.1 CO_2 emissions of fossil-fired power plants in Germany (as of 2015) [5, 6]

Source	CO_2-emissions [million t/a]	CO_2-conc. in flue gas [vol.-%]	Load profile
Lignite	157	11–12	Base load
Hard coal	92	14	Mid load
Natural gas	42	2–5	Mid/peak load
Mineral oil	4	13	
Other (waste, fuel gas, etc.)	14		

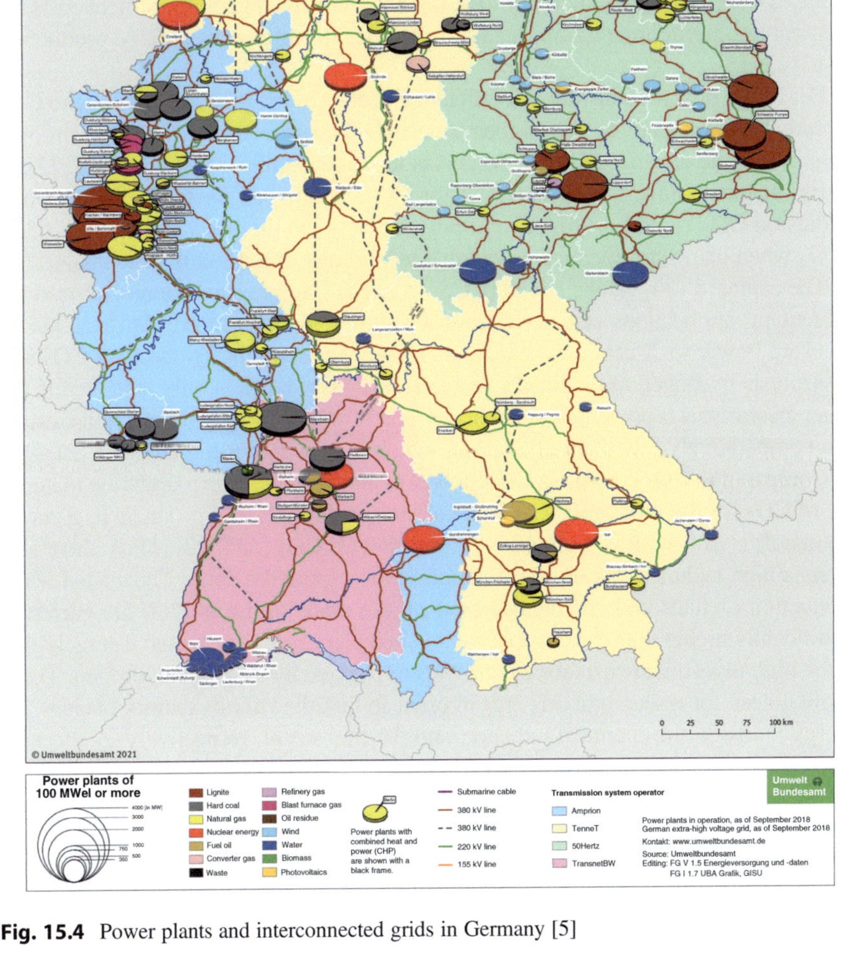

Fig. 15.4 Power plants and interconnected grids in Germany [5]

In the case of the IGCC and oxyfuel processes described, physical scrubbing, e.g., with methanol, is advantageous because the processes are carried out at much higher pressures and at larger CO_2 concentrations in the gas. This reduces the cost of CO_2 capture compared to flue gas scrubbing.

15.4 Economic Analysis of CO_2 Utilization

By separating CO_2 from the flue gas of the power plant, its net electrical efficiency decreases due to the additional thermal (especially for scrubber regeneration) and electrical (e.g., for CO_2 compression) energy consumed by the plant itself. As a result, for the same amount of fuel, the electrical output decreases compared to power plants without CO_2 capture. The loss of net efficiency for capture of 90% of CO_2 is about 8–15% points for aqueous amine solutions (including compression of captured CO_2) [11]. The efficiency loss, together with the higher investment costs for CO_2 capture and compression, leads to significantly higher electricity production costs [12, 13].

For IGCC power plants, the net efficiency decreases by 7–11% points due to CO_2 capture [14–16]. Due to the higher investment costs compared to conventional technology, the electricity production costs here are significantly higher [15, 17].

Power plants have a high potential for CO_2 capture (cf. Sect. 15.2). Economic studies on CO_2 capture in fossil-fired power plants put the costs at about 30–65 €/ t CO_2 when using commercially available technology [13, 18–20].

The captured CO_2 can be stored (CCS—Carbon Capture and Storage) or used for subsequent applications (CCU—Carbon Capture and Utilization). An example of CCU are Power-to-X concepts, which are mostly based on the synthesis of CO_2 and electrolysis hydrogen to chemicals or fuels. CO_2 avoidance is only achieved if the water electrolysis is powered by a majority of renewable energy sources. This results in a comparatively low plant utilization, which in turn causes higher costs [21].

Compared to CO_2 emissions from power plant flue gases, the worldwide material use of CO_2 for the production of chemical intermediates or end products (e.g., urea, methanol) is low at only about 100–160 million t/a CO_2 [3, 4, 22]. Thus, there still exists a high technical potential for coupling the captured carbon. The costs for the production of fuels and chemicals based on the material use of CO_2 are currently significantly higher than the costs of established processes [23], which is largely due to the high investment costs for electrolysis to provide the required hydrogen. There is still a need for research in this area in order to link the various sectors, such as the energy industry, the chemical industry, and the fuel sector, respectively. A successful interaction of renewable energies and fossil-fired power plants with CO_2 capture and subsequent CO_2-based syntheses can offer promising solutions for the production of chemicals and fuels as well as the long-term chemical storage of electrical energy. This would have particular advantages in regions where energy production and chemistry are in close proximity and infrastructure for transporting feedstock and products is already in place.

References

1. IPCC (2005) In: Metz B, Davidson O, de Coninck H, Loos M, Meyer L (eds) Carbon dioxide capture and storage. Cambridge University Press, New York, p 442 ff
2. de Vries GJ, Ferrarini B (2017) What accounts for the growth of carbon dioxide emissions in advanced and emerging economies? The role of consumption, technology and global supply chain participation. Ecol Econ 132:213–223. https://doi.org/10.1016/j.ecolecon.2016.11.001
3. Behr A, Neuberg S (2009) Mögliche Nutzungen von Kohlendioxid in der chemischen Industrie. Erdöl Erdgas Kohle 125(10):367–374
4. Olah GA, Goeppert A, Prakash GKS (2009) Chemical recycling of carbon dioxide to methanol and dimethyl ether: from greenhouse gas to renewable, environmentally carbon neutral fuels and synthetic hydrocarbons. J Org Chem 74(2):487–498. https://doi.org/10.1021/jo801260f
5. Umweltbundesamt (2017, 2018) Energiebedingte Treibhausgas-Emissionen. http://www.umweltbundesamt.de/daten/energiebereitstellung-verbrauch/energiebedingte-emissionen#textpart-1 http://www.umweltbundesamt.de/daten/energiebereitstellung-verbrauch/energiebedingte-emissionen#textpart-3 http://www.umweltbundesamt.de/sites/default/files/medien/372/bilder/dateien/de_kraftwerkskarte_2018.de
6. Fleige M (2015) Direkte Methanisierung von CO_2 aus dem Rauchgas konventioneller Kraftwerke. Springer, Heidelberg
7. Notz R, Tönnies I, Scheffknecht G, Hasse H (2010) CO_2-Abtrennung für fossil befeuerte Kraftwerke. CO_2 capture for fossil fuel fired power plants. Chem Ing Tech 82(10):1639–1653. https://doi.org/10.1002/cite.201000006
8. Bhown AS (2014) Status and analysis of next generation post-combustion CO_2 capture technologies. Energy Procedia 63:542–549. https://doi.org/10.1016/j.egypro.2014.11.059
9. Birnbaum U, Bongartz R, Linssen J, Markewitz S, Vögele S (2010) Fossil basierte Kraftwerkstechnologien, Wärmetransport, Brennstoffzellen. In: Wietschel M, Arens M, Dötsch C, Herkel S, Krewitt W, Markewitz MD, Scheufen M (eds) Energietechnologien 2050 – Schwerpunkte für Forschung und Entwicklung. Fraunhofer-Verlag, Stuttgart
10. EPRI (2012) CO_2 Capture Technologies. Post combustion capture. Global CCS Institute, Canberra
11. Global CCS Institute (2014) The global status of CCS: 2014. Global CCS Institute, Melbourne
12. Yu J, Wang S (2015) Modeling analysis of energy requirement in aqueous ammonia based CO_2 capture process. Int J Greenh Gas Control 43:33–45. https://doi.org/10.1016/j.ijggc.2015.10.010
13. Goto K, Yogo K, Higashii T (2013) A review of efficiency penalty in a coal-fired power plant with post-combustion CO_2 capture. Appl Energy 111:710–720. https://doi.org/10.1016/j.apenergy.2013.05.020
14. Tola V, Pettinau A (2014) Power generation plants with carbon capture and storage: a techno-economic comparison between coal combustion and gasification technologies. Appl Energy 113:1461–1474. https://doi.org/10.1016/j.apenergy.2013.09.007
15. Lockwood T (2017) A comparative review of next-generation carbon capture technologies for coal-fired power plant. Energy Procedia 114:2658–2670. https://doi.org/10.1016/j.egypro.2017.03.1850
16. Martelli E, Kreutz T, Consonni S (2009) Comparison of coal IGCC with and without CO_2 capture and storage: shell gasification with standard vs. partial water quench. Energy Procedia 1(1):607–614. https://doi.org/10.1016/j.egypro.2009.01.080
17. Gräbner M, Morstein OV, Rappold D, Günster W, Beysel G, Meyer B (2010) Constructability study on a German reference IGCC power plant with and without CO_2-capture for hard coal and lignite. Energy Convers Manag 51(11):2179–2187. https://doi.org/10.1016/j.enconman.2010.03.011
18. Mancuso L (2014) Advances in gasification plants for low carbon power and hydrogen co-production. New horizons in gasification. Foster Wheeler Italiana, Rotterdam

19. Wolfersdorf C, Meyer B (2017) The current status and future prospects for IGCC systems. In: Wang T, Stiegel GJ (eds) Integrated gasification combined cycle (IGCC) technologies. Woodhead Publishing, Cambridge, pp 847–889. https://doi.org/10.1016/B978-0-08-100167-7.00024-X
20. Schiffer H-W, Thielemann T (2017) 20 Jahre CCS – Erfolge einer Technologie im Wartestand. Energiewirtschaftliche Tagesfragen 67(1/2):40–46
21. Molina CT, Bouallou C (2017) Comparison of different CO2 recovery processes in their optimum operating conditions from a pulverized coal power plant. Energy Procedia 114: 1360–1365. https://doi.org/10.1016/j.egypro.2017.03.1257
22. EPEA – Internationale Umweltforschung GmbH (2016) Report – Evaluation zur Nutzung von Kohlendioxid (CO2) als Rohstoff in der Emscher-Lippe-Region – Erstellung einer Potentialanalyse. Koch T, Scheelhaase T, Jonas N, Hungsberg M, Osswald D (Eds.) http://www.emscher-lippe.de/wp-content/uploads/2017/05/2017-05-03_Report-CO2-Potentialanalyse_final.pdf
23. Wolfersdorf C, Forman C, Keller F, Meyer B (2017b) CO2-to-X and coal-to-X concepts in pulverized coal combustion power plants. Energy Procedia 114:7171–7185. https://doi.org/10.1016/j.egypro.2017.03.1821

Utilization of C1 Gas Streams from Chemical Processes

16

Kerstin Schwarze-Benning, Hans-Jürgen Körner,
and Görge Deerberg

Abstract

The chemical industry is one of the most energy-intensive sectors of the manufacturing industry and is thus also one of the largest emitters of CO_2 in this sector. The ammonia, petrochemical and carbon black processes alone are responsible for 80% of the chemical industry's CO_2 emissions. The suitability of this emission as a raw material for subsequent recycling varies greatly, as only 2% of these CO_2 streams are highly pure. They are used as CO_2 industrial gas, e.g. in the food industry. Other emission streams would have to be purified before further utilization. Since some of the emitting plants are very large in volume, they are point sources with particular utilization potential. Sites with such chemical plants may therefore be particularly suitable for the material use of CO_2.

Keywords

Chemical industry · CO_2 · Carbon Capture and Utilization (CCU) · Ammonia · Petrochemical · Carbon black

16.1 C1 Gas Streams from Chemical Processes

The most important C1 source for chemical processes besides syngas is CO_2, which accounts for 80% of German greenhouse gas emissions from the chemical industry. Germany, as well as the international community, is obliged as a signatory to the United Nations Framework Convention on Climate Change (UNFCCC; [1]) to

K. Schwarze-Benning · H.-J. Körner · G. Deerberg (✉)
Fraunhofer UMSICHT - Fraunhofer Institute for Environmental, Safety and Energy Technology, Oberhausen, Germany
e-mail: kerstin.schwarze-benning@umsicht.fraunhofer.de;
hans-juergen.koerner@umsicht.fraunhofer.de; goerge.deerberg@umsicht.fraunhofer.de

© The Author(s), under exclusive license to Springer Nature Switzerland AG 2023   219
M. Kircher, T. Schwarz (eds.), *CO2 and CO as Feedstock*, Circular Economy and Sustainability, https://doi.org/10.1007/978-3-031-27811-2_16

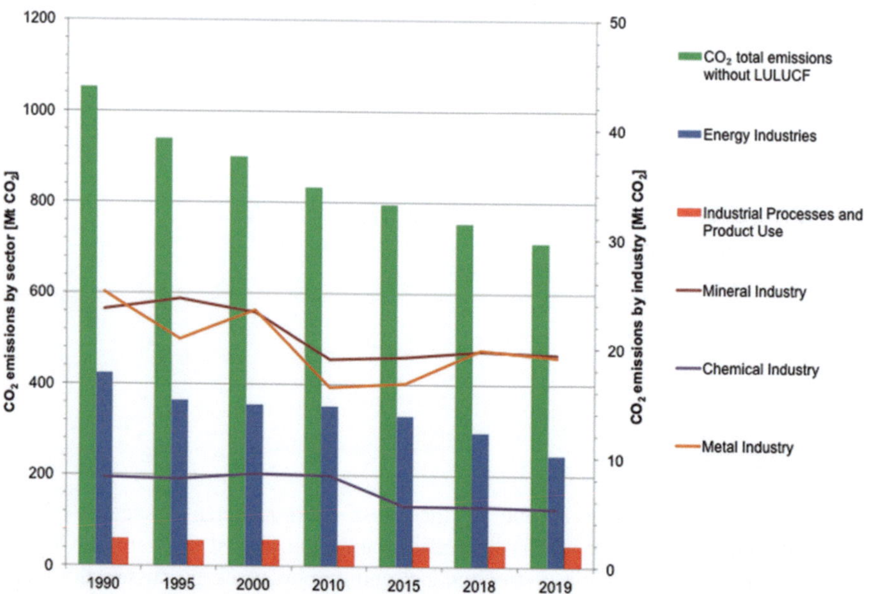

Fig. 16.1 CO_2 emissions in Germany by sector and industry

compile inventories on national greenhouse gas emissions. The collection of emissions data, which is carried out according to the rules of the Intergovernmental Panel On Climate Change (IPCC; [1]), is used to verify the agreed GHG emissions reduction targets under the Kyoto Protocol. In addition, since 2013, chemical industry production facilities have been subject to emissions trading in Europe, which aims to reduce GHG emissions from the energy sector and energy-intensive industries. To this end, the Emissions Trading Authority publishes an annual report on the emissions of all plants subject to emissions trading in the participating industries. The data used here are taken from these two sources, unless otherwise indicated [2, 3].

In Germany, the chemical industry is one of the most important industrial sectors. With achieved sales of EUR 198.3 billion in 2019, it ranks third among manufacturing industries. Alongside cement production and the iron and steel industry, it is one of the most energy-intensive industrial sectors, which also makes it one of the main emitters of greenhouse gases in the manufacturing sector nationally and internationally. The entire German industrial sector emitted approximately 45.9 million metric tons of CO_2 in 2019 due to processes [1, 5].

In the chemical-pharmaceutical and petrochemical industries, CO_2 emissions are both energy-related (39.5 Mt. CO_2) from process firing and own electricity generation, and process-related in the manufacture of chemical products. The process-related emissions here result from the material use of fossil raw materials, with process-related emissions of 5.3 Mt. CO_2 accounting for a share of 11.5% of the industrial sector. In Fig. 16.1, the CO_2 emissions of the energy sector, the industrial sector, and the three manufacturing sectors are shown for selected years between 1990 and 2019 in Germany [1].

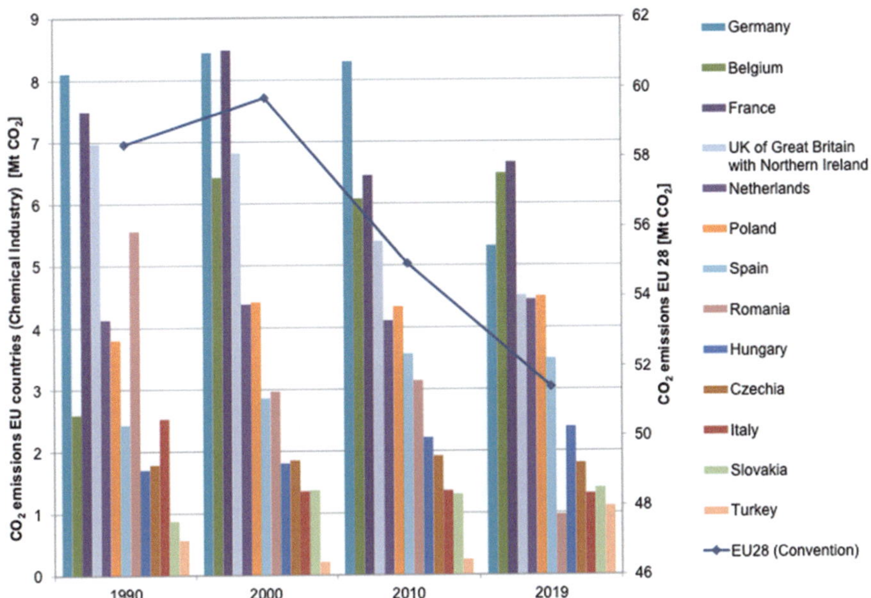

Fig. 16.2 CO_2 emissions of selected European countries in 2015 [4]

In the 28 EU countries, process-related carbon dioxide emissions from the chemical industry were 51.4 Mt. CO_2 in 2019. Belgium, France, Germany, Poland, the United Kingdom, and the Netherlands together accounted for 32 Mt. CO_2, or about 62% of the emissions [2], (Figs. 16.2 and 16.3).

Globally, China, Europe, India and the USA are among the largest emitters of CO_2. The United Nations publishes the climate-relevant emissions for the industrialized countries as well as for the emerging economies. According to this, CO_2 emissions in 2019 were around 38 Gt, with China alone accounting for around 30% of emissions. This is partly due to production volumes and partly due to the use of coal as a low-cost fossil feedstock for petrochemical syntheses [1] Chemical and petrochemical production accounts for around 2.2% of global CO_2 emissions [2]. The Fig. 16.4 shows CO_2 emissions by main products for the largest emitters worldwide..

In the last decade, efforts have been intensified to incorporate the climate-damaging flue gas CO_2 as a raw material in industrial production processes and to use it as a carbon source. In this context, the focus falls on large point sources with total emissions of more than 100,000 t CO_2 per year. Most of these emission sources worldwide come from the fossil energy sector (76%) and the industrial plants of the cement and iron and steel industries (22%). For the chemical industry in Germany, such point sources of energy- and process-related emissions are found in the large chemical industry parks of the petrochemical and fertilizer industries in North Rhine-Westphalia, in the Rhine-Main area, and in the Central German chemical triangle in Saxony-Anhalt (Fig. 16.5). The distribution of large CO_2 point sources in

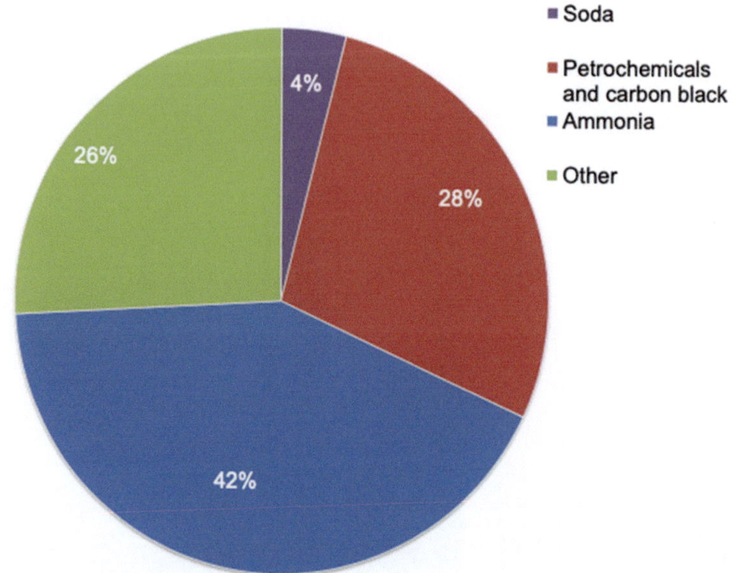

Fig. 16.3 Share of soda ash, petrochemical/carbon black and ammonia products in total CO_2 emissions from European industry in 2019

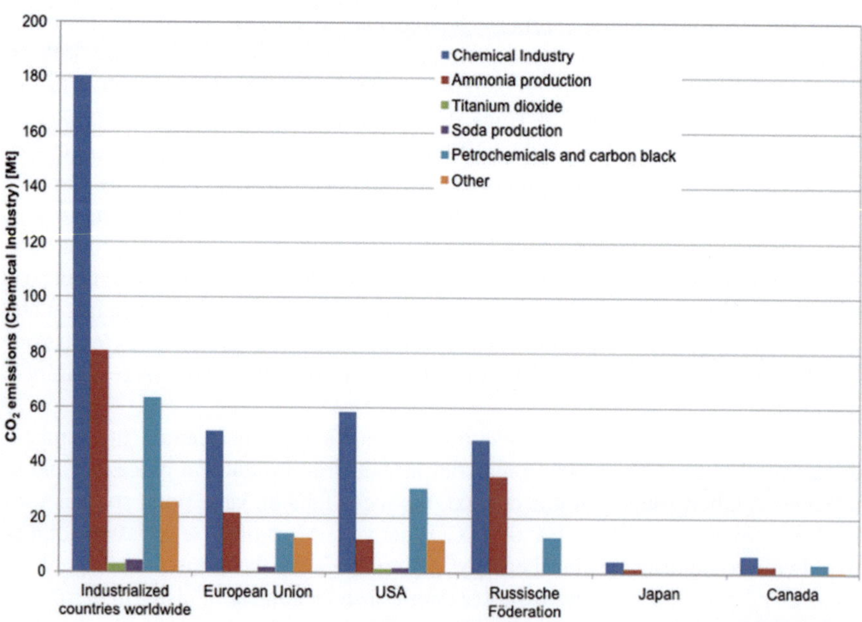

Fig. 16.4 Global CO_2 emissions for chemical industry in 2019

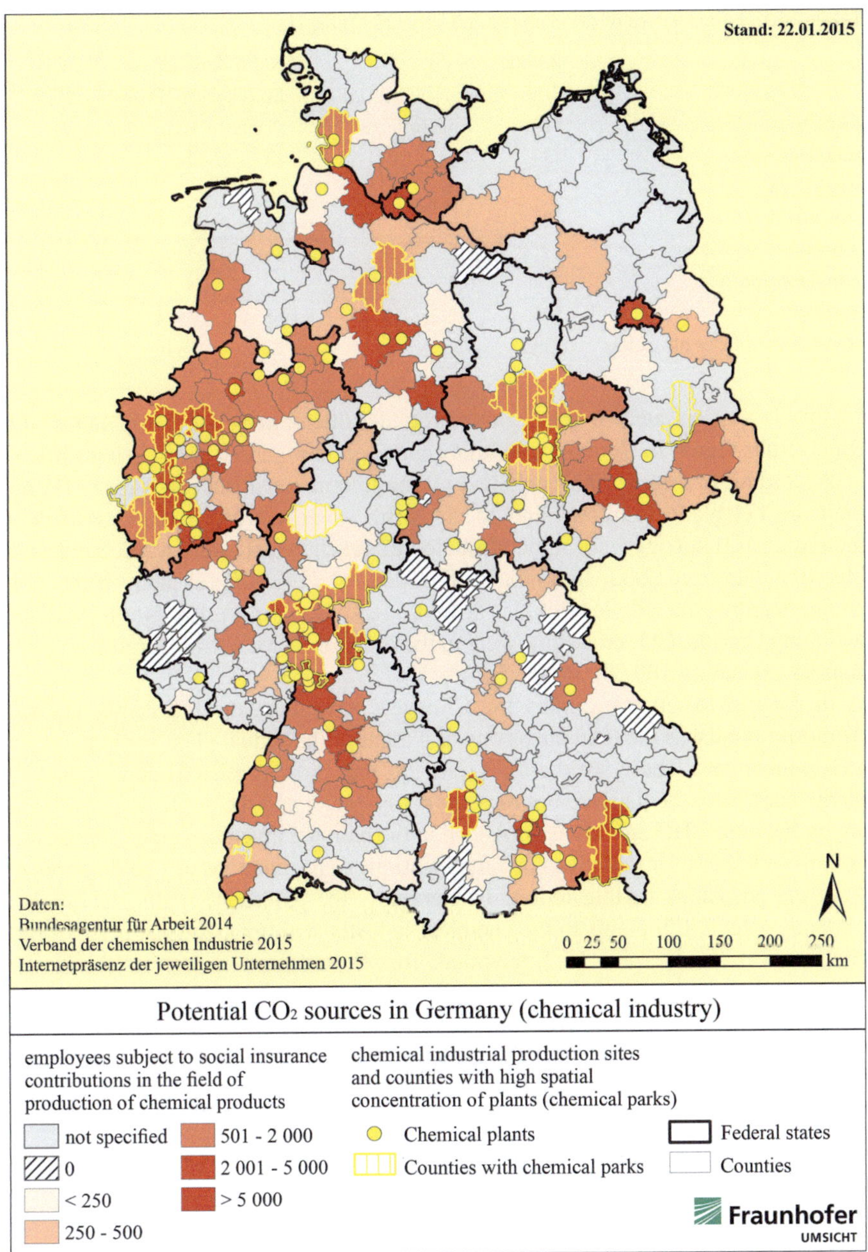

Fig. 16.5 Potential CO$_2$ sources at chemical industry sites

Table 16.1 Point sources of the chemical industry (2002) [6, 7]

Sector	Europe		North-America incl. Canada	
	Emissions > 100.000 t/a		Emissions > 100.000 t/a	
	Plants	Emissions [Mt/a]	Plants	Emissions [Mt/a]
Ethylene	48	48	43	72
Hydrogen	18	6	40	9
Ammonia	42	17	25	9
Ethanol			35	7
Ethylenoxide	3	0,4	8	1
Energy	795	1033	1156	2544
Total emissions of all sectors	1352	1535	2082	3846

Europe is concentrated in the regions around Rotterdam in the Netherlands, the Rhine-Ruhr region in Germany, and the central-eastern part of Great Britain [6].

According to information published by the International Energy Agency (IEA), there are 111 plants in Europe and 151 in North America with annual emissions of more than 100 kt CO_2 that can be attributed to the chemical industry. According to literature data from 2002, emissions in Europe total 71.4 Mt/a for plants producing ammonia, ethylene, ethylene oxide, or hydrogen (Table 16.1). For North America including Canada, CO_2 emissions were in the order of 98 Mt./a from large stationary sources exceeding 100,000 t/y [6, 7].

In contrast to reporting under the Climate Convention, the Emissions Trading Authority publishes the total emissions of a plant or individual plant components in CO_2 equivalents. These are the total (energy- and process-related) emissions of a plant. Therefore, the data are not directly comparable, and the published emissions are higher than GHG emissions published in the national inventory report. However, both the number of installations and the locations of installations can be determined from the emissions trading report. In Germany, the emissions trading obligation currently affects 189 plants in the chemical industry with total emissions of 18 Mt. CO_2 equivalents. In Table 16.2, the plants for selected products are listed with the corresponding CO_2 emission equivalents.

A closer look at the data listed so far shows that process-related CO_2 emissions are generated in particular in the production of ammonia, organic base chemicals (ethylene, ethylene oxide, industrial carbon black) in the petrochemical industry and in soda ash production. In many processes, the high CO_2 emissions can be traced back to the use of hydrogen in synthesis, 90% of whose production is still based on fossil raw materials.

16.2 Petrochemistry

In the petrochemical industry, natural gas and fractions of crude oil are used to produce basic organic chemicals, which are further processed into a variety of intermediate and end products. The most important process here is steam cracking,

Table 16.2 Chemical industry installations subject to emissions trading and emissions in metric tons of CO_2 equivalent

Industrial carbon black (14290)			2016	2019
Orion Engineered Carbons GmbH	Gas soot plant	Cologne	43,449	46,494
Orion Engineered Carbons GmbH	Flame soot plant	Cologne	3418	2348
KG Deutsche Gasrußwerke	Furnace carbon black	Dortmund	279,199	290,757
Orion Engineered Carbons GmbH	Furnace carbon black	Cologne	279,441	239.,856
Ammonia (14615)			2016	2019
SKW Stickstoffwerke Piesteritz GmbH	Ammonia plant 1	Wittenberg	1,248,503	1,083,692
SKW Stickstoffwerke Piesteritz GmbH	Ammonia plant 2	Wittenberg	1,105,510	1,302,923
BASF SE	Ammonia plant	Ludwigshafen	893,240	869,523
BASF SE	Ammonia plant	Ludwigshafen	648.607	595.114
INEOS Köln GmbH	Ammonia plant	Cologne	652,151	512,092
Herstellung organischer Grundchemikalien (Ethylen, Ethylenoxid)				
Ethylene (14215)			2016	2019
Basell Polyolefine GmbH	Petrochemical plant	Münchsmünster	431,132	391,633
Basell Polyolefine GmbH	Ethylene plant OM4	Wesseling	332,279	361,063
Basell Polyolefine GmbH	Ethylene plant OM6	Wesseling	997,851	849,690
BASF	Steam cracker I	Ludwigshafen	285,163	214,279
BASF	Steam cracker II	Ludwigshafen	532,725	580,296
INEOS	Cracker	Cologne	849,948	766.650
DOW Olefinverbund GmbH	Ethylene plant	Böhlen	1,164,797	767,255
DOW Olefinverbund GmbH	Polyethylene plant	Schopkau	29,836	27,409
Ethylenoxid (14616)			2016	2019
INEOS Köln GmbH	Ethylene oxide plant	Cologne	12,800	12,470
BASF	Ethylene oxide plant	Ludwigshafen	100,818	98,311
Sasol Germany	Ethylene oxide plant	Marl	73,808	52,438
Production of hydrogen and synthesis gas (14611)			2016	2015
Covestro AG	Brunsbüttel reformer plant	Brunsbüttel	115,121	114,304
Linde Gas	Burghausen plant	Burghausen	61,014	48,398
Linde Gas	Dormagen plant	Dormagen	24,527	33,214
Linde Gas	Leuna plant, SR1,2	Leuna	420,379	441,069
Linde Gas		Leuna	298,643	233,678

(continued)

Table 16.2 (continued)

	Leuna plant, unit 824			
Evonik Operations GmbH	Hydrogen plant	Marl	203,068	198,740
Evonik Operations GmbH	Synthesis gas plant	Marl	25,876	20,418
Air Liquide	Synthesis gas plant	Oberhausen	63,799	52,792
Nippon Gases Deutschland GmbH	CO plant	Dormagen	5694	6782
Air Liquide	CO plant	Stade	14,387	12,401
BASF Schwarzheide	Steam reforming plant	Schwarzheide	39,838	38,796
BASF SE	Synthesis gas plant	Ludwigshafen	171,352	156,903
BASF SE	Hydrogen plant	Ludwigshafen	277,029	258,691
Linde Gas	Harburg	Hamburg	52,823	58,193
Air Liquide	SMR DOR III	Dormagen	41,369	40,462
Soda (14641)			2016	2019
Solvay	Sodium carbonate production	Bernburg	165,735	207,036
CIECH	Soda plant Staßfurt	Staßfurt	171,196	148,880
Solvay	Soda plant	Rheinberg	154,493	124,472
BASF	Sodium carboxylate/Soda	Ludwigshafen	77,266	76,235
BASF	Salmiak	Ludwigshafen	781	766

in which hydrocarbons are broken down in the presence of steam and at high temperatures. The most common steam cracker products from ethane, naphtha, hydrowax, gasoil and liquefied petroleum gas include ethylene, propylene, C4 cut, isoprene in the gas phase and aromatics in the liquid phase. Steam reforming of refinery gases can also yield CO and hydrogen, which are subsequently required for the production of ammonia, acetic acid, and methanol, as well as for hydrogenation processes. Figure 16.6 shows the variety of synthesis routes of organic base chemicals from petroleum.

A large amount of heat energy is required for the processes of distillation, conversion (cracking, coking, reforming) as well as for the post-treatment and refining of products. CO_2 emissions occur both in the generation of heat and process steam and in the production of hydrogen for reforming hydrocarbons and in the burning of petroleum coke in the catalytic cracker. In Germany, the plants producing inorganic basic chemicals emitted a total of 2.3 Mt. CO_2, with energy-related emissions accounting for the largest share. In addition to the amount of CO_2 emitted, the concentration of the gas in the waste gas stream is also an important parameter for the economic use of the CO_2 waste gas. Table 16.3 shows the proportions of CO_2 emissions and CO_2 concentrations in the respective waste gas streams of the subprocesses of a typical refinery.

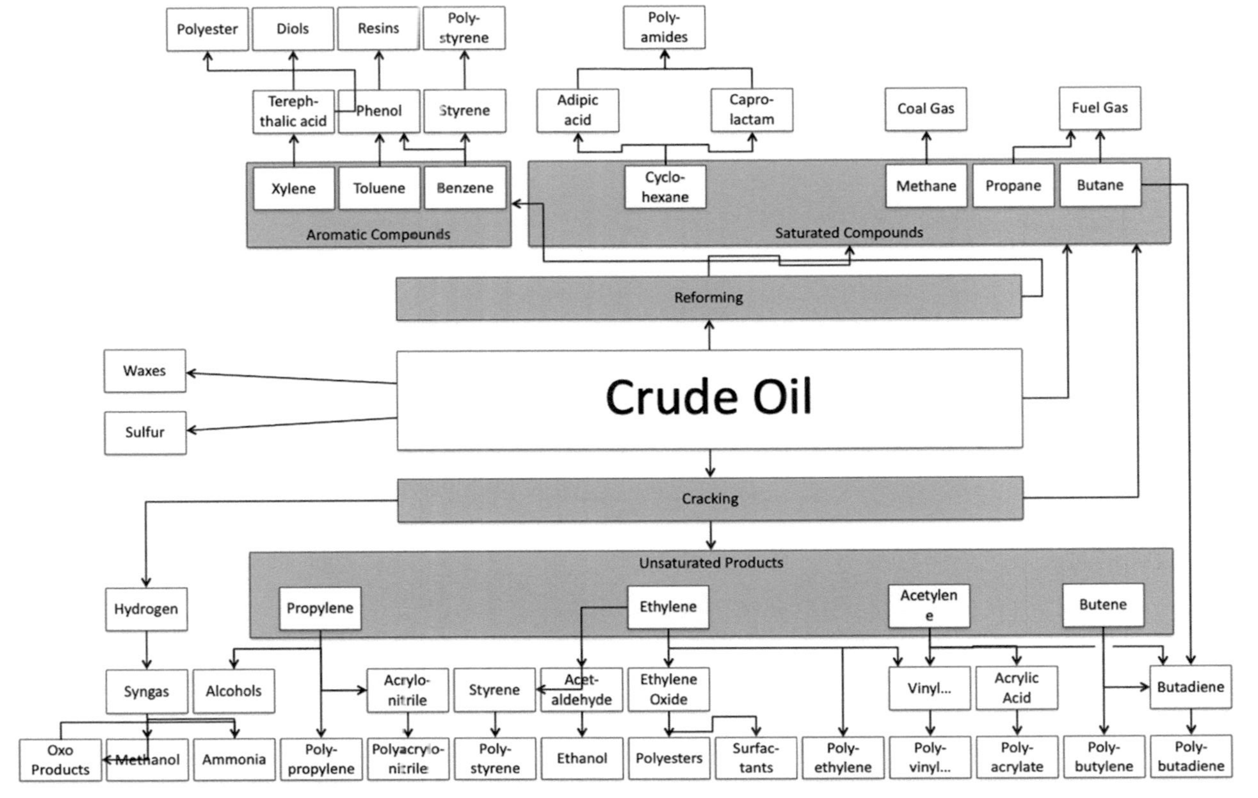

Fig. 16.6 Important products of the petrochemical industry (after Römpp [8])

Table 16.3 CO_2 emissions of a refinery

Emission source	Description	Share of CO_2-Emissions	CO_2-concentration (Vol.) in the exhaust gas flow
Process furnaces	Heat generation by combustion of fossil fuels for distillation columns and reactors	30–60%	8–10%
Steam generator	Process steam generation by burning fossil fuels	20–50%	4–15%
Catalytic crackers	Burning off petroleum coke	20–50%	10–20%
Hydrogen production	Reforming hydrocarbons to H_2 and CO_2	5–20%	20–99%

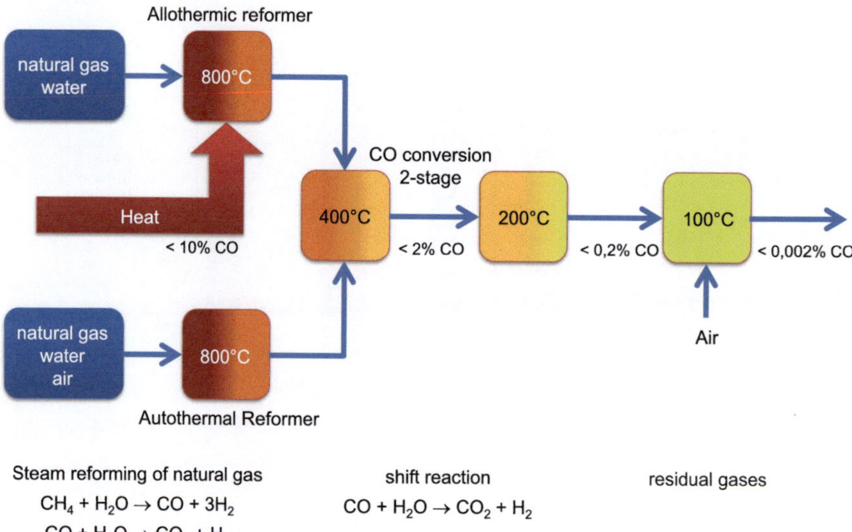

Fig. 16.7 Production routes of hydrogen from fossil raw materials (after Energy Agency NRW [9])

16.3 Hydrogen Production

All hydrogen production processes are based on the splitting of water (Chap. 3). The energy for splitting water can be supplied as chemical energy, heat, or electrical energy (Fig. 16.7). Worldwide, 96% of the hydrogen required is produced from fossil raw materials and up to 4% via electrolysis. Among these, steam reforming is the most widely used process for producing H_2-rich synthesis gas from light hydrocarbons (natural gas, liquefied petroleum gas, naphtha) and steam. The first process stage produces a hydrogen-rich gas mixture over a nickel catalyst (reforming reactor). The resulting mixture contains a high proportion of CO, which is converted into CO_2 together with steam in two downstream catalytic converters (shift reactors). A subsequent gas purification stage removes unreacted CO. Finally, the hydrogen is purified from further components in a pressure swing adsorption.

16.4 Production of Ammonia

Ammonia is the starting material for many nitrogen-containing chemical products, in particular nitric acid, nitrogen-containing mineral fertilizers and urea resins. The industrial production of ammonia is carried out by reacting hydrogen with nitrogen according to the Haber-Bosch process, which can essentially be divided into the three steps of synthesis gas production, compression, and ammonia synthesis (Fig. 16.8). Carbon dioxide is produced during the generation of hydrogen.

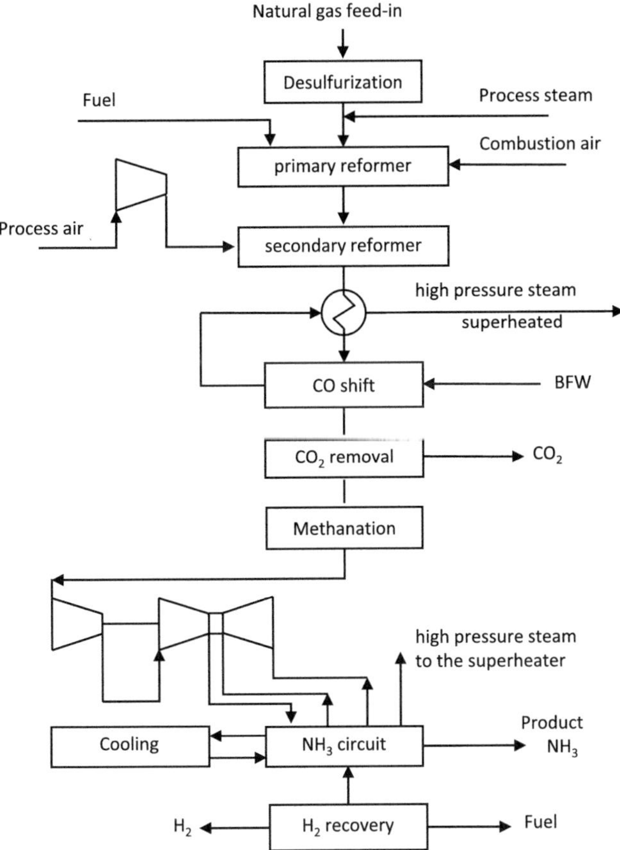

Fig. 16.8 Ammonia synthesis according to Haber-Bosch

Table 16.4 Specific CO_2 emissions according to calculations by Otto [10]

	Consumption per kg NH_3	Specific CO_2-emissions	Emissions g_{CO2}/ kg NH_3
Natural gas as a raw material	0.354 kg	Natural gas upstream 245,29 g_{CO2}/kg NH_3	86.8
Heat	3327 kWh_{th}	Gas burner 235 g/kWh_{th} incl. Gas supply	781.8
Power	0.06 kWh_{el}	Electricity mix 606 g_{CO2}/kWh	36.4
Process emissions			970
Total			1875

The level of CO_2 emissions is based on the fossil raw materials used for synthesis gas production (natural gas, heavy oil, coal). In the work by Alexander Otto, average operating data for ammonia synthesis at ThyssenKrupp Uhde were used to determine the CO_2 emissions. The specific CO_2 emissions for the provision of electricity and heat together with the consumption data are listed in Table 16.4. For energy and raw material supply, emissions of 904 gCO_2/kg NH_3 as well as process-related emissions of 970 g CO_2/kg NH_3 occur as a byproduct of steam reforming [10].

Asia currently accounts for 61% of global ammonia production, with coal being the raw material base there (Fig. 16.9).

In Germany, ammonia is produced at four sites (Table 16.5). In addition to the steam reforming process with natural gas, partial oxidation with heavy oil (emission factor 42.5 t CO_2/t NH_3) is also carried out, which produces significantly higher CO_2 emissions [1]. The CO_2 generated during ammonia production occurs with a purity of 99% and 75% of it is already captured and used for the production of other products, such as urea (Ad-Blue, fertilizer), methanol and as industrial gas. Due to international reporting regulations, the capture and reuse of CO_2 is taken into account in the emissions of downstream products.

16.5 Industrial Carbon Black

Industrial carbon blacks are produced by incomplete combustion of gaseous or liquid hydrocarbons. Coal tar oils (anthracene oils) or pyrolysis and crack oils from petroleum refineries are used. A hot gas of 1200–1800 °C is generated in a combustion chamber by burning natural gas or oil. A carbon black feedstock, mostly coal- and petroleum-derived carbon black oils, is injected into this hot gas (Fig. 16.10). The carbon black is formed by imperfect combustion and thermal cracking (pyrolysis) of the carbon black raw material. By adjusting the process parameters of oil loading and residence time, the desired property profile is obtained. After a certain residence time, the process gas mixture is abruptly cooled by water injection (quenching) and the carbon black is separated in bag filters. The furnace process (emission factor 2.62 t CO_2/t industrial carbon black) produces 90% of industrial carbon blacks, with the remaining 10% of production coming from the

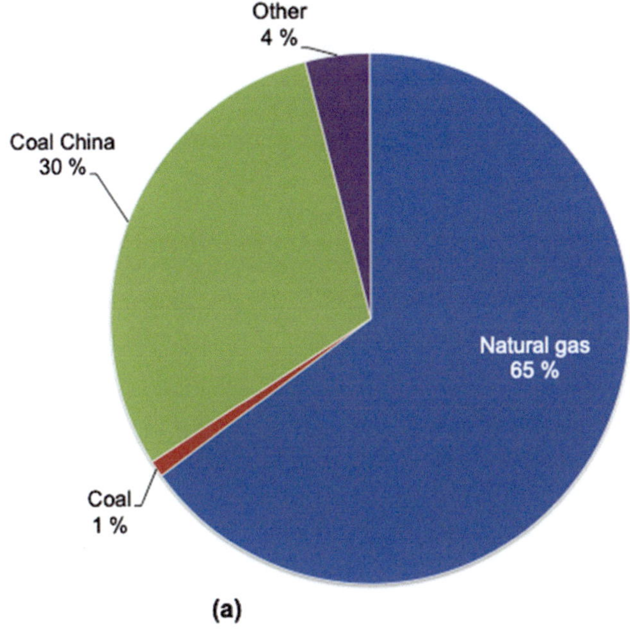

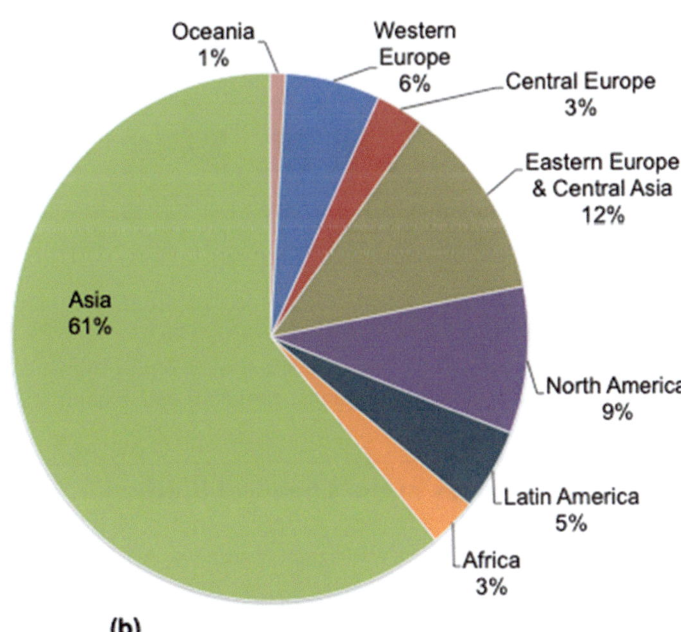

Fig. 16.9 (a) Raw material basis of ammonia production; (b) Global production of ammonia

Table 16.5 Producers of ammonia in Germany

Company	Product	Number of plants	Capacity	Location
Yara	Ammonia, urea	Not available	Not available	Brunsbüttel
BASF SE	Ammonia	4	1.5 Mio. t	Ludwigshafen
INEOS	Ammonia	Not available	Not available	Cologne
SKW Piesteritz	Ammonia	2	3300 t/d	Wittenberg
	Urea	3	4150 t/d0	Wittenberg

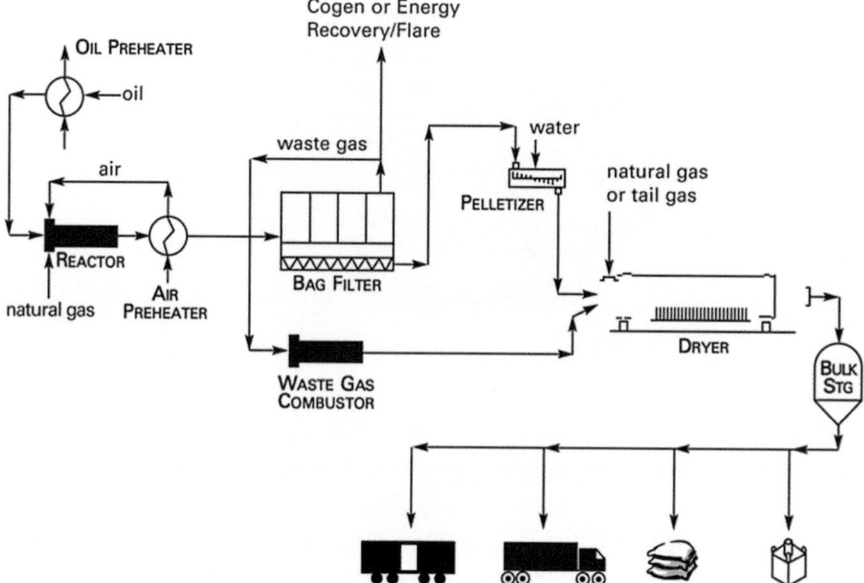

Fig. 16.10 Carbon black production

flame carbon black and gas carbon black processes. Industrial carbon blacks are mainly used as reinforcing fillers for elastomers, of which about 65% are used in the tire industry and another 25% for the production of technical rubber articles. Smaller quantities go into the production of paints, varnishes and coating materials [8].

16.6   Carbon Sources in the Chemical Industry

Coal, natural gas and crude oil as well as renewable raw materials represent the main raw material base of the chemical industry. In 2015, the chemical industry used around 17.3 million metric tons of fossil raw materials as materials. Naphtha and other petroleum derivatives accounted for 75% of this total, used to produce basic chemicals from these material streams (Fig. 16.11). In Germany and Europe, naphtha dominates as a raw material; in other parts of the world, ethane and propane tend

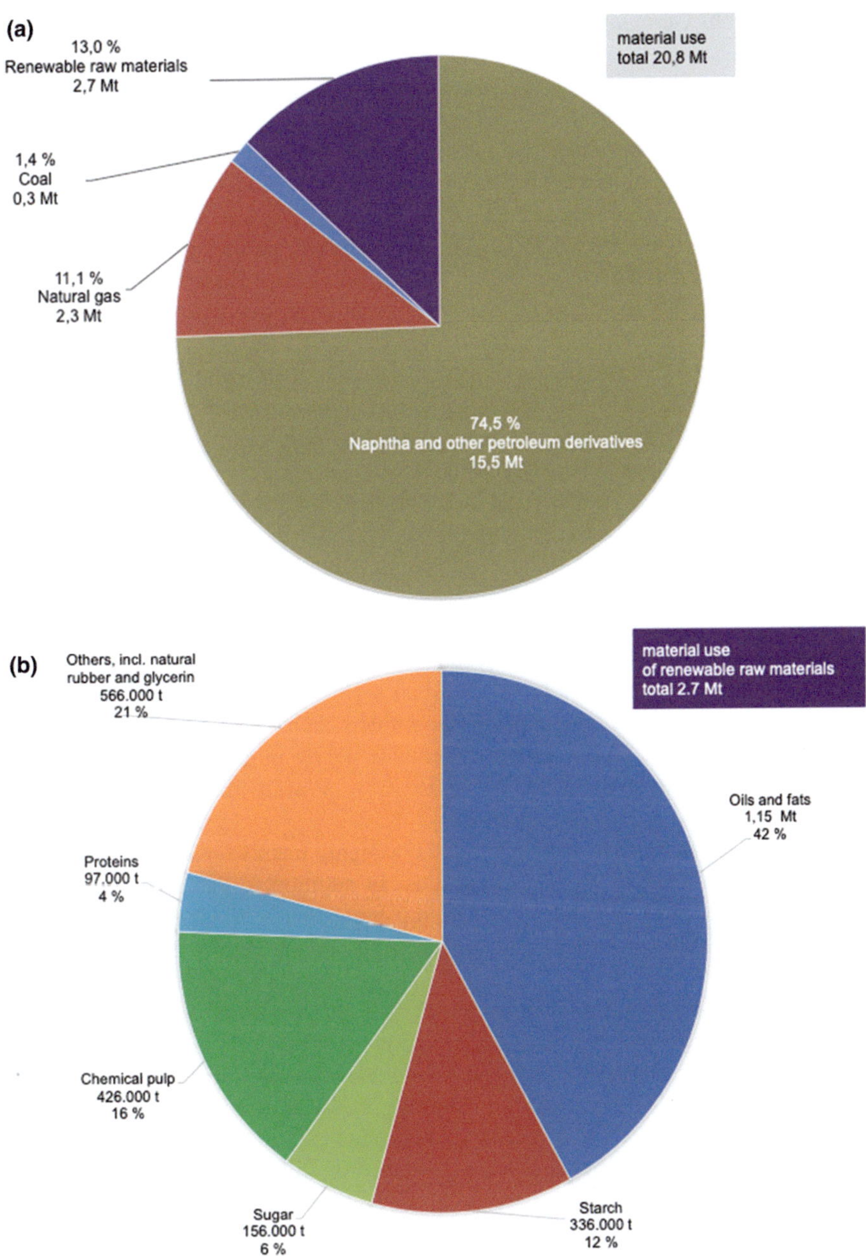

Fig. 16.11 Raw material base of organic chemistry and material use of renewable raw materials in Germany 2015 (according to VCI [11])

to dominate. In Germany, about 15% of mineral oil is used as a material by the chemical industry.

Natural gas represents a further raw material base with a share of approx. 11%. Due to the natural gas deposits, natural gas is considered as an alternative raw material base. However, the technologies to produce ethylene, propylene, C4 olefins, benzene and xylenes from methane, the main component of natural gas, are currently not yet economically feasible.

Another source of raw materials in the chemical industry is renewable raw materials derived from biomass, which now account for 13% of the raw material base. Renewable raw materials have become established wherever there are technical and economic advantages over fossil feedstocks. Fats and oils are mainly used in oleochemistry, e.g. for the production of surfactants. Starch and sugar are used in various applications; as raw materials for fermentation processes, they are the basis for industrial biotechnology. Cellulose (chemical pulp) is mainly used for the production of man-made fibers (viscose).

The use of CO_2 as a raw material in the chemical industry has been the focus of science, politics and industry for some time. Research into capture and geological storage (carbon capture and storage) and capture and utilization as a raw material (carbon capture and recycling [CCR] or carbon capture and utilization [CCU]) are the subject of numerous research projects [12]. CCR measures are aimed at using CO_2 as an industrial gas or as a C1 building block for chemical conversions. Capture and utilization are economically viable when high concentrations and purities are present in the flue gas stream. High-purity emitters include ammonia synthesis and hydrogen production. These processes produce nearly pure CO_2 as a byproduct, which can be captured for about 35 EUR/t CO_2. Worldwide, only about 2% of industrial emitters supply high-purity CO_2.

Since CO_2 is very inert, auxiliary agents or highly reactive reactants are usually required to involve it in chemical reactions. In addition to the use of additional energy, catalysis is a key factor in developing energetically efficient processes. The possible conversion processes and potential utilization pathways are shown in Fig. 16.12.

Today, around 180 million metric tons of CO_2 per year are already used as a raw material in the chemical industry. A distinction must be made between direct use, which includes use as carbon dioxide in beverages, as an industrial gas, as a fertilizer in greenhouses, and in packaging to improve the shelf life of foodstuffs. As an industrial gas, CO_2 finds application in enhanced oil/gas recovery (EOR/EGR), where more natural gas and crude oil is produced by injecting CO_2 into geological reservoirs. In the chemical industry, CO_2 already serves as a feedstock in a few industrial processes such as the production of urea, methanol, salicylic acid, inorganic and organic carbonates, and polycarbonates. In Table 16.6, the production volume of the respective substances is listed.

Numerous new synthesis routes based on CO_2 are currently being investigated and developed at laboratory, pilot or demonstration scale. As part of A. Otto's dissertation, 123 CO_2 utilization reactions were investigated that could show the potential to contribute to the reduction of CO_2. A large proportion of these reactions

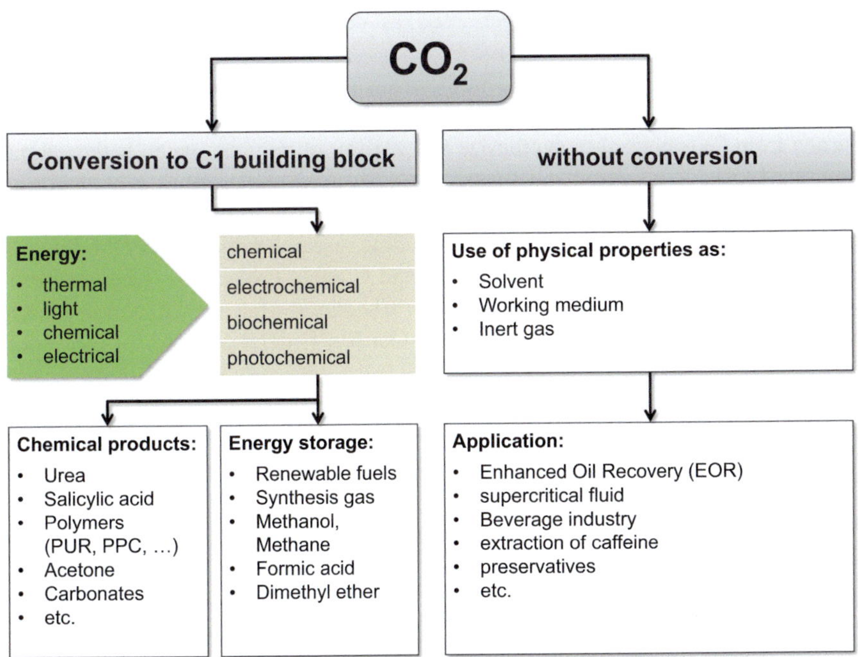

Fig. 16.12 CO_2 utilization pathways with and without conversion (after Otto [10])

Table 16.6 Production volume as well as the bound mass of CO_2 per kilogram of product and in total per substance (after Otto [10])

Material	Produced using CO_2 [Mt/a] (valid year)	Mass CO_2 per kilogram product	CO_2 bound in the product [Mt/a]
Urea	150 (2010)	0.73	109.5
Methanol	4.4 (2007)	1.37	6
Methanol	0.004 (2014)	1.37	0.00548
Salicylic acid	0.17 (2011)	0.318	0.054
Organic carbonates	0.1 (2009)	0.43–0.49	0.043–0.049
Bisphenol A polycarbonate	0.6 (2009)	0.17	0.102
Polypropylene carbonate	0.07 (2010)	0.43	0.03

of bulk and fine chemicals have only been synthesized on a laboratory scale. Based on a proprietary evaluation matrix, Otto identified six bulk chemicals (formic acid, oxalic acid, formaldehyde, methanol, urea and dimethyl ether) with high CO_2 reduction potential and high value added. For methanol and dimethyl ether, he conducted a comparison between conventional and CO_2-based processes. Through the comparison, it was found that the CO_2-based processes only lead to a reduction of CO_2 emissions if the hydrogen used as coreactant comes from renewable sources [10].

Since June 2016, Covestro AG in Dormagen has been operating a production plant for the manufacture of polyurethane foams with a 20% CO_2 content. The carbon from CO_2 is used to synthesize a new form of polyols, which is the central building block of polyurethane foam. The CO_2 used comes from a neighboring chemical company, where it is a waste product [13]. The implementation in industrial production was preceded by a long-standing research project under the funding measures "Technologies for Sustainability and Climate Protection—Chemical Processes and Material Utilization of CO_2" and "CO_2Plus—Material Utilization of CO_2 to Broaden the Raw Material Base" of the German Federal Ministry of Education and Research (BMBF).

16.7   Conclusion

The chemical use of carbon dioxide is a key approach to reducing CO_2 emissions in the process industry. If one considers the potential emission reductions through the use of renewable energy in the energy sector, these naturally turn out to be significantly higher. Significant emissions occur from ammonia synthesis and petrochemical (ethylene, ethylene oxide) syntheses, with carbon dioxide from ammonia synthesis often recycled in downstream urea synthesis. The challenges in CO_2 conversion to chemical feedstocks are in the area of catalysis to convert the unreactive CO_2. The material utilization of CO_2 is a strategically important project, which in the long term can lead to new manufacturing processes by combining it with CO_2-free methods of hydrogen production.

References

1. UNFCCC Greenhouse Gas Inventory Data – Detailed data by Party. United Nations Framework Convention on Climate Change (Eds.). http://di.unfccc.int/detailed_data_by_party. Accessed 30 May 2022
2. DEHSt (2020) Treibhausgasemissionen 2019 – Emissionshandelspflichtige stationäre Anlagen und Luftverkehr in Deutschland (VET-Bericht). Deutsche Emissionshandelsstelle im Umweltbundesamt (Eds.). https://www.dehst.de/SharedDocs/downloads/DE/publikationen/VET-Bericht-2019.pdf. Zugegriffen: 30 Mai 2022
3. UBA (2017) Berichterstattung unter der Klimarahmenkonvention der Vereinten Nationen und dem Kyoto-Protokoll 2017. Nationaler Inventarbericht zum Deutschen Treibhausgasinventar 1990–2015. Climate Change 13/2017, Dessau-Roßlau, ISSN 1862–4359
4. EEA (2017) Data viewer on greenhouse gas emissions and removals, sent by countries to UNFCCC and the EU Greenhouse Gas Monitoring Mechanism (EU Member States) (EEA greenhouse gas – data viewer). European Environment Agency (Eds.). http://www.eea.europa.eu/ data-and-maps/data/data-viewers/greenhouse-gases-viewer. Accessed 28 Sept 2017
5. VCI (2021) Chemiewirtschaft in Zahlen 2021. Verband der chemischen Industrie e. V. (Eds.) https://www.vci.de/vci/downloads-vci/publikation/chemiewirtschaft-in-zahlen-print.pdf. Accessed 30 May 2022

6. IEA (2005) Building the cost curves for CO_2 Storage: European Sector. International Energy Agency (Eds.) Cheltenham, United Kingdom (IEA GHG Report, 2005/2)

7. IEA (2005) Building the cost curves for CO_2 Storage: North American sector. International Energy Agency (Eds.) Cheltenham, United Kingdom (IEA GHG Report, 2005/3)

8. Römpp (2017) Datenbank Römpp Online. Chemie-Enzyklopädie von Thieme

9. Energieagentur.NRW (2013) Wasserstoff – Schlüssel zur Energiewende. http://www. energieagentur.nrw/brennstoffzelle/netzwerk-brennstoffzelle/ erzeugung?mm=Wasserstoff#ts. Accessed 28 Sept 2017

10. Otto A (2015) Chemische, verfahrenstechnische und ökonomische Bewertung von Kohlendioxid als Rohstoff in der chemischen Industrie. Dissertation. RWTH Aachen

11. VCI (2017) Rohstoffbasis der chemischen Industrie. Verband der chemischen Industrie e. V. (Eds.) https://www.vci.de/vci/downloads-vci/top-thema/daten-fakten-rohstoffbasis-der-chemischen-industrie-de.pdf. Accessed 28 Sept 2017

12. Olfe-Kräutlein B, Naims H, Bruhn T, Lorente Lafuente AM (2016) CO2 als Wertstoff: Herausforderungen und Potenziale für die Gesellschaft. IASS Study. https://doi.org/10.2312/iass.2016.025

13. Covestro AG (2016) Premiere für neuen Rohstoff. Covestro startet industrielle Kunststoff-Herstellung mit CO_2. Pressemitteilung vom 17 Juni 2016. http://presse.covestro.de/news.nsf/id/aazc76-premiere-fuer-neuen-rohstoff. Accessed 28 Sept 2017

Utilization of C1 Gas Streams from Bioprocesses Including Biogas Plants

17

Bettina Sayder, Ute Merrettig-Bruns, and Görge Deerberg

Abstract

CO_2 is produced as an end product in various industrial processes based on biological or biotechnological processes. Especially the production of bioethanol is to be highlighted in this context, as bioethanol is the most produced biotech product worldwide and also causes the most process-related CO_2 emissions. Since CO_2 is produced in high purity and in large quantities during bioethanol production, it has already been used for some time in the food industry (e.g. for carbonating beverages), so CO_2 emissions are already being saved here. In addition, CO_2 is also a component of the biogas produced in sewage treatment plants and biogas plants, but it is not suitable as a substrate for CO_2-based processes because of the relatively small proportion in this gas mixture. Other processes, such as the production of yeast, glutamate or lysine, also do not appear to be economical sources of CO_2 to enter the circular carbon economy as a reactant.

Keywords

Biotechnological processes · CO_2 · Bioethanol · Food industry · Biogas plant

Relevant C1 gas flows are to be expected in the context of biological processes exclusively in the form of CO_2. Major CO_2 emissions result from white and gray biotechnology plants (especially biogas). The other areas of biotechnology (Table 17.1) can be excluded for a more in-depth consideration.

B. Sayder · U. Merrettig-Bruns · G. Deerberg (✉)
Fraunhofer Institute for Environmental, Safety and Energy Technology UMSICHT, Oberhausen, Germany
e-mail: bettina.sayder@umsicht.fraunhofer.de; goerge.deerberg@umsicht.fraunhofer.de

239

Table 17.1 Classification of biotechnology (BT) application areas in Germany

Green BT	Plant biotechnology
White BT	Industrial biotechnology
Red BT	Use in medicine and pharmaceuticals; medical biotechnology
Gray BT	Use in waste management
Brown BT	Environmental biotechnology
Blue BT	Biotechnological utilization of marine resources
Yellow BT	Biotechnological utilization of insects

Table 17.2 Production of biotechnological products (fermentative production) with $\geq$0.1 Mt/a worldwide [1]

Product	World annual production [Mt/a]
Bioethanol	70
Biogas	6.6
L-glutamate	2.5
Feed / baking yeast	1.9
L-lysine	1.5
Citric acid	1
Lactic acid	0.6
L-threonine	0.2
Acetic acid	0.2
Polylactic acid	0.1
Gluconic acid	0.1
Ascorbic acid	0.1

About 10% of the dedicated biotechnology companies in Germany state white biotechnology as their main area of activity [1]. In addition, there are other biotechnologically active companies (e.g. pharmaceutical and chemical companies, seed manufacturers), which are also active in the field of detergents and cleaning agents, antibiotics, vitamins, hormones, food industry and enzymes.

In biotechnology, CO_2 is produced as a by-product from fermentative processes. In purely enzymatic processes, such as the production of glucose-fructose syrup, no CO_2 is formed. The production figures for biotechnological products with fermentative production worldwide are shown in Table 17.2. With regard to the estimation of their C1 potential, biotechnological processes with production quantities greater than 500 kt/a are considered in more detail below.

While no CO_2 is released during the production of lactic acid, which takes place under anaerobic conditions by means of homofermentative lactic acid fermentation, CO_2 is produced as a by-product during the biotechnological production of the following products:

– Bioethanol
– Biogas
– Feed/baking yeast
– Amino acids: glutamate, lysine
– Citric acid

Table 17.3 Carbohydrate conversion rates for selected fermentation products

Product	Carbohydrate conversion rate [%]
Bioethanol	48
Feed−/baking yeast	67
Glutamic acid	75
Lysine	71
Citric acid	76
Lactic acid	85

Usually, carbohydrate-containing substrates such as molasses and starch hydrolysates are used as feedstock for fermentations. The substrate/product conversion rate gives an indication of the specific raw material requirement and, if the underlying metabolism is known, of the formation of biomass and by-products. Table 17.3 shows the carbohydrate conversion rates for selected fermentation products.

Within the products shown, the conversion rate for bioethanol is the lowest at 48%. In addition to ethanol, CO_2 and also biomass are produced as a result of metabolism. The conversion rates for yeast, amino acids and citric acid are much higher at approx. 60–80% and suggest a lower formation of the by-product CO_2. So far, CO_2 has not been captured and utilized in these products. The quantities of CO_2 formed and the potential for possible utilization are listed or estimated below for the products bioethanol, biogas, yeast, amino acids and citric acid.

17.1 Bioethanol

In Germany, bioethanol production was around 0.65 MT in 2019 (Fig. 17.1). Bioethanol is produced during the fermentation of feed grains or industrial beets, in which 1 mol of glucose is converted to 2 mol of ethanol and 2 mol of CO_2. The production of one ton of ethanol thus generates about 0.95 tons of CO_2. In 2016, bioethanol production in Germany thus generated 0.62 Mt of CO_2. An overview of the production sites in Germany is shown in Fig. 17.2.

The CO_2 released during ethanol fermentation is obtained in high purity and used in Germany in the beverage and food industries. The estimated amounts of CO_2 used here are 2.9 Mt/a (carbonation of beverages) and 8.2 Mt/a (food packaging) [45].

In Europe, France, Germany, Poland and Hungary are among the largest producers of bioethanol. In 2019, about 5.6 Mt of bioethanol was produced in the European Union [5], which corresponds to a CO_2 amount of about 5.3 Mt. According to the Agency for Renewable Resources (FNR), the production of bioethanol worldwide is over 70 Mt [6]. The USA (bioethanol from corn) and Brazil (bioethanol from sugar cane) are the two largest producers.

According to Alonso-Moreno et al. [7], 12.7 Mt of CO_2 is generated globally by the wine and spirits industry. The authors conclude that the best solution for capturing and using CO_2 from alcoholic fermentation is to produce Na2CO3 using 50% caustic soda. This approach is expected to reduce CO_2 emissions from current

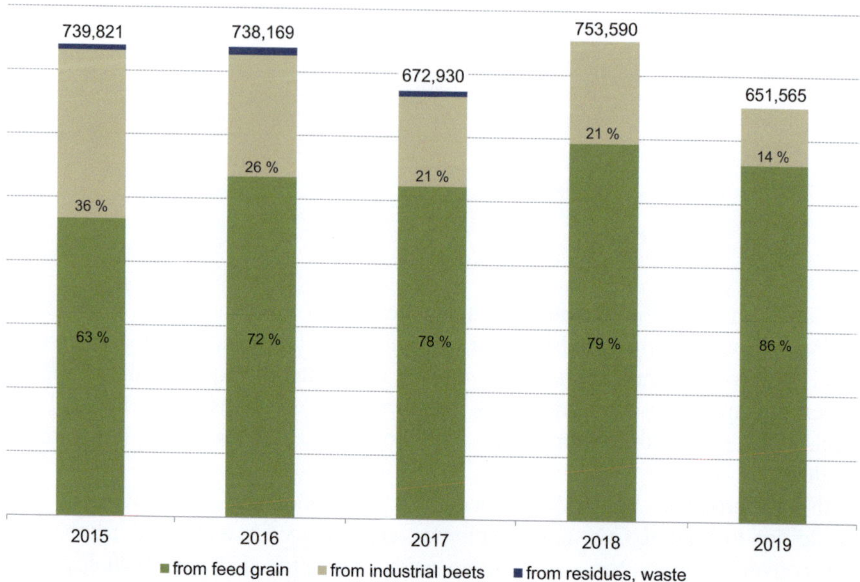

Fig. 17.1 Bioethanol production in Germany in tons according to market data 2019, Bundesverband der deutschen Bioethanolwirtschaft e. V. [3]

Na2CO3 production by a factor of 3–4 [7]. CO_2 can also find application, even in smaller quantities, in the following areas: Fruit and perishable food storage (CA storage), as dry ice, as an extinguishing agent, and as a shielding gas in welding [8].

17.2 Biogas

When considering electricity generation in Germany, CO_2 emissions occur in the area of biogas, sewage gas, and landfill gas. The share of biogas, biomethane, sewage gas and landfill gas in total electricity generation from renewable sources in 2019 was 11.9%, 1.1%, 0.6% and 0.1%, respectively [9]. Due to the small share in the total electricity generation and the associated small amount of CO_2, as well as the decentralized distribution of the corresponding plants, sewage gas and landfill gas are not further considered as significant C1 sources.

In 2019, about 9500 biogas plants were operated in Germany [9], including 216 plants for biomethane production (injection into the natural gas grid) [10] with a total installed electrical capacity of 5.3 GW [9]. About 80% of the biogas plants in Germany have an installed electrical capacity of less than 0.5 MW [11]. In the field of biomethane plants, there are currently 16 larger plants in Germany, whose gas production corresponds to an installed electrical capacity of 3.8–20 MW [12].

In Europe, about 13,800 biogas plants were in operation in 2019. About 70% of the plants are located in Germany. Based on data from 2016, other countries with a

Production capacities in tons/year

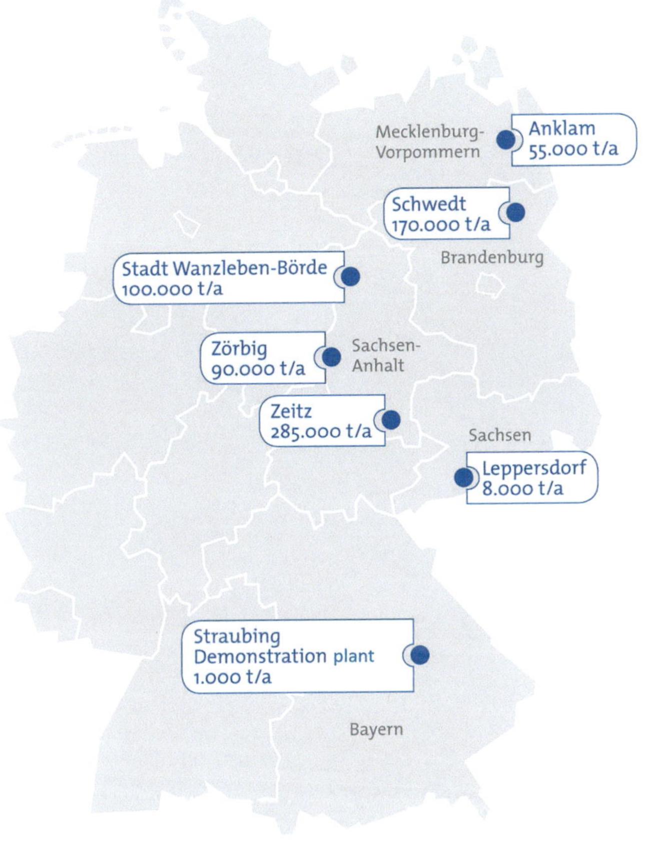

Fig. 17.2 Bioethanol plants in Germany [4]

significant number of biogas plants are Italy (1020), Czech Republic (500), Austria (360), Netherlands (220), Switzerland (140) and France (140) [13, 14]. The average plant size worldwide is 0.5 MW of installed electrical capacity, and 0.4 MW in Germany [11]. The three countries with the highest average installed electrical capacity are Belgium (2 MW), Romania (1.9 MW), and China (1.75 MW). In Belgium, about one third of the total biogas plants are in the range of 2–7.5 MW [11].

Based on the data of the installed electrical capacity of all biogas plants in Germany (5.3 GW; [9]) and the feed-in capacity of biomethane plants with a standard volume flow of 146,000 m^3/a [13], a total emission of 14 Mt CO_2 per

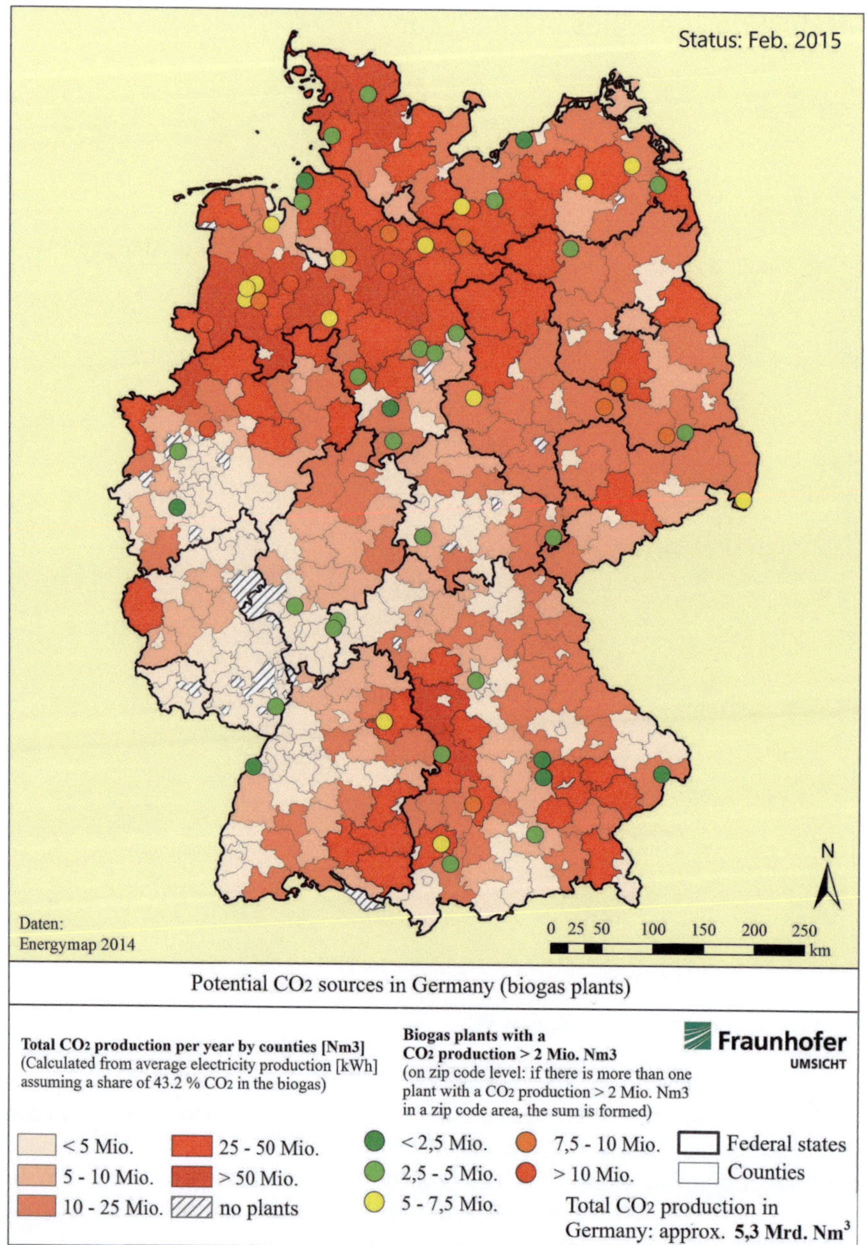

Fig. 17.3 Illustration of biogas plants in Germany according to potential CO_2 sources

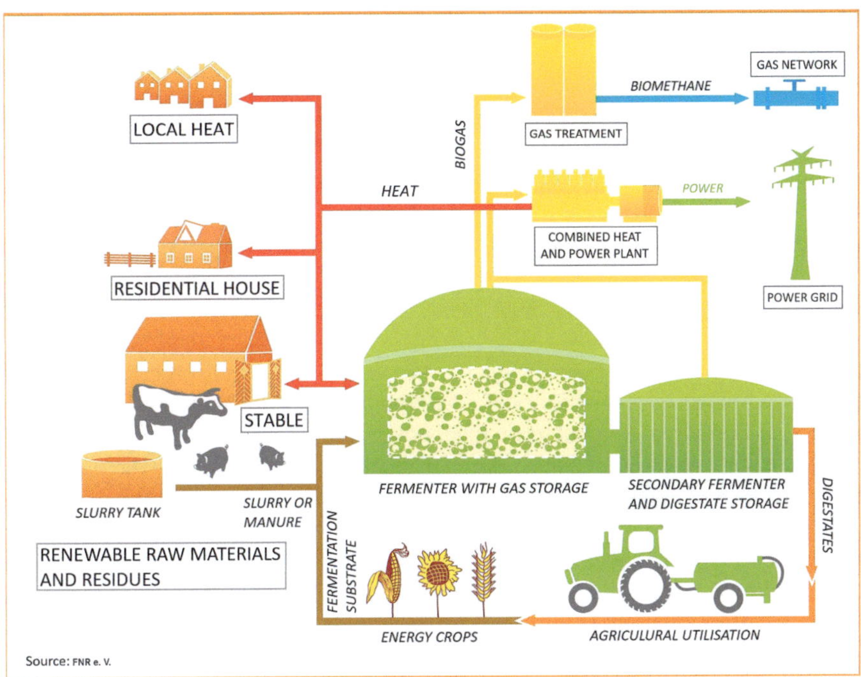

Fig. 17.4 Schematic of an agricultural biogas plant [16]. © Fachagentur Nachwachsende Rohstoffe

Table 17.4 Average composition of biogas [15]

Component	Concentration
Methan (CH_4)	50–75 vol.-%
Carbon dioxide (CO_2)	25–45 vol.-%
Water vapour (H_2O)	2–7 vol.-%
Hydrogen sulfide (H_2S)	20 20.000 ppm
Oxygen (O_2)	< 2 vol.-%
Nitrogen (N_2)	< 2 vol.-%
Ammonia (NH_3)	< 2 vol.-%
Hydrogen (H_2)	< 2 vol.-%
Trace gases	< 2 vol.-%

year can be calculated. The calculated value is based on a purely material consideration and includes both the CO_2 flows from biogas plants in which the biogas is converted to electricity and CO_2 flows from biomethane plants in which the biogas is upgraded to biomethane and fed into the natural gas grid. The plot in Fig. 17.3 illustrates the decentralized distribution of biogas plants in Germany.

CO_2 streams are generated in the biogas process, firstly during gas upgrading to biomethane and secondly during use in the combined heat and power plant (Fig. 17.4). The carbon sources used for biogas production are exclusively biogenic.

Mainly renewable resources, such as corn silage, grass silage, or cereal whole plant silage, as well as farm manure, e.g., cattle manure, pig manure, or dry chicken manure, are used as substrates [15].

The composition of biogas can be seen in Table 17.4. According to a study by the Institute for Advanced Sustainability Studies (IASS), the capture costs for CO_2 from the bioenergy sector are 26 €/t CO_2 [47]. However, this does not explain which technologies the bioenergy sector includes exactly. It can therefore be assumed that this value cannot be adopted without restriction for the biogas sector.

In biomethane plants, after desulfurization and drying, the biogas is upgraded for injection into the natural gas grid using different techniques: Pressurized water scrubbing (DWW), pressure swing adsorption (PSA), amine scrubbing, organic-physical scrubbing (PEG scrubbing) or membrane separation processes [17]. In pressurized water scrubbing and PEG scrubbing, CO_2 is desorbed via air injection and is thus present in a diluted form. In all other treatment processes, CO_2 is released into the environment in an almost pure form. About 60% of biomethane plants use PSA, amine scrubbing, or membrane separation processes [18]. The total possible amount of CO_2 from biomethane plants in Germany is 1.1 Mt/a. The much larger amount of CO_2 (9.5 Mt/a) comes from biogas plants that use the biogas in a combined heat and power plant. However, since these plants usually operate with excess air ("lean mixture") to reduce NO_x emissions, the CO_2 streams are present in a highly diluted form. Utilization of the resulting CO_2 streams does not currently take place.

In this context, biological methanation (power-to-gas) can play a significant role in the future energy system [19]. In this process, CO_2 from biogas, biomethane, or wastewater treatment plants is biologically converted with hydrogen from the electrolysis of electricity from renewable sources into storable biomethane. Plants for biological methanation on a demonstration scale are operated by the company Microbenergy GmbH at a biogas plant in Allendorf (Eder) and by the company Electrochaea GmbH as part of the BioCat project ("power-to-gas via biological catalysis", funded by Energienet.dk, ForskEL) at a wastewater treatment plant in Denmark.

Another possibility for the use of smaller, decentrally produced quantities of CO_2 is its use in greenhouses [20]. In this case, the CO_2 produced by heating the greenhouses with natural gas furnaces is mixed directly into the greenhouse air without further purification [21].

17.3 Biomass: Yeast

Yeast, in the form of baker's yeast, is one of the most important raw materials for the baking industry. Of the nearly 2 MT of yeast produced worldwide, Europe produces a total of about 1 MT per year (Confederation of EU Yeast Producers—COFALEC), of which about 200,000 tons are produced in Germany.

Yeast is produced by a strictly aerobic fermentation process using molasses as substrate. To achieve the highest possible yield of biomass from the substrate, ethanol formation is largely suppressed by suitable fermentation conditions. This includes intensive aeration and strict limitation of molasses (sugar) in the feed [22]. With good process control, a yield of up to 54 g yeast solids per 100 g sugar is achieved. The conversion rate for yeast of 67% given in Table 10.11 therefore appears to be rather high. Due to the aerobic respiration process, the other half of the sugar is converted into CO_2. In the highly aerated process, the CO_2 concentration in the off-gas stream is less than 5%, so separation appears uneconomical and has not been carried out so far [23].

17.4   Amino Acids

The amino acids with the largest production volumes worldwide are L-glutamic acid and L-lysine. The former is mainly used as an additive (flavor enhancer) for human nutrition, while the essential amino acid L-lysine is a feed additive for livestock. Glutamic acid and the resulting monosodium glutamate are produced to more than 90% in Asia [2] and only about 60,000 tons per year in Europe. Annual consumption in Germany is about 6000 tons, all of which is imported. The production of L-lysine also takes place mainly in Asia; out of 1.5 MT annually, only about 100,000 tons are produced in Europe. No lysine production takes place in Germany.

Both amino acids are produced biotechnologically by fermentation with bacteria such as Corynebacterium glutamicum [24]. Molasses and starch hydrolysates serve as substrates. These are exclusively aerobic processes that require intensive aeration during fermentation. CO_2 is produced during amino acid synthesis and respiratory metabolism, some of which is consumed again during amino acid metabolism. In the fermentative production of glutamic acid, the supply of the substrate is controlled by the CO_2 concentration in the exhaust air to minimize inhibition and by-product formation [22]. Due to the high conversion rate of more than 70% (Table 10.11) and taking into account that part of the substrate is converted into biomass, a lower formation of CO_2 is expected compared to yeast production. Due to the intensive aeration, the CO_2 in the exhaust air accumulates in a highly diluted form; a separation would probably not be economical.

17.5   Citric Acid

The tricarboxylic acid citric acid has a wide range of applications, e.g. as an acidifier in the food industry and, due to its good complex formation with divalent cations, for decalcification. Worldwide, 1 MT of citric acid are produced annually, of which more than half is produced in China [2] and about 300,000 tons in Europe. There is currently no production facility in Germany.

Citric acid is produced exclusively biotechnologically by fermentation of molasses and other sugar-containing substrates with the fungus Aspergillus niger. This is an aerobic process that requires intensive aeration, especially during the multiplication phase of the fungus. Part of the carbon dioxide formed during respiration metabolism is consumed by means of carboxylation in citric acid metabolism; therefore, a relatively high carbohydrate conversion rate of 76% results (Table 10.11). A portion of the substrate not converted to citric acid is also used for biomass formation. Similar to yeast production and amino acid productions, a small amount of CO_2 can be expected in the exhaust air.

References

1. BIOCOM AG (2017) The German Biotechnology Sector 2017. http://biotechnologie.de/statistics_articles. Accessed Sept 2017
2. ECO SYS GmbH (2011) Die Wettbewerbsfähigkeit der Bundesrepublik Deutschland als Standort für die Fermentationsindustrie im internationalen Vergleich. Fachagentur Nachwachsende Rohstoffe e. V. (Ed.) https://www.iwbio.de/fileadmin/Publikationen/IWBio-Artikel/Ind_Biotech_22003310.pdf
3. Bundesverband der deutschen Bioethanolwirtschaft e. V. (2017) Die deutsche Bioethanolwirtschaft in Zahlen. https://www.bdbe.de/daten/marktdaten-deutschland. Accessed 20 May 2022
4. Bundesverband der deutschen Bioethanolwirtschaft e. V. (2015) Bioethanol – effiziente Bioökonomie. https://www.bdbe.de/application/files/6114/3566/7565/BDBe_Broschuere_Bioethanol_2015.pdf. Accessed 8 Sept 2017
5. Bundesverband der deutschen Bioethanolwirtschaft e. V. (2017) Bioethanol weltweit. https://www.bdbe.de/daten/bioethanol-weltweit. Accessed 8 Sept 2017
6. FNR (2017) Bioethanol weltweit. https://biokraftstoffe.fnr.de/kraftstoffe/bioethanol/. Accessed 8 Sept 2017
7. Alonso-Moreno C, García-Yuste S (2016) Environmental potential of the use of CO_2 from alcoholic fermentation processes. The CO_2-AFP strategy. Sci Total Environ 568:319–326. https://doi.org/10.1016/j.scitotenv.2016.05.220
8. Weidner E, Pollak S (2015) Einsatz und Verwendung von CO_2. In: Fischedick M, Görner K, Thomeczek M (eds) CO_2: Abtrennung, Speicherung, Nutzung. Springer, Berlin/Heidelberg, pp 93–110
9. Fachverband Biogas (2021) Basisdaten Bioenergie Deutschland 2021. https://bioenergie.fnr.de/bioenergie. Accessed 11 Nov 2022
10. dena – Deutsche Energie-Agentur (2015) biogaspartner – gemeinsam einspeisen. Biogaseinspeisung in Deutschland und Europa – Markt, Technik und Akteure. https://shop.dena.de/fileadmin/denashop/media/Downloads_Dateien/erneuerbare/5026_biogaspartner_gemeinsam_einspeisen_dt_REG.pdf. Accessed 8 Sept 2022
11. ecoprog GmbH (2012) Biogas to Energy. Der weltweite Markt für Biogasanlagen 2012/2013
12. Biogaspartner (2021) Einspeiseatlas Deutschland. https://www.biogaspartner.de/einspeiseatlas/. Accessed 20 May 2022
13. Brijder M, Dumont M, Blume A (2014) Contribution greengasgrids project to development in biomethane markets. Final report greengasgrids project. Deutsche Energie-Agentur (Ed.) http://www.greengasgrids.eu/fileadmin/greengas/media/Downloads/Other_Downloads/D5.3_Publishable_Report.pdf
14. EurObserv'ER (2014) Biogas barometer 2014. Marktbericht. https://www.eurobserv-er.org/biogas-barometer-2014/. Accessed 8 Sep. 2022
15. FNR (2016) Basisdaten Bioenergie Deutschland 2016. Fachagentur Nachwachsende Rohstoffe e. V. (Ed.) http://www.bioenergyfarm.eu/wp-content/uploads/2015/09/Basisdaten_Bioenergie_FNR_2016.pdf. Accessed 20 May 2022
16. FNR (2017) Schema einer landwirtschaftlichen Biogasanlage. https://mediathek.fnr.de/grafiken/pressegrafiken/bioenergie/schema-einer-landwirtschaftlichen-biogasanlage.html. Accessed 8 Sept 2017

17. FNR/Kuratorium für Technik und Bauwesen in der Landwirtschaft (2013) Faustzahlen Biogas. 3. Ausgabe
18. FNR (2014) Leitfaden Biogasaufbereitung und -einspeisung. 5. Auflage. Adler P, Billig E, Brosowski A, Daniel-Gromke J, Falke I, Fischer E (Eds.) Gülzow-Prüzen: Fachagentur für Nachwachsende Rohstoffe e. V
19. Götz M, Lefebvre J, Mörs F, McDaniel Koch A, Graf F, Bajohr S, Reimert R, Kolb T (2016) Renewable power-to-gas. A technological and economic review. Renew Energy 85:1371–1390. https://doi.org/10.1016/j.renene.2015.07.066
20. EPEA – Internationale Umweltforschung GmbH (2016) Report – Evaluation zur Nutzung von Kohlendioxid (CO_2) als Rohstoff in der Emscher-Lippe-Region - Erstellung einer Potentialanalyse. Koch T, Scheelhaase T, Jonas N, Hungsberg M, Osswald D (Eds.) http://www.emscher-lippe.de/wp-content/uploads/2017/05/2017-05-03_Report-CO$_2$-Potentialanalyse_final.pdf. Accessed 8 Sept 2017
21. DBG (2010) Feuerlöscher oder Klimakiller. Kohlendioxid CO_2 – Facetten eines Moleküls. Deutsche Bunsen-Gesellschaft für Physikalische Chemie (Hrsg) https://www.gdch.de/fileadmin/downloads/Publikationen/Weitere_Publikationen/PDF/co2.pdf
22. Schmid RD (2016) Taschenatlas der Biotechnologie und Gentechnik, 3rd edn. Wiley-VCH, Weinheim
23. Pollmann E (2017) Persönliche Mitteilung. Versuchsanstalt der Hefeindustrie e. V., Berlin
24. Becker J, Wittmann C (2012) Bio-based production of chemicals, materials and fuels – Corynebacterium glutamicum as versatile cell factory. Curr Opin Biotechnol 23:631–640. https://doi.org/10.1016/j.copbio.2011.11.012

Utilization of Residuals and C1 Gas Streams: Organic Waste, Sludge and Agricultural Residuals

18

Thomas Bayer

Abstract

The utilization of residual materials and organic waste, sewage sludge and agricultural residues leads to C1 gas streams such as CO, CO_2 through gasification and pyrolysis. Methane and CO_2 are the result of anaerobic digestion of the mentioned residues and wastes. Conversion processes such as Fischer-Tropsch of CO, methanation of CO_2 and other Carbon Capture and Utilization (CCU) processes enable the industrial utilization of C1 gas streams. The chapter highlights the necessary technologies, the quantitative dimensions of the possible gas streams and the legal framework.

Keywords

Carbon Capture and Utilization (CCU) · Organic waste · Sludge · Agricultural residuals · CO · CO_2 · Conversion processes · Fischer-Tropsch · Anaerobic digestion · Methanation · Gasification · Pyrolysis

18.1 Introduction

Today, the raw material supply of organic chemistry is mainly based on the fossil raw materials oil, gas and coal, which are also used for energy supply. Renewable raw materials (fats, oils, starch, sugar and cellulose) account for a share in the 10% range of the total demand and are used for specialties or in biotechnological processes. The figures for the European and German chemical industry can be found in Table 18.1.

T. Bayer (✉)
Provadis School of International Management and Technology AG, Frankfurt am Main, Germany
e-mail: thomas.bayer@provadis-hochschule.de

Table 18.1 Raw materials of the chemical industry in Europe 2013 [1] and Germany 2015 [2] for organic chemistry

Raw material requirements	Europe		Germany	
	Amount [Mt]	Share [%]	Amount [Mt]	Share [%]
Fossil raw materials (oil, gas, coal)	71	89,9	17,3	87,4
Renewable raw materials (fats, oils, cellulose, starch, sugar, natural rubber, glycerin, bioethanol)	8	10,1	2,5	12,6
Total	79	100,0	19,8	100,0

Organic chemistry is thus dependent on fossil fuels to a considerable extent. How could alternative C1-based gas streams be used, and what are the sources? Here, C1 sources are understood to be methane, CO, and CO_2, which are produced by various processes (e.g., CO_2 in cement production, Chap. 14) or are produced by recycling organic compounds (e.g., methane in biogas plants).

Methane is the simplest hydrocarbon compound and the main component of natural gas and biogas. It is used on a large scale for the production of methanol (via syngas) and for the production of chlorination products.

Carbon monoxide (CO) is formed during the incomplete oxidation of carbon and its compounds and is a component of synthesis gas (mixture of CO and hydrogen H_2; Chap. 5), the raw material for syntheses of basic chemicals. It is produced by gasification of coal, petroleum, or natural gas. However, biomasses can also be converted to synthesis gas. The chemical industry has long used processes such as steam reforming or coal gasification to generate carbon monoxide and hydrogen as starting components for the synthesis of organic compounds. Today, the raw materials are largely the fossil fuels natural gas and coal. In China, for example, coal serves as a raw material source for organic chemicals to about 50% [3], and in the USA, the availability of shale gas has led to a new flourishing in basic chemistry [4].

Steam Reforming

Synthesis gas (mixture of carbon monoxide and hydrogen) can be produced from natural gas (main component is methane) by reaction with steam at temperatures of 800–900 °C and a pressure of about 25–30 bar on a nickel catalyst:

$$CH_4 + H_2O + heat \rightarrow CO + 3\,H_2 \tag{18.1}$$

The heat required for this endothermic process is provided by partial oxidation of methane:

$$CH_4 + \tfrac{1}{2}\,O_2 \rightarrow CO + 2\,H_2 + heat \tag{18.2}$$

Carbon dioxide comes into consideration as a further source due to the discussions on climate change (2-degree target). Today, CO_2 is already used in large quantities for the synthesis of urea (worldwide about 110 MT per year), using here CO_2 produced during the generation of the hydrogen necessary for the synthesis of the precursor ammonia [5]. Realistically, however, it should be seen that the use of CO_2 as a feedstock for organic chemistry cannot solve the problem of CO_2 emissions alone. According to estimates, the demand of chemistry is only about 0.6% of anthropogenic CO_2 emissions [6]. Since CO_2, as the end product of the oxidation of carbon-containing compounds, is thermodynamically a very inert compound, its use as a raw material always requires an energy carrier as a reducing agent, usually hydrogen. Only if this is produced from sustainable energy sources can it be ensured that no more CO_2 is emitted than is chemically used. However, the use of CO_2 with the help of hydrogen produced from renewable electricity can help reduce the German chemical industry's dependence on fossil raw materials. However, this is only realistic if renewable electricity can be produced cheaply and the price of fossil fuels rises. On new uses for CO_2, research and industry are developing joint efforts, especially in Germany [5]. The use of organic waste as a raw material for the chemical industry plays practically no role today.

> **Coal Gasification**
> Water gas is produced from coal in an endothermic process by adding steam:
>
> $$C + H_2O + heat \rightarrow CO + H_2 \qquad (18.3)$$
>
> The required heat is provided by partial combustion of the coal.

In this chapter, the availability of carbon-containing compounds as raw materials will be considered. The focus is on Germany. Resources that are already used for energy purposes will also be considered. The extent to which a switch to feedstock utilization is practicable will be discussed here.

18.2 Raw Material Sources

18.2.1 Organic Waste

In the Federal Republic of Germany, the Act on the Promotion of the Circular Economy and Ensuring the Environmentally Sound Management of Waste (Closed Substance Cycle Waste Management Act—KrWG) [7] is the central federal law regulating German waste legislation. The legislator sees the purpose of this law as "...to promote the circular economy to conserve natural resources and to ensure the protection of people and the environment in the generation and management of waste, and in particular to promote recycling and other material recovery of waste." Section 6 of the KrWG establishes a five-level waste hierarchy as a hierarchy of

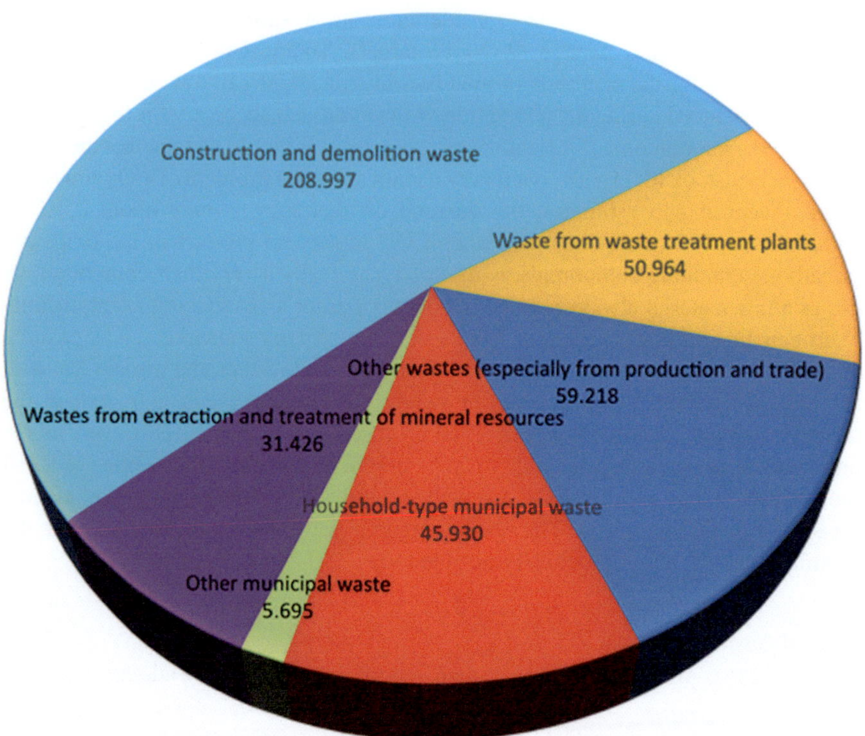

Fig. 18.1 Waste produced in Germany in 2015 in tons

measures for the prevention and management of waste. According to this, waste is to be: (1) avoided, (2) reused, (3) recycled, (4) recovered (especially energetically) or (5) disposed of. By closing loops and using the material potential of waste in particular, valuable primary raw materials can be replaced and thus conserved. Quota requirements are intended to promote the recycling of individual wastes.

Commercial municipal waste is commercial and industrial waste. They are similar to waste from private households due to their nature or composition. They can be collected separately from or together with household waste as so-called "business waste". The Commercial Waste Ordinance [8] regulates the handling of this waste and sets requirements for high-quality recycling of this waste.

A total of 402,229,000 tons of waste were generated in Germany in 2015 according to the waste balance sheet of the Federal Statistical Office [9]. The composition is shown Fig. 18.1.

Construction and demolition waste (mainly excavated soil) and waste from the extraction and treatment of mineral resources (mainly tailings from mining) account for around 60% of the waste generated and are not eligible as C1 sources. Of interest are typical household municipal waste, which includes waste from the organic waste garbage can, biodegradable garden and park waste (including cemetery waste) and

Table 18.2 Organic waste on household municipal waste in metric tons in Germany 2015 [9]

Waste	Amount	Disposal	Energy recovery	Material recovery
Waste from the organic waste garbage can	4.232.000	0	85.000	4.147.000
Biodegradable garden and park waste (incl. cemetery waste)	5.771.000	3.000	145.000	5.623.000
Biodegradable kitchen and canteen waste	928.000	6.000	47.000	875.000
Market waste	60.000	8.000	3.000	50.000
Total	10.991.000	17.000	280.000	10.695.000

mixed packaging/valuable materials, as well as other municipal waste, and here in particular biodegradable kitchen and canteen waste and market waste.

The legislator understands waste from private households to mean the residues generated in private households during private living. The organic waste collected in many municipalities via the "organic waste garbage can" is therefore likely to be of interest for the recycling envisaged here. According to the Federal Statistical Office, approximately 4.23 MT were generated in 2015 [9]. From the waste balance of the Federal Statistical Office 2015, the organic waste shown in Table 18.2 (biodegradable garden and park waste incl. cemetery waste; biodegradable kitchen and canteen waste and market waste) from household-type municipal waste can be considered as potential C1 sources. Mixtures of materials are not considered here. Furthermore, the current disposal or recovery routes are also indicated. The recovery takes place energetically or materially. For example, separately collected biowaste is used to produce compost, which is used as fertilizer or for soil improvement.

The total potential is just under 11 Mt, of which >97% is already recycled today.

In the case of packaging waste, the plastic fractions can be regarded as a C1 source, which today go to thermal recycling and are not reused as materials. Here, only the quantity that was collected separately is considered, since only here can sufficient enrichment be assumed, which is not the case with plastic fractions in household waste. This quantity was 842,000 tons in 2015 [10].

18.2.2 Sewage Sludge

Furthermore, sewage sludge from wastewater treatment plants can also be considered organic material and thus a potential C1 source. It must be taken into account that sewage sludges always have a high water content. The dry matter content of pumpable sludges is between 5 and 10%, of mechanically dewatered sewage sludges usually a good 30%. These are then already solid. In Germany, 1.8 Mt (dry matter) was produced in 2015, of which 64% was incinerated [11]. Thermal disposal is increasingly on the rise, as the application of municipal sewage sludge as fertilizer is to be abandoned due to the pollutants it contains. There will also be a need for new processes in view of the phosphorus recycling from sewage sludge demanded by legislation in some years. This is where recycling processes for C1 generation could come in, if they also allow phosphorus compounds to be reused as fertilizer. Thus, a possible potential from organic waste, the plastic fraction from packaging, and sewage sludge from wastewater treatment is about 13.6 Mt per year under the assumptions made here. On the one hand, this quantity still includes impurities and water, but on the other hand, it can be significantly higher if further organic waste types are collected separatly. This could also be previously exported waste or appropriately pre-sorted imported waste.

18.2.3 Agricultural Residues

In principle, all organic compounds synthesized by nature can be considered for use as a carbon source for organic chemistry. Due to the limited availability of arable land on earth, the principle of cascade utilization should be applied. The principle states that food should come first, followed by the use of polymers (cellulose, starch) as well as fats for the production of chemical products.

The long-term average (1999–2007) of agricultural residues (Table 18.3) is approx. 215 Mt of fresh mass; in relation to dry mass, this still results in 63 Mt/a [12].

Agricultural residues are used in different ways today. Rapeseed press cake is used as cattle feed, liquid and solid manure as fertilizer and in biogas plants, harvest residues such as from root crops, rapeseed straw and grain maize straw for humus formation. Cereal straw is also used as bedding for livestock.

In terms of quantity, cereal straw is the most interesting. If the quantities required for humus formation are considered, the sustainable straw potential is 7–11 Mt [12], which can be used for the synthesis of chemical compounds.

Today, wood is not only used for material purposes (sawmill industry, paper production, wood-based materials), but about 24% is also used for energy purposes. These approximately 12 Mt of firewood (dry) could also be used as a C1 source.

This results in a potential of 32–37 Mt of organic material from the above categories. This is significantly more than what is currently used as renewable raw materials by the chemical industry. However, these materials vary greatly in composition, water content and energy content. They accumulate decentrally (straw) or in centers (municipal wastewater treatment plants) and are already being used or recycled today.

Table 18.3 Amount of agricultural residues in Mt dry matter per year [12]

Cereal straw	25,8
Rape straw	6,4
Grain corn straw	3,1
Crop residues root crops	3,1
Rapeseed press cake	2,7
Liquid manure	12,2
Solid manure	7,5
Total	61

18.3 Conversion Processes

Due to the diverse composition of these wastes, conversion processes are required (after separation of individually usable streams) for material utilization, which generate standardized C1 components from the wastes for the construction of organic chemicals. Both biological (fermentation) and thermal processes (gasification) can be used for this purpose. In a further step, chemical base materials can be produced, also via biological processes (fermentation) or classic chemical processes (e.g. Fischer-Tropsch synthesis, methanol synthesis). It must be ensured that the quality of the basic chemicals produced in this way is comparable to today's products, so that there is acceptance for their use. Figure 18.2 provides a schematic overview of what such a value chain might look like. However, it does not claim to be complete. For reasons of clarity, the necessary pretreatment and purification steps as well as procedures for the removal of residues from the processes and conceivable recirculations have not been shown. The processes are endothermic and exothermic, so that energy (usually heat) must be introduced or can be used in some cases. These couplings are also only indicated here.

Classically, the required energy is generated via fossil energy sources or extracted from the energy-containing raw materials by combustion. If the focus is to be on the extensive utilization of the carbon contained, energy supply from regenerative sources is required. For the utilization of CO and CO_2, hydrogen is also required, which can be produced by the electrolysis of water using regeneratively generated electricity.

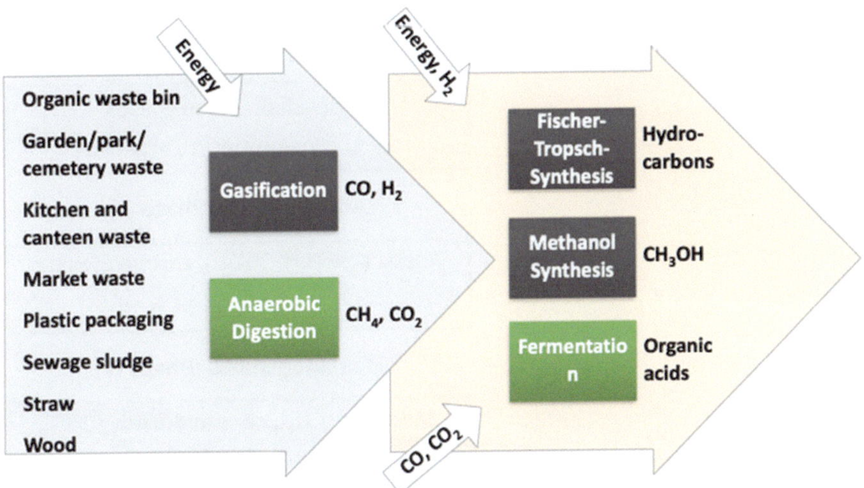

Fig. 18.2 Recycling chain from waste to chemical raw materials

Fischer-Tropsch Synthesis
Hydrocarbons are produced from synthesis gas with the aid of catalysts at temperatures of 160–300 °C and pressures of up to 25 bar. The synthesis gas can be produced from various C sources (e.g. coal, natural gas, biomass).

$$n\,CO + (2n+1)\,H2 \rightarrow CnH(2n+2) + n\,H2O + heat \tag{18.4}$$

The hydrocarbons produced can be used as fuel or chemical feedstock.

18.3.1 Anaerobic Digestion

Anaerobic (occurring in the absence of oxygen) fermentation of complex, organic substances is carried out by different microorganisms in four substeps (Fig. 18.3).

In the first stage, carbohydrates, fats and proteins are broken down into simple molecules (sugars, fatty acids and amino acids) (hydrolysis). In the next stage,

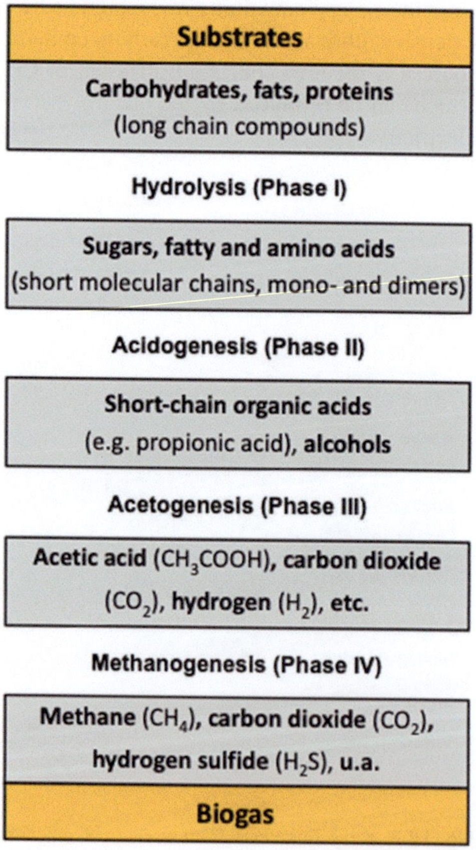

Fig. 18.3 Schematic process of anaerobic digestion

organic acids, alcohols, hydrogen and carbon dioxide are formed (acidogenesis). In a third phase, the organic acids and alcohols are broken down into acetate (salt of acetic acid) (acetogenesis). In the last phase, methane and carbon dioxide are formed from acetate and hydrogen by methane-forming archaea (methanogenesis). The resulting biogas also contains some other components, e.g. hydrogen sulfide (H_2S) and ammonia (NH_3). Depending on the further use of the biogas, complex purification steps may therefore be necessary. For example, sulfur components act as catalyst poisons. For the chemical processes used to synthesize organic compounds, the sulfur contents should be in the ppb range so as not to reduce the lifetimeof catalysts too much.

All four phases can proceed in a single stage in a reaction vessel or in two stages. The process depends on a good interaction of the different microorganisms. Therefore, an adjustment is necessary if the substrate composition is changed. Under optimal conditions, a stable pH value in the weakly alkaline (7.5–8) range is established. The methane content is between 50 and 70%, depending on the type of substrate. The time required for the process is in the range of several weeks. The processes can be carried out in batch or continuous mode. Fermentation can be carried out in the mesophilic (32–45 °C) or thermophilic range (50–60 °C). The mesophilic mode leads to more stable processes, but takes longer. The fermentation residues (non-degradable organic compounds and inorganic substances) must be disposed. In industrial plants, this is usually done by incineration with utilization of the heat generated. In agricultural biogas plants, which are operated with renewable raw materials, the residues are usually as fertilizers on the fields.

Today, the utilization of the gas produced in biogas plants focuses on energy recovery (Renewable Energy Sources Act (EEG)-subsidization of electricity generation). However, the biogas (50–70% methane, 30–50% carbon dioxide) or a biomethane (natural gas quality) produced after processing can also serve as a raw material for the chemical industry. With the expiry of the EEG subsidy (20 years per plant), the provision of raw materials for chemical production could possibly result in a new business model for plant operators. The process-related CO_2 produced during fermentation could also be used as a C1 source. This would require a methanation plant. This also applies to CO_2 generated during the production of bioethanol.

Biogas plants that utilize fermentable organic waste are today also considered state of the art on an industrial scale. Figure 18.4 shows the biogas plant at the Höchst Industrial Park in Frankfurt am Main as an example. Here, industrial sewage sludge and further organic wastes, such as residues from biotechnological production plants, are fermented together to produce biogas, which is used to generate electricity and heat and to produce biomethane. After mechanical dewatering, the digestate is burned in a sewage sludge incinerator. The heat released in the process is used to generate steam and fed into the industrial park's plant network.

Anaerobic digestion thus represents a method of producing C1 components (in this case CO_2 and CH_4) from (biodegradable) organic waste as a raw material for the chemical industry. To ensure a uniform quality (natural gas standard) and to

Fig. 18.4 Biogas plant in Industriepark Höchst, Frankfurt am Main: 2 × 10,800 m^3 digesters (blue), biogas production 40,000 m^3/d. © Infraserv GmbH & Co. Höchst KG

enable transport to the chemical sites, biogas upgrading plants already connected to the natural gas grid could be used.

In Germany, about 180 plants for the production of biomethane from biogas were already in operation in 2015, with more under construction. The Fachagentur Nachwachsende Rohstoffe e. V [13]. estimates the feed-in capacity in 2016 to be 122,400 Nm^3/h. Assuming an operating time of >8000 h/a, this capacity results in a calculated production volume of 1 billion m^3/a. This corresponds to more than 1% of Germany's natural-gas consumption of 80.5 billion Nm^3 in 2016 [14].

Industriepark Höchst also operates a biogas upgrading plant that feeds the biomethane produced into the public natural-gas grid (Fig. 18.5). Here, the biogas produced in the biogas plant shown above is upgraded.

This biomethane, which is used today as fuel and to generate electricity and heat, could also be used for materials. The volume could replace around one-third of the chemical industry's material natural gas consumption of around 3 billion cubic meters. However, since it is considerably more expensive than natural gas, this is not currently happening.

Agricultural biogas plants would first have to be supplemented with upgrading or methanation plants and connected to the natural gas distribution network. This would further increase the costs.

Fig. 18.5 Biogas upgrading plant of Infranova Bioerdgas GmbH in Industriepark Höchst, Frankfurt am Main, capacity: 80,000 MWh/a of biomethane. © Infraserv GmbH & Co. Höchst KG

Methanation

Methane can be produced from biogas by adding hydrogen by chemical or biological means (Sabatier reaction). If renewable electricity is used to produce the hydrogen, the process is sustainable and reduces CO_2 emissions:

$$CO_2 + 4\,H_2 \rightarrow CH_4 + 2\,H_2O + \text{heat} \tag{18.5}$$

Separation of CO_2 from exhaust streams is energy intensive and inversely proportional to concentration. Therefore, recovery should focus on point sources with high fractions of CO_2 in the exhaust gas. Biogas plants are certainly a good place to start, especially if separation in the form of biogas upgrading plants already exists. Here, methane is separated to be able to feed it into the natural gas grid. This produces a highly concentrated CO_2 stream of up to 99%.

18.3.2 Gasification

For the classic gasification processes, the biomass must be in dry form. This requires an energetically expensive drying of the normally moist biomasses. However, processes are also being developed that can work with moist biomass.

Gasification processes are designed to generate energy-rich carbon-containing gases from mainly solid, but also liquid or pasty organic feedstocks. This requires the introduction of air, oxygen, water vapor or carbon dioxide as gasification agents. Since the gasification reactions are endothermic, energy (heat) must be introduced into the process. When air or oxygen is used, the energy is generated during gasification by partial oxidation of the organic substances in the process (autothermal gasification). When steam or carbon dioxide is used, the energy must be supplied from outside by heating (allothermic gasification) [15].

The best-known process here is certainly wood gasification, which was used to power automobiles, especially after World War II. However, these processes are critical from an environmental point of view and are also not economically competitive at current oil prices.

Gasification of organic residues has regained importance as a waste treatment method and in the course of the use of renewable raw materials. In recent decades, numerous pilot and demonstration plants for gasification have been built around the world, but these have not made it to large-scale implementation. Two projects from Germany can be cited as examples. In Spreetal, Saxony (Sekundärrohstoff-Verwertungszentrum Schwarze Pumpe GmbH), a plant (throughput up to 30 tons per hour) was built to produce methanol and electricity from waste. Input materials were pelletized household waste, contaminated wood, tar sludge and mixed plastics blended with coal. Operations ceased in 2007 for economic reasons, and the gasifiers were dismantled [16]. The Thermoselect process, developed in Switzerland between 1985 and 1992, was built in a demonstration plant (110 tons per day) in Italy, and the process was further developed and operated as a disposal facility until 1999. A larger plant (750 tons per day) was built in Karlsruhe and operated from 1999. This plant was not able to reach the planned capacity and was finally shut down in 2004 [16].

The possibility of gasification of dried sewage sludge has also been tested by Kopf Syngas since 2010 at the wastewater treatment plant in Mannheim. The pilot plant has a capacity of 5000 tons per year for dried sewage sludge.

Since organic residues usually do not have a high energy content per mass, compression is required for decentralized materials. This can be done by pyrolysis. The resulting pyrolysis oils can then be gasified in a large-scale central plant.

> **Pyrolysis**
> Thermal decomposition of organic compounds breaks large molecules into smaller ones at high temperatures (200–900 °C). A pyrolysis oil as well as gas and solid are formed. The composition depends on the temperature, residence time, and composition of the organic material. In contrast to gasification or combustion, no additional oxygen is added.

A funded demonstration plant (BioLiq plant, www.bioliq.de) has been built at the Karlsruhe Institute of Technology (KIT) [17], which produces an oil from agricultural residues (straw) via upstream decentralized pyrolysis, which is then gasified in a central plant. The resulting syngas can be used for energy or materials. Solid residues (ash) and tar-like substances are also produced, which must be disposed of in an environmentally safe manner.

Unfortunately, there is little published data on operating experience, energy yields and costs. This may also be related to the fact that gasification technology faces fundamental challenges related to the quality of waste processing (removal of impurities), high tar and dust contents (complex gas cleaning), and high maintenance costs, which have not led to any commercially successful implementation to date.

Alternatively, gasification can also be carried out in supercritical water (hydrothermal gasification). Here, a mixture of hydrogen, carbon dioxide and methane is produced at temperatures > 374 °C and pressures > 220 bar [18]. The VERENA plant (Experimental plant for the energetic use of agricultural materials) with a throughput of 100 kg/d is located in Karlsruhe [19].

Methane formation can be promoted using catalysts. The possibility of also using moist feedstock such as sewage sludge is advantageous. Energy-intensive drying of the biomass is therefore not necessary. Furthermore, the inorganic salts can be separated due to their lower solubility in supercritical water. The resulting gas can be freed from CO_2 via water scrubbing and a high methane content can be achieved. A disadvantage is the material problems caused by the considerable corrosiveness of supercritical water. The process is currently being tested.

18.4 Conclusion

Since direct landfilling of organic waste has no longer been permitted since 2005, numerous plants for biological or thermal treatment have been built since then, and the separate collection of waste (e.g. "bio garbage can") has been further developed.

In Germany, 68 waste incineration plants and 35 substitute fuel incineration plants are operated for thermal treatment, with capacities of 20 and 5.4 MT, respectively. There are 912 composting plants and 1439 anaerobic digestion plants (including combined anaerobic digestion and composting plants) for the recycling of biowaste. The quantities added were 7.37 and 5.5 Mt, respectively [20]. From this, it can be concluded that collection and logistics work. However, this is not true for straw. Here, decentralized plants for pyrolysis would be required.

In principle, the material flows considered could also be diverted to new material recycling plants to be built. However, this is likely to meet with considerable resistance from the operators of existing plants, as these would then no longer be needed. An entry into a largely feedstock-based use as C1 sources would therefore certainly have to be approached in the long term and supported by legal framework conditions. Economically, a certainly more costly feedstock use is not possible at the current prices for fossil fuels.

With the increasing amount of regeneratively produced electricity, the use of CO_2 via the integration of regeneratively produced hydrogen should also be of interest. This could come, for example, from plants that continue to use waste streams even if they are used as materials (e.g., plants to produce bioethanol, cement plants). It is to be hoped that the trend towards lower costs in the production of regenerative electricity will continue and that an entry into the use of organic residues in material utilization will also become economically interesting, even if many technical problems still need to be solved. Therefore, the existing need for research and development for the technologies discussed should also be further supported by European and national funds.

References

1. Bazzanella A, Ausfelder F (2017) Low carbon energy and feedstock for the European chemical industry. Dechema, Frankfurt am Main
2. VCI (2017) Daten und Fakten. Verband der Chemischen Industrie, Frankfurt am Main. http://www.vci.de. Zugegriffen: 8 Jan 2018
3. Xie K, Li W, Zhao W (2010) Coal chemical industry and its sustainable development in China. Energy 35:4349–4355. https://doi.org/10.1016/j.energy.2009.05.029
4. Handelsblatt (2014) Durch Gas-Boom in USA: Linde und BASF setzen auf neue Geschäfte. http://www.handelsblatt.com/unternehmen/industrie/durch-gas-boom-in-usa-linde-und-basf-setzen-auf-neue-geschaefte-/9985408.html. Zugegriffen: 12 Jan 2018
5. Bazzanella A, Krämer D (2017) Technologien für Nachhaltigkeit und Klimaschutz – Chemische Prozesse und stoffliche Verwertung von CO_2. Dechema, Frankfurt am Main. http://dechema.de/dechema_media/CO2_Buch_Online-p-20003330.pdf.   Zugegriffen: 12 Jan 2018
6. Ausfelder F, Bazzanella A (2008) Verwertung und Speicherung von CO_2. Dechema, Frankfurt am Main. https://dechema.de/dechema_media/diskussionco2-view_image-1-called_by-dechema-original_site-dechema_eV-original_page-124930.pdf. Zugegriffen: 12 Jan 2018
7. KrWG (2012) Gesetz zur Förderung der Kreislaufwirtschaft und Sicherung der umweltverträglichen Beseitigung von Abfällen. http://www.gesetze-im-internet.de/krwg/index.html. Zugegriffen: 12 Jan 2018
8. GewAbfV (2017) Verordnung über die Bewirtschaftung von gewerblichen Siedlungsabfällen und von bestimmten Bau- und Abbruchabfällen. https://www.gesetze-im-internet.de/gewabfv_2017/BJNR089600017.html. Zugegriffen: 12 Jan 2018
9. Destatis (2017) Abfallbilanz 2015 des Statistischen Bundesamtes. https://www.destatis.de/DE/Publikationen/Thematisch/UmweltstatistischeErhebungen/Abfallwirtschaft/Abfallbilanz.html. Zugegriffen: 3 Nov 2017
10. Schüler K (2017) Aufkommen und Verwertung von Verpackungsabfällen in Deutschland im Jahr 2015. Umweltbundesamt, Dessau-Rosslau. https://www.umweltbundesamt.de/sites/default/files/medien/1410/publikationen/2017-11-29_texte_106-2017_verpackungsabfaelle-2015.pdf. Zugegriffen: 13 Jan 2018
11. Destatis (2016) Pressemitteilung Nr. 446 vom 12. Dezember 2016. 64% des Klärschlamms wurden 2015 verbrannt. https://www.destatis.de/DE/PresseService/Presse/Pressemitteilungen/2016/12/PD16_446_32214.html. Zugegriffen: 12 Jan 2018
12. Zeller V, Thrän D, Zeymer M, Bürzle B, Adler P, Ponitka J, Postel J, Müller-Langer F, Rönsch S, Gröngröft A, Kirsten C, Weller N, Schenker M, Wedwitschka H, Vetter A, Weiser C, Wagner B, Deumelandt P, Reinicke F, Henneberg K, Wiegmann K (2012) Basisinformationen für eine nachhaltige Nutzung von landwirtschaftlichen Reststoffen zur Bioenergiebereitstellung. DBFZ Report Nr. 13. Deutsches Biomasseforschungszentrum,

Leipzig. https://www.dbfz.de/fileadmin/user_upload/Referenzen/DBFZ_Reports/DBFZ_Report_13.pdf. Zugegriffen: 12 Jan 2018

13. FNR (2016) Anlagen zur Biomethan-Produktion. https://mediathek.fnr.de/grafiken/daten-und-fakten/bioenergie/biogas/anlagen-zur-biomethan-produktion.html. Zugegriffen: 12 Jan 2018

14. Statista (2018) Erdgasverbrauch in Deutschland in den Jahren von 1980 bis 2016. https://de.statista.com/statistik/daten/studie/41033/umfrage/deutschland%2D%2D-erdgasverbrauch-in-milliarden-kubikmeter/. Zugegriffen: 10 Jan 2018

15. Quicker P, Neuerburg F, Noël Y, Huras A, Eyssen RG, Seifert H, Vehlow J, Thomé-Kozmiensky KJ (2017) Sachstand zu den alternativen Verfahren für die thermische Entsorgung von Abfällen. Umweltbundesamt, Dessau-Rosslau. https://www.umweltbundesamt.de/publikationen/sachstand-zu-den-alternativen-verfahren-fuer-die. Zugegriffen: 11 Jan 2018

16. Gleis M (2011) Pyrolyse und Vergasung. In: Thomé-Kozmiensky KJ, Beckmann M (eds) Energie aus Abfall – Band 8. TK Verlag Karl-Thomé-Kozmiensky, Neuruppin, pp 438–465

17. Dahmen N, Dinjus E (2010) Synthetische Chemieprodukte und Kraftstoffe aus Biomasse Volume. Chem Ing Tech 82:1147–1152. https://doi.org/10.1002/cite.201000082

18. Möbius A, Boukis N, Sauer J (2013) Gasification of biomass in supercritical water (SCWG). In: Mendez-Vilas A (ed) Materials and processes for energy: communicating current research and technological developments. Formatex Research Center, Badajoz, pp 264–268

19. Boukis N, Dinjus E (2008) Wasserstoff- und Methanerzeugung aus nasser Biomasse. http://www.ikft.kit.edu/downloads/boukis-flyer-verena.pdf. Zugegriffen: 3 Nov 2017

20. Jaron A, Walter N (2016) Abfallwirtschaft in Deutschland 2016 Fakten, Daten, Grafiken. BMUB, Berlin. https://www.bmub.bund.de/fileadmin/Daten_BMU/Pools/Broschueren/abfallwirtschaft_2016.pdf. Zugegriffen: 3 Nov 2017

Utilization of Residuals and C1 Gas Streams: Pyrolysis Process of Concord Blue

Stefan Burmester and Johannes Booz

Abstract

The chemical industry can in principle use C1 material streams as a carbon source. Suitable feedstocks for the production of these gases include organic waste, sewage sludge and agricultural residues. These are substances of diverse compositions that would have to be standardized into an industrially suitable C1 gas prior to industrial material use in the chemical industry. Suitable processes include fermentation to biogas (methane) and gasification to synthesis gas (CO). In terms of volume, these raw materials and processes could offer a carbon volume in Germany that corresponds to a considerable extent to the needs of the German chemical industry. Under the current framework conditions, however, C1 gases produced in this way are not competitive compared with established fossil carbon sources.

Keywords

Organic waste · Sewage sludge · Agricultural residues · Chemical industry · Biogas · Methane · Gasification · Synthesis gas (CO)

The CBR® patented by the Concord Blue group of companies operates according to the staged reforming process and thus, according to Sect. 5.4.1, corresponds to a multi-stage fixed-bed gasifier in which the pyrolysis of the feedstock and the reforming of the pyrolysis gases take place spatially separated from each other. The required thermal energy for the thermal conversion processes is provided by an inert heat transfer medium (aluminum oxide spheres, d = 15 mm), which is circulated as a moving bed. Figure 19.1 represents the basic process schematically:

S. Burmester (✉) · J. Booz
Concord Blue Engineering GmbH, Herten, Germany
e-mail: sb@concordblue.de

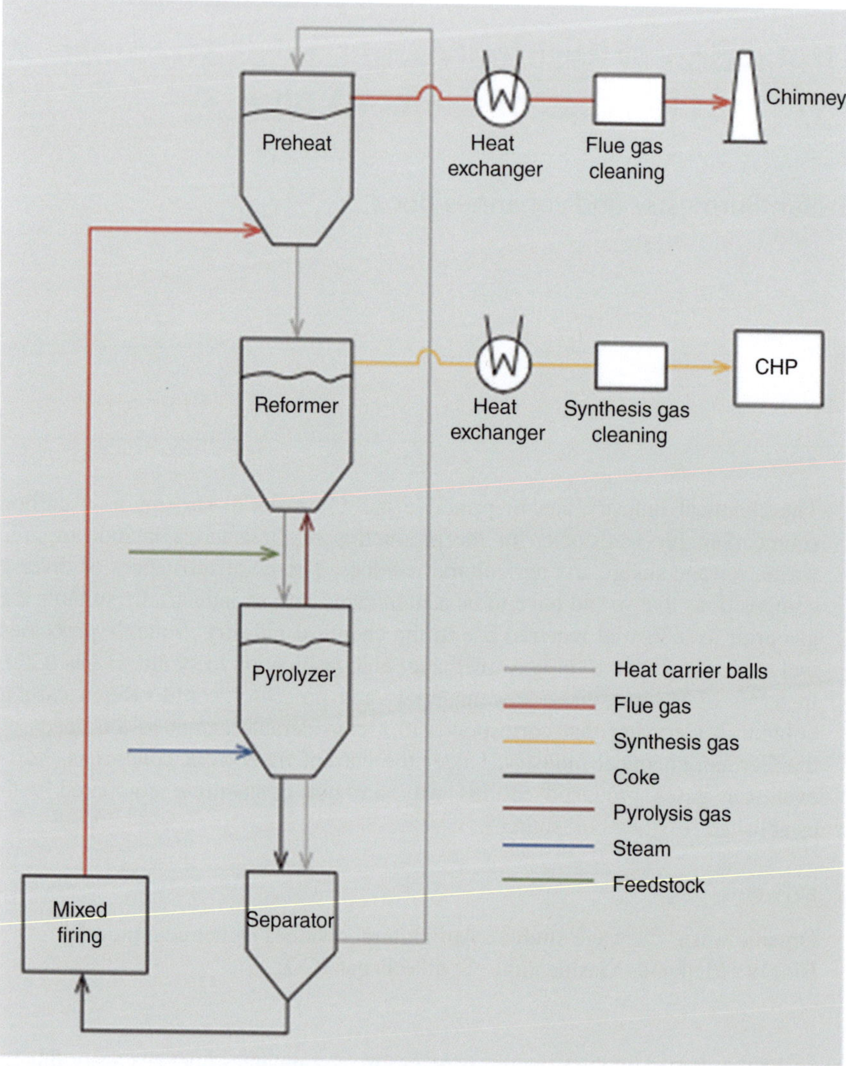

Fig. 19.1 Schematic representation of the Concord Blue Reformer (CBR®)

The core process of this gasification technology consists of three reactors arranged one above the other. The feedstock is fed together with the heat transfer medium into the lowest of the three main reactors, the pyrolyzer. In the pyrolyzer, heat is transferred between the heat transfer medium and the feedstock, so that the feedstock, which is usually already pre-dried, is pyrolyzed (thermally decomposed) at temperatures between 600 and 800 °C. This thermal decomposition process is called pyrolysis. During this thermal decomposition process, most of the feedstock is converted into a volatile gas phase ("pyrolysis gas"), while a solid residue ("pyrolysis coke") remains in the reactor. The convective heat transfer between heat transfer medium and feedstock is supported by the addition of preheated steam. This leaves the pyrolyzer at the reactor head together with the pyrolysis gas, while the pyrolysis coke is discharged at the bottom of the reactor together with the heat transfer medium.

Pyrolysis coke and heat transfer medium are metered via control elements into a screen unit (separator), where both material streams are separated from each other. The heat transfer medium, which is still hot at approx. 550 °C, is conveyed by a conveyor system to the upper end of the CBR$^{®}$ tower, where the heat transfer medium preheater is located. Ultimately, this reactor corresponds to a heat exchanger in which the heat transfer medium is heated to a target temperature of 1050 °C with the aid of a flue gas stream generated in a mixed firing system. For this purpose, the hot flue gas is introduced into the lower conical section of the preheater at approx. 1065 °C and flows through the reactor in countercurrent to the gravity-driven heat transfer fluidized bed. The flue gas exits the reactor at the top of the reactor head at temperatures of 550 to 650 °C and is then first fed to one or more heat exchangers to utilize the available residual heat energy, before being cleaned and discharged to the environment via a stack.

The hot flue gas is produced in a mixed furnace in which, in conventional CBR$^{®}$ operation, the pyrolysis coke produced in the process is burned. For this purpose, after leaving the pyrolyzer and being separated from the heat transfer medium, it must first be screened and ground to the particle size required for the burner's requirements. The combustion air required for complete combustion is preheated using the above-mentioned residual heat energy in the flue gas and added in stages to the combustion chamber of the mixed firing system.

The heat transfer media brought to the target temperature with the aid of the hot flue gas are metered from the preheater into the reformer via control elements. In this reactor, the mixture of pyrolysis gas and steam generated in the pyrolyzer is heated to up to 950 °C. In the reformer, a water vapor reaction takes place. In the process, steam reforming takes place in the reformer, during which the long-chain hydrocarbons (tars) in the pyrolysis gas are thermally broken down and decomposed into short-chain molecules.

After the CBR$^{®}$, the hot, reformed synthesis gas is cooled to approx. 500 °C in a heat exchanger using the available thermal energy and fed to a subsequent synthesis gas purification stage, in which unwanted pollutant gases (e.g., H2S, NH3) and any residual tar content that may still be present are separated using rapeseed methyl ester (RME) (Table 19.1). The tar-loaded RME is fed to the mixed firing system as an additional fuel and thus utilized for energy generation. The purified synthesis gas is then available for energy recovery in a combined heat and power plant (CHP) or for material use in further synthesis steps.

In an alternative mode of operation of the CBR$^{®}$, the pyrolysis coke produced in the pyrolyzer is not used energetically in the mixed firing system after separation from the heat transfer medium, but is discharged from the process. Initial laboratory analyses have shown that the pyrolysis coke has a high carbon content and good adsorption properties. In particular, when biomasses are used, this coke is available

Table 19.1 Typical composition of CBR®-synthesis gas

Component	Vol.%
H_2	32.9
CH_4	5.6
C_2H_4	0.3
C_xH_x	0.1
CO	19.0
CO_2	11.7
H_2O	30.1
H_2S	<0.1
COS	<0.1
NH_3	0.1
N_2	0.1
HCl	<0.1

as biochar for material recycling, e.g. as a soil improver in agriculture and horticulture, as animal feed, as a raw material for the production of activated carbon, and for many other applications. Appropriate certification as biochar by the European Biochar Foundation as well as by the International Biochar Initiative is in progress.

The functionality of the process for a wide range of different feedstocks (wood chips, shrub cuttings, household waste, sewage sludge) has been demonstrated in a pilot plant in Herten (Germany). The plant (Fig. 11.7, left) had a nominal capacity of 1 MWth at a throughput of about 200 kg/h and was operated from 2002 to 2005. Subsequently, the technology was successfully used in several commercial plants in India and Japan for both gasification of municipal waste with energy recovery and production of hydrogen.

In 2013, Concord Blue entered into a strategic partnership with Lockheed Martin, under which Lockheed Martin assumes liability or warranty beginning with plant design, including process and plant engineering, and ending with plant performance during initial operation. Together, an initial demonstration plant (Fig. 19.2, right) was built at the Owego site (USA) and successfully commissioned in 2016. Currently, Concord Blue and Lockheed Martin are working on a follow-up project for the Herten site. This commercial plant will utilize approximately 60,000 tons/year of wood chips at a rated capacity of 25 MWth and generate 5.8 MWel using gas engines. Commissioning is scheduled for the end of 2019.

Fig. 19.2 Left: Pilot plant in Herten, Germany (2002–2005), right: demonstration plant in Owego, USA (2016–present)

Utilization of Residuals and C1 Gas Streams: CO$_2$ Sources in Agriculture

Michael Binder

Abstract

Low emission farming can significantly reduce greenhouse gas emissions from livestock production. Inevitably, excrement and other waste can be fermented into biogas. The resulting biogas is usually used directly on site for energy, but can also be upgraded to pure biomethane. Associated with this is the capture of CO$_2$. With methane and CO$_2$, waste from animal husbandry thus leads to two C1 gases, both of which can also be utilized as materials. At the same time, the carbon footprint of animal husbandry is improved.

Keywords

Livestock production · Biogas · Bio-methane · Carbon Capture and Utilization (CCU)

20.1 Introduction

Agricultural animal production plays a significant role in the context of current global environmental problems. The ever-increasing demand for plant-based raw materials for both renewable energy and food/animal production is shaping entire landscapes, affecting natural areas and thereby also contributing to climate change on the one hand. On the other hand, technological progress is also a key factor in the development of global agricultural food production. Improved productivity and the use of concentrated feed can make a significant contribution to reducing the impact on the climate [1]. For this reason, the Nutrition & Care GmbH segment of Evonik Industries AG has long been engaged in analyzing the positive effects of modern

M. Binder (✉)
Evonik Industries AG, Essen, Germany
e-mail: michael.binder@evonik.com

technologies in agricultural production through life cycle analyses—so-called life cycle assessments—from "cradle to grave." In addition to process efficiency in production, the focus is also on optimized feed utilization through the use of feed additives and thus the efficient use of nutrients. This has been shown to reduce discharges to air, water and soil [2–4]. Based on this more than 15 years of scientific experience in life cycle assessment of agricultural production systems, work has now also been underway for some time on a forward-looking approach to summarize and further reduce, in a holistic approach, the environmental emissions and waste streams that inevitably occur in agricultural production. This is the Low Emission Farm. The success of the development of this concept, with cycles as closed as possible, is analyzed and visualized in depth according to the specifications of the international standard of the ISO family 14040–14044. So far, this has mainly evaluated the ecological impact of nitrogen and phosphorus inputs to air, soil and water [5, 6].

20.2   The Low Emission Farm

Low Emission Farming, or LEF for short, is a future-oriented concept for more sustainability in agriculture. The aim is to combine all the expertise of the chemical industry and modern agricultural production. In the long term, this should enable agricultural food production with virtually closed nutrient cycles. At the same time, the ecological footprint that agriculture inevitably leaves behind is to be reduced to the best possible minimum.

To achieve this, the concept for more sustainable agriculture with nutrient cycles that are as closed as possible comprises three important elements:

Nutrient management, which is the provision of optimal feed mixtures that meet demand and conserve resources;

Recyclables management, whereby organic waste from livestock production, such as liquid manure or slurry, is combined in biogas plants;

Emissions management, where raw biogas is upgraded to highly purified biogas and, further compressed or liquefied, used as a renewable energy source to replace fossil energy sources.

The LEF concept is the consistent further development of a life cycle assessment of the agricultural production chain from individual product life cycle assessments to a holistic approach for recording and measuring the holistic ecological footprint. By creating new back-integrated production concepts based on existing technologies, nutrient cycles, especially nitrogen, carbon, and phosphorus, are closed as far as possible. Thus, as far as possible, all organic side streams from feed production and animal fattening should be collected in a biogas plant. Other waste streams can come from the food processing industry (such as mills, large bakeries, slaughterhouses) or from canteens, catering companies, or restaurants. Individual solutions depend very much on regional conditions. Furthermore, the concept provides for consistent optimization towards highly purified and compressed biogas (biomethane; CH_4). This ensures maximum efficiency and the greatest flexibility in use. By way of example, it should be mentioned here that part of the biogas produced in this way can of course be used for own use in agricultural production, while the remaining surplus can also serve as an alternative for feeding directly into the public natural gas grid. Further options are shown in Fig. 20.1, e.g., filling in small containers such as gas cylinders for decentralized use without infrastructure or else as an ecologically sensible diesel substitute as a renewable energy source in mobility. However, these approaches are not described further in this chapter. Ultimately, the remaining digestate, further processed by advanced methods, can additionally be used as a high-quality nitrogen and/or phosphate fertilizer.

At present, a further use of the CO_2 produced during the purification of the raw biomethane to high-purity biomethane is not yet considered (Chap. 8).

Figures 20.2 and 20.3 impressively demonstrate the effect of the stepwise integration of the various elements of the LEF concept using the example of a representative pig production in Europe. By consistently controlling or optimizing nutrient use (especially nitrogen and phosphorus), the contribution to both the greenhouse effect and eutrophication was significantly reduced, recognizable from the first two bars of conventional production without protein reduction with amino acids (pig con, Figs. 20.2 and 20.3) and nutrient-adapted animal production with amino acid supplementation (pig AA, Figs. 20.2 and 20.3). Using the biogas in the combined heat and power plant and as a substitute for fossil diesel fuel provides another slight improvement.

20.3 The Importance of Biogas

As described at the beginning, the concept of biogas production through fermentation and subsequent conversion into electricity in combined heat and power plants or feed-in as biomethane into the gas grids represents a significant contribution to the utilization of biowaste from households, municipalities and agriculture. An alternative is the targeted production of CO_2-neutral energy from energy crops, the so-called renewable raw materials. However, this approach plays no role in the low-emission farm concept. The LEF concept consistently concentrates on residual or secondary streams so as not to compete with other material streams or usable arable land for direct agricultural food production. Due to the large number of possible feedstocks (substrates) and their easy regional availability, the potential energy available from biogas amounts to 417 petajoules (PJ) in Germany alone, which corresponds to 3% of the total primary energy consumption in Germany (13,878 PJ in 2007). Most of the biogas produced or used in this way, around 85%, is produced in agriculture. Additional gas volumes come from municipal and commercial biowaste, from landfill gases, and from sewage sludge treatment [7].

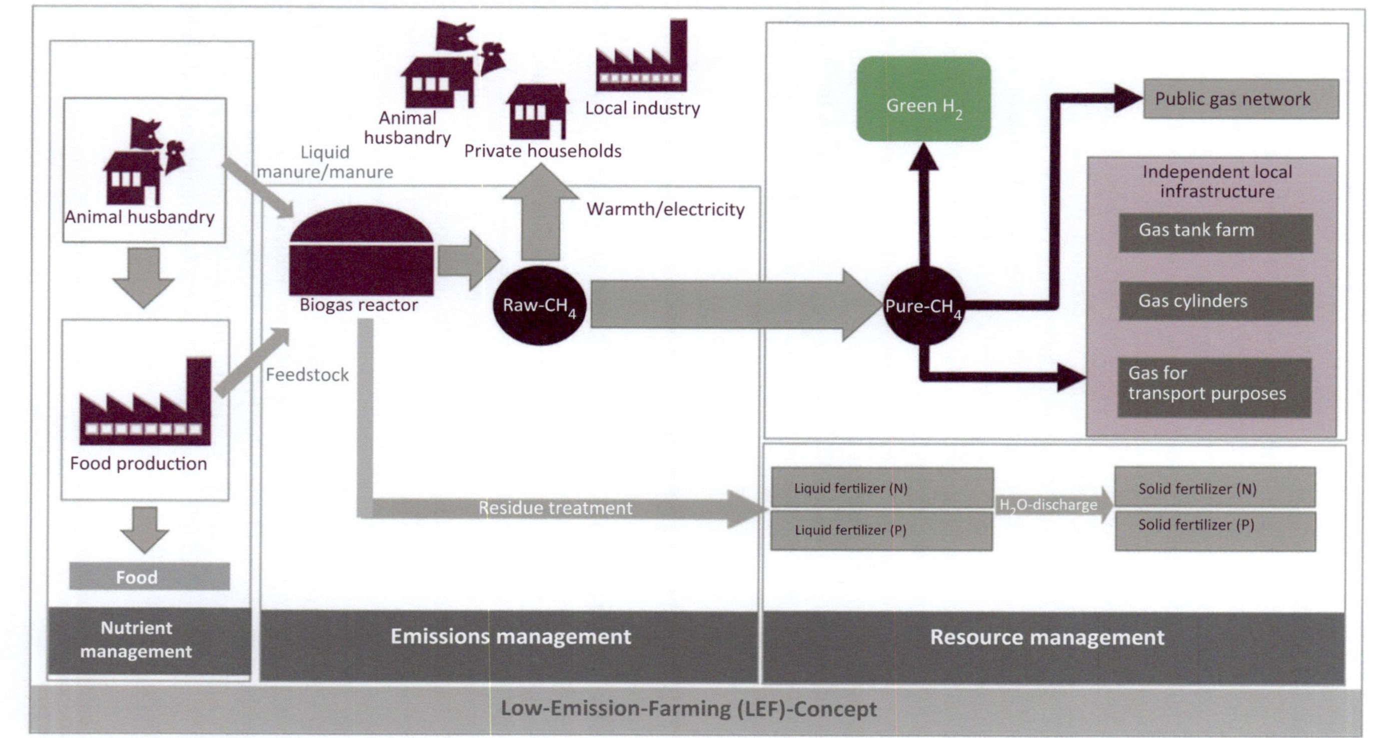

Fig. 20.1 LEF—the future concept of integrated agriculture to reduce the ecological footprint [6]. © Evonik

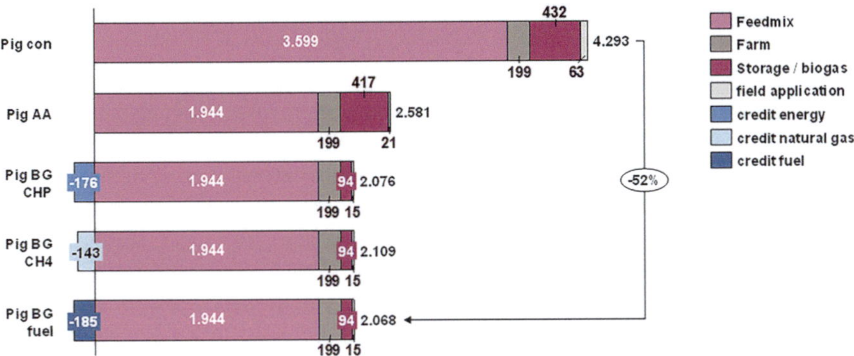

Fig. 20.2 Change in greenhouse gas potential GWP [kg CO$_2$e] using the example of European pig production through stepwise implementation of the LEF concept per 1000 kg live weight (LW) [6]. Pig con: pig production without amino acids; Pig AA: pig production with amino acids; Pig BG CHP: pig production with amino acids and biogas production; Pig BG CH$_4$: pig production with amino acids and biogas production with purification; Pig BG fuel: pig production with amino acids and biogas production with purification and fossil diesel replacement credits. © Evonik

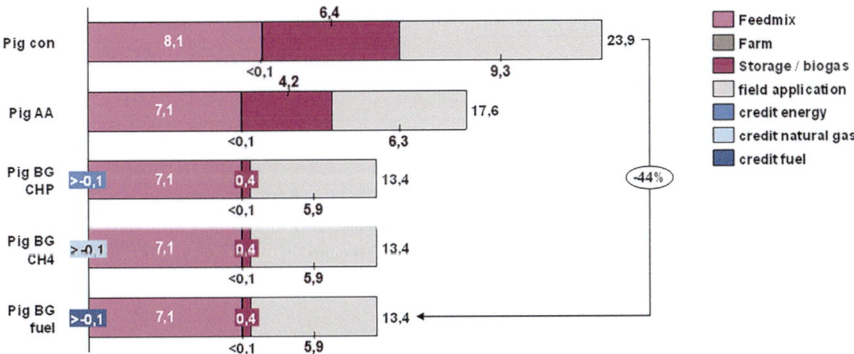

Fig. 20.3 Change in eutrophication potential EP [kg PO4e] using the example of European pig production through stepwise implementation of the LEF concept per 1000 kg live weight (LW) [6]. Pig con: pig production without amino acids; Pig AA: pig production with amino acids; Pig BG CHP: pig production with amino acids and biogas production; Pig BG CH$_4$: pig production with amino acids and biogas production with purification; Pig BG fuel: pig production with amino acids and biogas production with purification and fossil diesel replacement credits. © Evonik

20.4 Biogas Production: The Biochemical Process

Biogas is generally produced whenever organic material is decomposed by microorganisms in the absence of oxygen. This is a process that is also considered a type of anaerobic digestion. Mainly observed in bogs, in the sediment of surface waters or in the rumen of ruminants, this biochemical process converts the organic material almost completely into biogas. Anaerobic degradation of organic matter can be divided into four degradation steps: Hydrolysis, acid formation, acetate formation, and methane formation (Fig. 20.4) [7].

In the first two sub-steps, the organic substances are liquefied and broken down. The actual conversion to methane then takes place in the last two degradation steps. Different microorganisms are involved in the individual sub-steps, resulting in different products. The end product of the fermentation is ultimately a combustible gas, known as biogas or biomethane. Table 20.1 shows a typical composition:

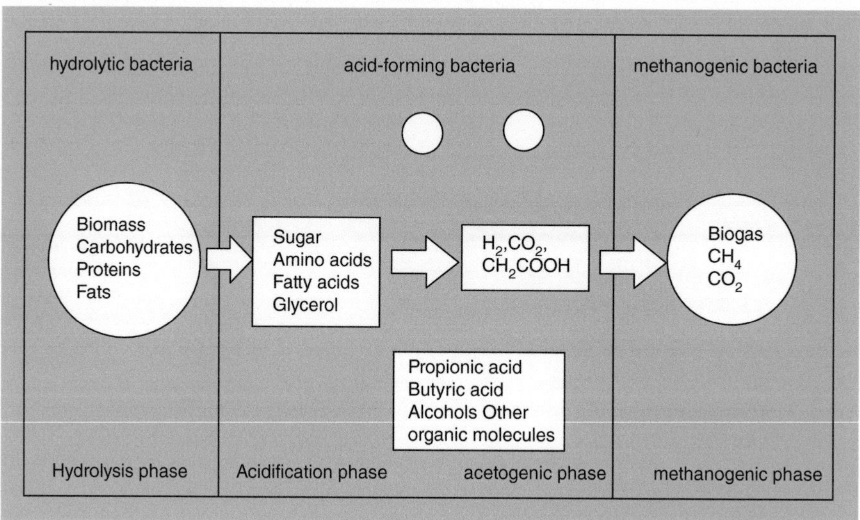

Fig. 20.4 Schematic representation of anaerobic metabolism (after [7, 8])

Table 20.1 Typical composition of biogas (biomethane, CH_4)

Component	Formula	Share [%]
Methan	CH_4	50–75
Carbon dioxide	CO_2	25–45
Water	H_2O	2–7
Oxygen	O_2	<2
Nitrogen	N_2	<2
Ammonia	NH_3	<1
Hydrogen sulfide	H_2S	<1

20.4.1 Biogas Upgrading: CH$_4$ Mass Flow

If biogas is not used directly in a cogeneration power plant, it is possible to purify it further by upgrading processes such as gas scrubbing (pressurized water scrubbing, pressureless amine scrubbing) or adsorption processes. In this process, undesirable components of the biogas (Table 20.1), but especially the CO$_2$, are separated. Further purification and concentration leads to an increase in the calorific value. Afterwards, the so-called "biomethane"can, as already described, either be used as fuel or fed directly into the public gas supply network.

The production of pure biomethane according to the above-mentioned processes results in a CO$_2$-rich off-gas, which normally still contains significant impurities of methane fractions (from $<0.1\%$ up to 4.5%, depending on the process). Thus, to reduce the methane emissions due to these impurities, further process steps are still required, such as additional thermal utilization of the residual off-gas in a lean gas boiler or off-gas cleaning processes such as a (catalytic) lean gas post-combustion [7]. Thus, a high-purity biomethane stream is achieved, but the accumulated or scrubbed CO$_2$ does not become efficiently usable as a carbon source.

Modern separation processes using polymer-based hollow fiber membranes eliminate the need for this extensive purification, drying, and conditioning. Hydrogen sulfide and ammonia are eliminated directly, for example to prevent corrosion in engines and downstream components such as heat exchangers, and essentially only two gas streams are produced, CH$_4$ and CO$_2$.

In recent years, various processes have become established on the market for the separation of CO$_2$ [9], which will not be discussed further here. In the majority of the processes, both the CO$_2$ and the contained trace gases are either adsorbed on porous materials or dissolved in water and washed out.

All of these processes have some serious drawbacks, as they require energy and also usually involve the use of auxiliary materials and chemicals. Waste and effluents are produced, which require costly treatment or disposal.

Improved flexibility, energy efficiency, and cost efficiency are made possible by the new membrane technology, with which Evonik, as a manufacturer of high-performance plastics, has had years of experience. Gas separation with the novel polymer membranes is based on the different size of the gas molecules and thus the different solubility in the polymer. In biogas, the CO$_2$ molecules are smaller than the CH$_4$ molecules of methane due to their spatial structure, which is why they can pass through the micropores of the membrane much faster. Thus, methane accumulates at the high-pressure side of the membrane, while water vapor, ammonia, hydrogen sulfide, and most of the CO$_2$ pass through the molecular sieve (Fig. 20.5).

20.4.1.1 Biogas Upgrading: CO$_2$ Mass Flow

As already described in detail, extensive steps for further purification, drying and conditioning are necessary to meet the specified technical requirements before biogas can be fed into the public grid.

The central step in the upgrading process, as already mentioned several times, is the effective separation of CO$_2$, hydrogen sulfide and ammonia. This is because CO$_2$

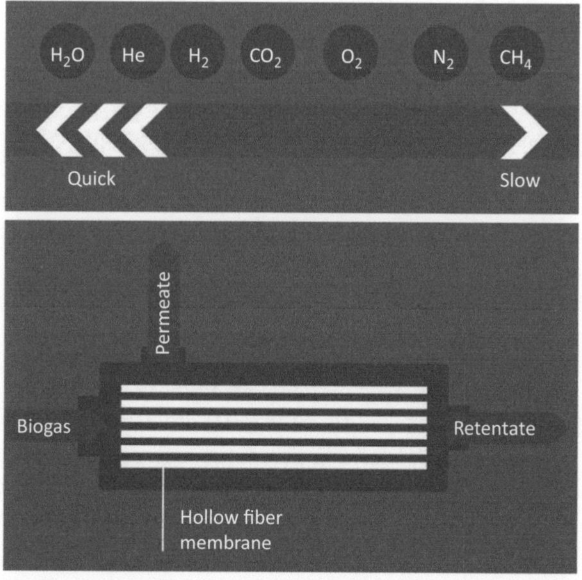

Fig. 20.5 Purification of biogas on a polymer membrane [9]. © Evonik

is not combustible and thus reduces the calorific value when either the biogas or the purified fractions are used further thermally. If the CO_2 isolated in this way is not either washed out or released back into the atmosphere after separation, as has been the case to date, it represents an interesting source of raw materials. While intensive work is generally being done on the utilization of CO_2 from other sources, such as waste gases from industrial plants and fossil-fired power plants, this CO_2 source has so far been rather ignored.

In Germany, more than 9000 biogas plants are in operation, most with energy use in combined heat and power plants (CHP) without efficient purification. Only a few plants operate purification using modern polymer membrane technology. There is great potential here to obtain relevant quantities of pure CO_2 by means of purification.

As mentioned before, using the methods of life cycle assessment, mass flows for ecologically relevant substances, such as nutrients, can be analyzed and presented in terms of their environmental impact [10]. So far, mainly nitrogen fluxes have been considered, but the same method can be used to assess carbon.

For example, the model used to analyze the greenhouse effect and eutrophication potential in Figs. 20.2 and 20.3 can also be used to capture the ratio of CO_2 produced when the LEF concept is implemented. This is illustrated below.

As Fig. 20.6 shows, the largest proportion of carbon that has been fixed by forage plants with the help of photosynthesis is first utilized by the animal itself, namely about 742 kg of a total of 1042 kg in the forage. A certain proportion is released directly back into the atmosphere via the animals' breath. This is not considered further in this balance here. Another part of about 300 kg remains in the excrements and then enters the biogas plant. Furthermore, it is assumed in this model that no

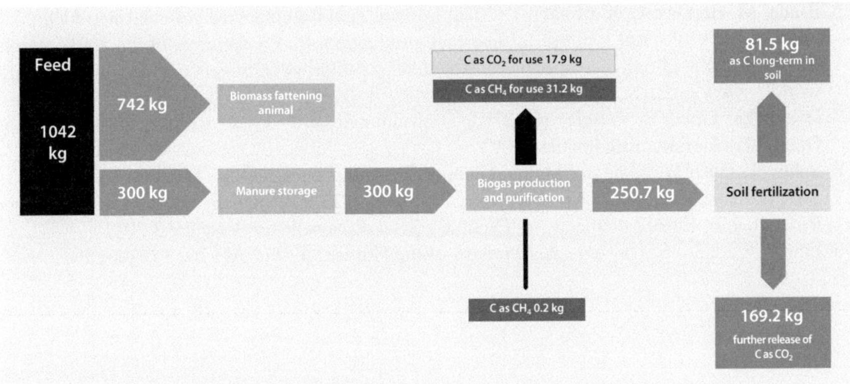

Fig. 20.6 Mass balance for carbon when the LEF concept is implemented (own calculations based on [6])

further carbon loss through CH$_4$ slip occurs due to appropriate technical measures during manure storage. Thus, about 28% of the carbon from the feed enters the biogas plant via animal excreta. Slightly more than 10% of this remaining carbon is converted to biogas (CH$_4$) by the microflora in the biogas reactor. The remaining carbon remains in the digestate and is thus an important ingredient for the further return of carbon to the field areas, thus preventing soil depletion of carbon.

Further purification of the raw biogas separates CO$_2$. This results in about 17.9 kg of sequestered carbon in the CO$_2$ stream. Since this CO$_2$ volume stream is free of other gases, is suitable to be used as feedstock in other production technologies. A small loss of carbon occurs due to methane slip in the biogas plant.

References

1. Griep W, Binder M (2017) Pork production's 'ecological burden' is decreasing thanks to more sustainability in animal feed practices. AMINONews® 21(4):13–24
2. Krauter J, Mosthaf F, Randy M, Cosgrove JA, Binder M, Kaufmann T, Scherer V (2017) What if …Evonik's animal nutrition contribution towards a more sustainable world, 1st edn ISBN 978-3-00-055744-6
3. Haasken C, Binder M (2015) Comparative life cycle assessment of MetAMINO®, Biolys®, ThreAMINO®, TrypAMINO® and ValAMINO® in broiler and pig production. Evonik Nutrition & Care GmbH
4. Kebreab E, Liedke A, Caro D, Deimling S, Binder M, Finkbeiner M (2016) Environmental impact of using specialty feed ingredients in swine and poultry production: a life cycle assessment. J Anim Sci 94:1–18. https://doi.org/10.2527/jas2016-9036
5. Binder M, Rother E, Kaufmann T (2015) Low emission farming – a significant step forward to improve the environmental impacts of livestock production. In: Scalbi S, Loprieno AD, Sposato P (eds) International conference on Life Cycle Assessment as reference methodology for assessing supply chains and supporting global sustainability challenges LCA FOR "FEEDING THE PLANET AND ENERGY FOR LIFE"

6. Binder M, Haasken C, Kaufmann T (2016) Low mission farming – a significant step forward to improve the ecological footprint of livestock production. In: Proceedings of the 2016 international nitrogen initiative conference, "solutions to improve nitrogen use efficiency for the world". www.ini2016.com
7. Daniel De Graaf D, Fendler R (2010) Umweltbundesamt Dessau-Roßlau, Biogasanbau in Deutschland. Hintergrundpapier
8. Sahm H (1981) Biologie der Methanbildung. Chem Ing Tech 53:854–863
9. Baumgarten G, Ungerank M, Schnitzer C, Kobus A (2011) Hochleistungspolymere erzeugen Biomethanin. Evonik elements 36, Quarterly Science Newsletter Ausgabe 3 2011
10. Leip A. Ledgard S. Uwizeye A. Members of the Nutrient TAG FAO 2017 in progress

Manfred Kircher

Abstract

Today, the energy sector, followed by the food and feed, and materials and chemicals sectors, are the largest consumers of fossil and biogenic carbon sources. The value chains of these product groups are largely linear, i.e. raw materials are processed into products and after use or disposal, the carbon contained is almost completely emitted into the atmosphere as CO_2. Point sources of highly concentrated CO_2 emission streams from the manufacturing sector that are suitable for carbon capture and utilization (CCU) or storage (CCS) are presented elsewhere in this book. This chapter discusses the potential for CO_2 emissions from waste-to-energy in the EU. Emissions from waste incineration still account for 9% of total EU CO_2 emissions. Their capture and recycling could meet the carbon needs of the European chemical industry, assuming sufficient energy supply for CCU processes. Utilizing the CO_2 emission from waste management would close the carbon cycle of the value chains leading to the wastes.

Keywords

Waste-to-energy · Chemical industry · Waste management · CO_2

21.1 Introduction

Today, not only fossil-based but also bio-based value chains are largely organized in a linear way. They use a raw material to manufacture products that through use (e.g. fuel combustion) or disposal (e.g. incineration and degradation of waste) emit

M. Kircher (✉)
KADIB, Frankfurt am Main, Hessen, Germany
e-mail: kircher@kadib.de

M. Kircher, T. Schwarz (eds.), *CO2 and CO as Feedstock*, Circular Economy and Sustainability, https://doi.org/10.1007/978-3-031-27811-2_21

CO_2 into the atmosphere. Some of this CO_2 emission is sequestered in biomass through the natural carbon cycle of photosynthesis; however, this is not recycling of CO_2 in the sense of value chains with closed carbon cycles.

This chapter examines the question of the extent to which CO_2 emissions generated during the production, use and disposal of carbon-containing products can be technically recycled, and whether carbon cycles can be closed in this way. A special focus is given to the CO_2 emissions from waste treatment.

To this end, the study first examines how much carbon from fossil and biogenic raw materials is consumed worldwide today for various applications. The value chains of these applications are then examined to determine the CO_2 emissions they lead to. This is followed by an analysis of which CO_2 emissions could in principle be suitable for recycling, how large their potential is in Europe in terms of volume, and what emission reductions this CO_2 recycling could lead to.

21.2 Consumption and Use of Energy and Carbon Sources

Table 21.1 shows the enormous volumes of fossil and renewable biogenic energy and carbon sources that the global economy turns over. In 2019, before the Corona Crisis, 15.9 bn t (billion tons) of fossil resources were consumed. Depending on their origin and quality, the carbon content of these fossil feedstocks varies, so only a range of carbon content of 10.7–12.9 bn t can be given for them. The resulting emission potential is 39.1–46.9 bn t of CO_2. This range is higher than the reported CO_2 emission from fossil sources of 36.7 bn t CO_2 (world, 2019) [1]. This can be

Table 21.1 Consumption and CO_2 emission potential of fossil and biogenic raw materials (global; 2019) (Production volume of coal, mineral oil, and natural gas [5]; share of carbon in coal [6], mineral oil [7], natural gas [8], biomass [9]; production of agricultural crop [10], forestry wood [11])

| C-Source | Global usage per year | | Carbon content | | CO_2-potential | Share |
		[b t]	[% per weight]	[b t]	[b t]	[%]
	Fossil feedstock					
Coal	8 bn t	8.0	60–75%	4.8–6.0	17.6–22.0	
Mineral oil	36.4 bn barrel	5.1	83–87%	4.2–4.4	15.3–15.8	
Natural gas	4014 bcm	2.8	60–90%	1.7–2.5	6.2–9.1	
Total fossil feedstock		*15.9*		*10.7–12.9*	*39.1–46.9*	*62–66%*
	Biomass					
Agricultural crop	9.4 bn t	9.4	40–55%	3.8–5.2	13.9–19.0	
Forestry wood	4.2 bn t	4.2	40–55%	1.7–2.3	6.2–8.4	
Total biomass	*13.6 bn t*	*13.6*		*5.5–7.5*	*20.1–28.4*	*36–32%*
Total				16.2–20.4	59.2–75.3	100%

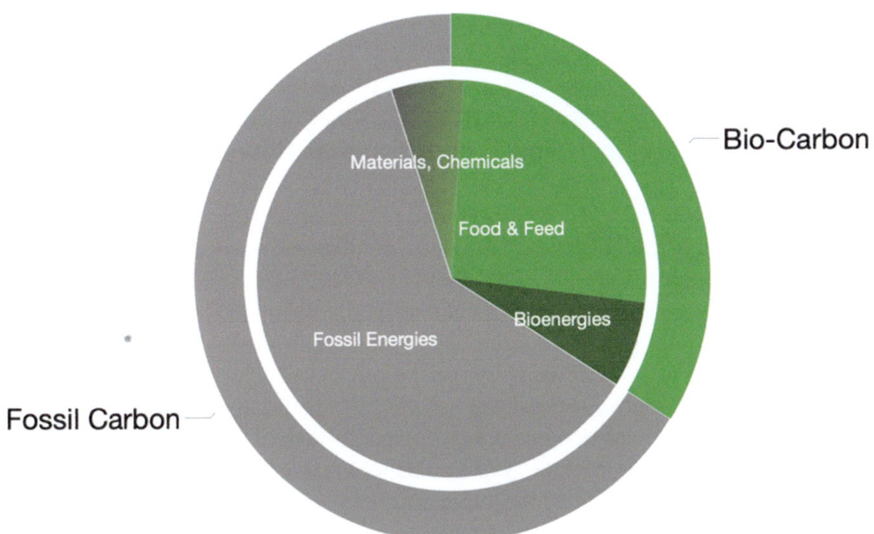

Fig. 21.1 Proportion of fossil and biogenic carbon consumption for different applications [13]

partially explained by the binding of part of the carbon in products (for example, globally 450 MT carbon in products of organic chemistry and polymers [2] (equivalent to 1,6 MT CO$_2$); 85 MT carbon in bitumen [3] (equivalent to 311 MT CO$_2$), CCS (carbon capture and storage) 39 MT CO$_2$ [4], waste recycling and disposal by landfill).

Biomass is also utilized as a source of energy and carbon and therefore must be included in the emission potential statistics. Agriculture and forestry produced 13.6 bn t of biomass in 2019 [12], with a carbon content of 40–55% (dw; dry weight), containing about 5.5–7.5 bn t of carbon, representing an emission potential of 20–28 bn t of CO$_2$. Although biomass and fossil feedstocks contribute about half by weight to feedstock supply, biomass contributes only about one-third to carbon supply because of its lower carbon density. Two thirds are supplied by fossil raw materials (Table 21.1).

These raw materials are consumed primarily for the generation of energy (heat, electricity, fuels), followed by the production of food and feed, chemicals, polymers and pharmaceuticals, and materials such as fibers and wood products (lumber, paper). The shares of fossil and biogenic raw materials for the production of these product groups are shown in Fig. 21.1.

When looking on carbon only, the largest consumer is the energy sector, which takes 67% of the total carbon (61% fossil and 6% biogenic origin), followed by the food and feed sectors (26% biogenic carbon) and 6% materials and chemicals (about 4% fossil and 2% biogenic carbon) [13]. The goal of climate neutrality, which was agreed in the Paris Climate Agreement (2015), forces the sectors that still use fossil raw materials today to switch to (1) carbon-free energy production and (2) renewable carbon sources. Renewable carbon sources include among other things (1) biomass, which is part of the natural carbon cycle, and (2) recycling of CO_2, which technically closes the carbon cycle of value chains.

21.3   Value Chains of Fossil and Biogenic Raw Materials

Today, the value chains of the sectors mentioned are largely linear, i.e. raw materials are processed into products and, after use or disposal, the carbon contained is predominantly emitted into the atmosphere as CO_2. This also applies to bio-based value chains, for which circularity is claimed in the literature with reference to the natural carbon cycle [14]. According to the European Commission recycling of waste is defined "as any recovery operation by which waste materials are reprocessed into products, materials or substances whether for the original or other purposes. It includes the reprocessing of organic material but does not include energy recovery and the reprocessing into materials that are to be used as fuels or for backfilling operations" [15]. Although the emission of CO_2 from the recycling of biogenic waste can be classified as climate-neutral in balance sheet terms, recycling of CO_2 through the natural carbon cycle does not really result in a closed value chain. In the following, process stages of exemplary value chains that emit CO_2 will be considered.

21.3.1 Production of Fossil Raw Materials

The production of fossil raw materials generates considerable amounts of greenhouse gases. For example, oil production produces natural gas (methane) as a by-product from geological deposits, which is flared at the extraction site for economic reasons. As a result, globally 400 MT of CO_2 are emitted at the point sources of flaring [16]. A comparison with Table 21.1 shows that this corresponds to about 2% of the total emission potential of the extracted natural gas.

Coal deposits also contain methane, which escapes from mines as mine gas. To prevent firedamp, the methane-containing mine gas must be emitted from the mine. Methane is either ventilated directly into the atmosphere or used to generate energy [17]. The latter technology emits CO_2 at the power plant as a point source.

21.3.2 Production of Biogenic Energy and Carbon Sources

Agriculture and forestry represent sectors that generate significant CO_2 emissions through the use of (still) fossil fuels and through tillage of soils. Soil tillage stimulates the metabolic activity of soil flora, which significantly increases their emissions. These CO_2 emissions account for 7.2% of total CO_2 emissions worldwide [18]. They occur over large areas.

21.3.3 Energy Recovery from Fossil and Biogenic Raw Materials

The energetic utilization of fossil raw materials (coal, crude oil, natural gas) leads to CO_2, which is predominantly emitted into the atmosphere, a small portion of which is deposited (Carbon Capture and Storage; CCS) and an also small portion of which is recycled (Carbon Capture and Utilization; CCU). Power and heat generation facilities represent large point sources of CO_2 emission. Examples for small point sources are heating systems for individual homes or individual stationary or mobile engines.

If biomass is energetically utilized directly by combustion or indirectly after transformation to biogas or biofuel, the contained carbon is also completely converted to CO_2 and emitted at point sources.

21.3.4 Material Utilization of Fossil and Biogenic Raw Materials

The material utilization of fossil and biogenic carbon sources also leads to CO_2 emissions. Other chapters in this book have already referred to the emission of CO_2 in chemical and biotechnological processes and the resulting point sources. In contrast, emissions from waste treatment have not been considered so far. Exemplary waste categories containing carbon will be considered. According to the EU classification, they are either completely biogenic (material flow 1), completely fossil (material flow 4) or of both biogenic and fossil origin (Table 21.2) [19].

These waste categories [20] and the type of waste treatment [21] (Table 21.3) are described first. For the fossil-based and biogenic organic wastes considered here, recovery operations R1 (use as energy source), R3 (recycling of organic substances) and R10 (land treatment) are particularly relevant. It should be noted that R3 (recycling) includes both mechanical recycling (e.g., waste paper to paper) and recovery to other products (e.g. bio-waste to compost and biogas). While in mechanical recycling the carbon contained in the waste remains fixed in the product, composting and the biodegradation of compost as well as the fermentation and combustion of biogas lead to only delayed complete emission of the carbon as CO_2. Only a small fraction of CO_2 is captured from biogas plants, e.g. for the carbonation of beverages [22], and from other sources to be used as chemical feedstock, e.g. urea.

Subsequently, the statistics of waste treatment of selected waste categories (EU28, 2018) [23] (Table 21.4) are evaluated to discuss the recovery potential of the associated CO_2 emission. Data for 2018 are the most recent available in Eurostat (accessed August 2022).

21.3.4.1 W11 Common Sludges (100% Biogenic)

Common sludges are waste water treatment sludges from municipal sewerage water and organic sludges from food preparation and processing. The majority of the sludges (64%) is recycled, which mainly means spreading on fields as fertilizer. There, the sewage sludge decomposes and emits CO_2 over large areas. The situation

Table 21.2 Correspondence of waste codes to the four material flows MF1 to MF4

| | | Feedstock | | | |
| | | MF 1
Biomass | MF2
Metal ores | MF3
Non-metallic minerals | MF 4
Fossil energy carriers |
Waste Category					
W11	Common sludges	100%			
W72	Paper and cardboard wastes	100%			
W75	Wood wastes	100%			
W91	Animal and mixed food wastes	100%			
W92	Vegetal wastes	100%			
W93	Animal faeces, urine and manure	100%			
W73	Rubber wastes				100%
W74	Plastics wastes				100%
W05	Health care and biological wastes	62%	1%	3%	35%
W76	Textile wastes	30%			70%
W101	Household and similar wastes	64%	7%	12%	16%
W102	Mixed and undifferentiated materials	31%	11%	9%	48%
W103	Sorting residues	50%	10%	11%	30%

is different for the energy recovery type of use, to which 16% of the sewage sludge (1.8 MT) is fed. These are point sources at large capacity plants. 11% is incinerated without energy recovery; whether these provide point sources is unknown. Another 9% is landfilled, with CO_2 emitted here presumably escaping over the entire landfill area.

21.3.4.2 W72 Paper and Cardboard Wastes (100% Biogenic)

Paper and cardboard waste from sorting and separate collection are almost completely recycled (98.5%). Apart from the emissions caused by the recycling process, the carbon of the waste remains bound in the recycled products.

21.3.4.3 W75 Wood Wastes (100% Biogenic)

This category includes wooden packaging; sawdust, shavings, cuttings; waste bark, cork and wood from production of pulp and paper; wood from construction and demolition of buildings; and separately collected wood waste. These materials are essentially burned or recycled in almost equal proportions for energy recovery (22 MT each). Energy recovery results in CO_2 emissions at facilities that may be point sources.

Table 21.3 Recovery operations pursuant to Annex I (D1-D12) and Annex II (R1-R11) of the Waste Statistics Regulation [24]

Code	Types of recovery operations
D1	Deposit ino or onto land (e.g. landfill etc.)
D2	Land treatment (e.g. biodegradation of liquid or sludgy discards in soils etc.)
D3	Deep injection (e.g. injection of pumpable discards into wells, salt domes or naturally occurring depositories, etc.)
D4	Surface impoundment (e.g. placement of liquid or sludgy discards into pits, ponds or lagoons, etc.)
D5	Specially engineered landfill (e.g. placement into lined discrete cells which are capped and isolated from one another and the environment, etc.)
D6	Release into a water body except seas, oceans
D7	Release into seas/oceans including sea-bed insertion
D10	Incineration on land
D12	Permanent storage (e.g. placement of containers in a mine, etc.)
R1	Use principally as a fuel or other means to generate energy
R2	Solvent reclamation/regeneration
R3	Recycling/reclamation of organic substances which are not used as solvents (including composting and other biological transformation processes)
R4	Recycling/reclamation of metals and metal compounds
R5	Recycling/reclamation of other inorganic materials
R6	Regeneration of acids and bases
R7	Recovery of components used for pollution abatement
R8	Recovery of components of catalysts
R9	Oil-refining or other reuses of oil
R10	Land treatment resulting in benefit to agriculture or ecological improvement
R11	Use of wastes obtained from any of the operation numbers R1 to R10

21.3.4.4 W91 Animal and Mixed Food Wastes (100% Biogenic)

This category includes animal waste of food preparation and products, incl. sludges from washing and cleaning, and mixed wastes of food preparation and products incl. Biodegradable kitchen/canteen wastes, and edible oils and fats. 87% are recycled (compost, biogas). Emissions from composting and biogas fermentation could theoretically be captured and recycled for larger plants. 2% is incinerated without or 6% with recovery of energy. Facilities with energy recovery can be assumed to be point sources (1.2 MT).

21.3.4.5 W92 Vegetable Wastes (100% Biogenic)

Vegetal waste from food preparation and products, including sludges from washing and cleaning is almost completely (94%) recycled by composting and biogas fermentation. 4.1% is used for energy recovery (2 MT); these facilities may be suitable point sources for CO$_2$ recycling.

Table 21.4 Waste management operations of selected waste categories (EU28; 2018)

Waste category		Volume [MT]	Disposal by landfill D5, D7, D12	Disposal by incineration D10	Recovery of energy R1	Recovery by recycling R2-R11
		Bio-based				
W11	Common sludges	11.49	8.5%	11.2%	15.9%	64.4%
W72	Paper and cardboard wastes	35.22	0.03%	0.0%	1.2%	98.5%
W75	Wood wastes	48.89	0.4%	4.4%	48.9%	46.3%
W91	Animal and mixed food wastes	20.28	1.5%	2.2%	6.2%	90.1%
W92	Vegetal wastes	47.89	1.6%	0.2%	4.1%	94.1%
W93	Animal faeces, urine and manure	10.6	1.4%	8.3%	6.1%	84.2%
		Fossil-based				
W73	Rubber wastes	2.85	0.3%	3.2%	33.5%	63.0%
W74	Plastic wastes	11.73	5.5%	0.5%	19.5%	74.5%
W74	Plastics wastes					
		Mixed bio- and fossil-based				
W05	Health care and biological wastes	0.74	21.6%	10.8%	66.2%	1.4%
W76	Textile wastes	1.69	8.9%	0.6%	12.4%	78.1%
W101	Household and similar wastes	119.38	34,1%	4.3%	46.9%	14.7%
W103	Sorting residues	96.84	48.0%	5.9%	32.7%	13.4%
	Total	407.26				

21.3.4.6 W93 Animal Faeces, Urine and Manure (100% Biogenic)

84% of slurry and manure including spoiled straw is recycled (fertilizer, biogas). 6.1% (0.6 MT) is recycled for energy. These CO_2 point sources may have potential for CO_2 recycling; another 8.1% (0.9 MT) is disposed of by incineration with no energy recovery to date.

21.3.4.7 W73 Rubber Wastes (100% Fossil)

63% of end-of-life wastes are recycled; the remainder is almost entirely recovered for energy (33.5%; 950 MT). The latter may represent point sources of CO_2.

21.3.4.8 W74 Plastic Wastes (100% Fossil)

Plastic packaging, plastic waste from plastic production and machining of plastics, and separately collected plastic waste is to 74% recycled. 19.5% (27 MT) is allocated to energy recovery and can potentially provide CO_2 from point sources.

21.3.4.9 W05 Health Care and Biological Wastes

This category comprises only biological waste from health care for animals and humans. 66% (0.49 MT) are used to generate energy in point source facilities.

21.3.4.10 W76 Textile Wastes

Textile and leather waste, textile packaging, fleshings and lime split from leather processing, worn clothes and used textiles, waste from fiber preparation and processing (man-made and natural fibers), waste tanned leather, and separately collected textile and leather waste is recycled to 78%. 12.4% (82.8 MT) is used for energy recovery with CO$_2$ emissions in plants that can be point sources.

21.3.4.11 W101 Household and Similar Wastes

Mixed municipal waste, bulky waste, street cleaning waste, kitchen waste, household equipment is mainly used for energy recovery with a share of 46.9% (56 MT). Another 4.3% (5.1 MT) is incinerated without energy recovery. CO$_2$ point sources can be expected for both types of equipment.

21.3.4.12 W103 Sorting Residues

Sorting residues from mechanical sorting processes for waste, such as screening, fluff-light fraction, premixed and combustible wastes from physico/chemical treatment of waste, combustible waste (refuse derived fuel), non composted fractions of biodegradable waste is used for energy recovery in plants with point sources for 32.7% (31.7 MT) and incinerated for 5.9% (5.7 MT).

21.4   Closing the Carbon Cycle with CO$_2$

In the previous section, it was shown that raw material production, generating energy and producing materials by transforming fossil and biogenic energy and carbon sources, and waste-disposal lead to CO$_2$ emissions. The ability to capture CO$_2$ for subsequent storage (CCS) or recycling (CCU) is determined primarily by whether CO$_2$ is emitted over large areas or from point sources. For areal emissions, it is technically not feasible to capture the CO$_2$ emission. The situation is different for CO$_2$ point sources, especially if they supply a CO$_2$ stream of high concentration. These are natural gas flaring plants, power and heat generation plants, steel and cement plants, biogas and fermentation plants, as presented in further chapters in this book. The focus of this chapter is on waste management, which today is the end point of linear value chains. Three of the four waste recovery methods presented for carbonaceous wastes result in CO$_2$ emission, namely landfilling, incineration and energy recovery. Landfills, where waste is slowly biodegraded in an uncontrolled manner with CO$_2$ emission, are no technically easily accessible sources of CO$_2$. Facilities for disposal by incineration and energy recovery, on the other hand, are point sources that can be expected to produce a CO$_2$ stream that can be captured without much effort.

The cost of CO_2 capture has recently been estimated at EUR 45–125/t CO_2 [25], which is in the range of the current price for CO_2 emission allowances of EUR 80/t CO_2 [26]. Currently, waste incineration is not subject to the European Emissions Trading Scheme, but the inclusion of these emissions from 2026 onwards was proposed by the EU Commission in June 2022 [27].

If this were to occur, the economics of CO_2 capture from waste incineration would take on a whole new look. As Table 21.5 shows, up to 270 MT of CO_2 (210 MT of fossil and mixed bio/fossil, 60 of biogenic origin) could be captured from point sources at waste incineration plants in the EU, which in 2018 was equivalent to about 9% of Europe's CO_2 emissions of 3.069 bn t CO_2 [28]. If the captured CO_2 were permanently stored (CCS), the value chains leading to the waste would not be closed, but CO_2 emissions to the atmosphere would at least be significantly reduced. The value chains would be closed if the captured CO_2 were used as a carbon source in CCU-processes. In fact, the German chemical industry is planning to use CO_2 as a raw material in the long term [29]. Today, the major emission sources of fossil energy generation come into question as CO_2 sources, but these are increasingly being replaced by emission-free energy sources. Therefore, CO_2 obtained by closing material cycles gains in importance. In principle, recycling must be given priority in waste management, but for non-recyclable waste, energy recovery combined with CCU is an option that also closes the material cycle. In this way, up to 74 MT of carbon could be tapped in Europe from waste incineration emissions, 88% thereof from plants with energy recovery (Table 21.5). Although this volume represents only 1–1.5% of the global production of biological carbon (cf. Table 21.1), the potential as a renewable carbon source for industry should not be underestimated. After all, these emissions could satisfy the current carbon demand of the European chemical industry, estimated at 63 MT (this estimate uses the European chemical industry's share of the world market (14.4%) [30] to estimate the European share of the global chemical industry's carbon consumption (450 MT)) [31]. This is, of course, contingent on the availability of sufficient energy or hydrogen to reduce CO_2, which is discussed elsewhere in this book.

Table 21.5 also shows that 23% of the emissions studied come from the incineration of biogenic wastes. The biogenic CO_2 to be captured here could also provide an acceptable carbon source for companies entering the market with bio-based products. In this way, CO_2 recycling could mitigate the land competition that otherwise exists for biomass from the production of food and raw materials for industrial purposes.

The by far largest single contributor is the incineration of waste category W101 (household and similar waste), which emits more than 40% of the CO_2 of total waste incineration. According to Table 21.5, emissions from disposal by energy recovery are calculated to be 94 Mt., which is fairly close to the reported emissions from waste to energy plants (95 MT; EU28; 2018) [32]. If these plants have to buy emission certificates, as proposed by the EU-Parliament [33], waste disposal by incineration will become significantly more expensive. CCU offers operators the option of avoiding this cost increase and closing value chains. Norway has been pricing these emissions for some time and the country's largest waste-to-energy plant in

Table 21.5 CO$_2$-point sources of waste treatment facilities by waste category (EU28, 2018) (The carbon content of waste categories was derived from the data in Table 21.2)

Waste category		Share of carbon [%]	Waste disposed by incineration (D10)			Waste disposed by recovery of energy (R1)			Share [%]
			Waste [kt]	Carbon content [kt]	CO$_2$-potential [kt]	Waste [kt]	Carbon content [kt]	CO$_2$-potential [kt]	
	Bio-based								
W11	Common sludges	50%	˙,290	645	2,361	1,830	915	3,349	1%
W72	Paper and cardboard wastes	50%	0	0	0	410	205	750	0%
W75	Wood wastes	50%	2,170	1,085	3,971	23,900	11,950	43,737	18%
W91	Animal and mixed food wastes	50%	450	225	824	1,250	625	2,288	1%
W92	Vegetal wastes	50%	110	55	201	1,950	975	3,569	1%
W93	Animal faeces, urine and manure	50%	350	425	1,556	630	315	1,153	0%
	Total bio-based		4,870	2,435	8,912	29,970	14,985	54,845	23%
	Fossil-based								
W73	Rubber wastes	85%	90	77	280	950	808	2,955	1%
W74	Plastic wastes	85%	60	51	187	2,290	1,947	7,124	3%
	Total fossil		150	128	467	3,240	2,754	10,080	4%
	Mixed bio- anc fossil-based								
W05	Health care and biological wastes	61%	810	494	1,808	570	348	1,273	1%
W76	Textile wastes	75%	10	8	27	210	158	576	0%
W101	Household and similar wastes	46%	5,170	2,378	8,704	56,000	25,760	94,282	39%
W102	Mixed and undifferentiated materials	56%	450	252	922	8,550	4,788	17,524	7%
W103	Sorting residues	51%	6,210	3,167	11,592	32,610	16,631	60,870	25%

(continued)

Table 21.5 (continued)

Waste category	Share of carbon [%]	Waste disposed by incineration (D10)			Waste disposed by recovery of energy (R1)			
		Waste [kt]	Carbon content [kt]	CO_2-potential [kt]	Waste [kt]	Carbon content [kt]	CO_2-potential [kt]	Share [%]
Total mixed fossil and bio-based		12,650	6,299	23,054	97,940	47,684	174,525	73%
Total		*17,670*	*8,861*	*32,433*	*131,150*	*65,423*	*239,449*	*100%*

Oslo is taking a step to capture its emissions. In March 2022, Oslo's mayor announced that from 2026, the plant's CO_2 emissions (400,000 t) will be captured and permanently stored geologically (CCS) [34].

References

1. World Economic forum (2021) Global CO2 emissions have been flat for a decade, new data reveals. https://www.weforum.org/agenda/2021/11/global-co2-emissions-fossil-fuels-new-data-reveals/. Accessed 3 Aug. 2022
2. Kähler F, Carus M, Porc O, vom Berg C (2021) Turning off the tap for fossil carbon: future prospects for a global chemical and derived material sector based on renewable carbon. Ind Biotechnol 17(5) https://www.liebertpub.com/doi/abs/10.1089/ind.2021.29261.fka?journalCode=ind. Accessed 3 Aug. 2022
3. Pyshyev S, Gunka V, Grytsenko Y, Bratychak M (2016) Polymer modified bitumen: review. Chem Chem Technol 10(4s):631–636. https://www.researchgate.net/publication/313332631_Polymer_Modified_Bitumen_Review/figures?lo=1. Accessed 3 Aug 2022
4. Global CCS Institute (2021) Global status of CCS 2020. https://www.globalccsinstitute.com/wp-content/uploads/2021/03/Global-Status-of-CCS-Report-English.pdf /. Accessed 3 Aug. 2022
5. IEA (2022) Global coal demand is set to return to its all-time high in 2022. https://www.iea.org/news/global-coal-demand-is-set-to-return-to-its-all-time-high-in-2022. Accessed 3 Aug. 2022
6. chemie.de (2022) Kohle. https://www.chemie.de/lexikon/Kohle.html#Zusammensetzung_einiger_Kohlensorten.07UNIQ4ef5016956377340-nowiki-00000003-QINU1.07UNIQ4ef501 6956377340-nowiki-00000004-QINU. Accessed 3 Aug 2022
7. Focus (2022) Woraus besteht Erdöl. https://praxistipps.focus.de/woraus-besteht-erdoel-das-stoffgemisch-und-seine-zusammensetzung_108759. Accessed 3 Aug 2022
8. CROFT (2022) Natural gas composition. https://www.croftsystems.net/oil-gas-blog/natural-gas-composition/. Accessed 3 Aug 2022
9. FAO (undated) Forests and climate change. https://www.fao.org/3/ac836e/AC836E03.htm. Accessed 3 Aug 2022
10. FAO (2020) FAO Statistical Yearbook 2021 - World Food and Agriculture. https://reliefweb.int/report/world/fao-statistical-yearbook-2021-world-food-and-agriculture. Accessed 3 Aug 2022
11. FAO (2019) Global forest products. https://www.fao.org/3/ca7415en/ca7415en.pdf. Accessed 3 Aug 2022
12. Heissenhuber A et al. (undated) Globale Landflächen und Biomasse nachhaltig und resssourcenschonend nutzen. Umweltbundesamt Dessau. https://www.umweltbundesamt.de/sites/default/files/medien/479/publikationen/globale_landflaechen_biomasse_bf_klein.pdf. Accessed 3 Aug. 2022
13. Manfred K (2020) Weg vom Öl. Springer, Berlin, Heidelberg
14. Lokesh K, Ladu L, Summerton L (2018) Bridging the gaps for a "circular" bioeconomy: selection criteria, bio-based value chain and stakeholder mapping. Sustainability 10:1695. https://doi.org/10.3390/su10061695
15. EC (2018) Directive 2008/98/EC of the European Parliament and of the Council of 19 November 2008 on waste and repealing certain Directives. https://eur-lex.europa.eu/legal-content/EN/ALL/?uri=CELEX%3A32008L0098. Accessed 3 Aug. 2022
16. Fuchsbriefe (2020) Gas-Abfackeln ist mit Abstand am billigsten. https://www.fuchsbriefe.de/gas-abfackeln-ist-mit-abstand-am-billigsten. Accessed 3 Aug 2022
17. IEA (2009) Energy sector methane - Recovery and Use. https://www.osti.gov/etdeweb/servlets/purl/21233584. Accessed 3 Aug 2022

18. Ritchie H, Roser M, Rosado P (2020) CO_2 and greenhouse gas emissions. OurWorldInData. https://ourworldindata.org/co2-and-other-greenhouse-gas-emissions. Accessed 3 Aug 2022

19. EC (undated) ANNEX - Correspondence of waste codes to the four material flows MF1 to MF4. https://ec.europa.eu/eurostat/documents/8105938/8465062/env_ac_cur_esms_MFA_correspon dence.pdf. Accessed 3 Aug 2022

20. EC (2010) Guidance on classification of waste according to EWC-Stat categories. https://ec. europa.eu/eurostat/documents/342366/351806/Guidance-on-EWCStat-categories-2010.pdf/0 e7cd3fc-c05c-47a7-818f-1c2421e55604. Accessed 3 Aug. 2022

21. Eurostat (2010) Manual on waste statistics. https://ec.europa.eu/eurostat/documents/3859598/ 5915865/KS-RA-10-011-EN.PDF/39cda22f-3449-4cf6-98a6-280193bf770c.   Accessed 3 Aug. 2022

22. Assunção LRC, Mendes PAS, Matos S, Borschiver S (2021) Technology roadmap of renewable natural gas: identifying trends for research and development to improve biogas upgrading technology management. Appl Energy 292:116849. https://doi.org/10.1016/j.apenergy.2021. 116849

23. Eurostat (2022) Treatment of waste by waste category, hazardousness and waste management operations. https://ec.europa.eu/eurostat/databrowser/view/ENV_WASTRT__custom_1 886870/default/table?lang=en. Accessed 3 Aug 2022

24. Eurostat (2010) Manual on waste statistics. https://ec.europa.eu/eurostat/documents/3859598/ 5915865/KS-RA-10-011-EN.PDF/39cda22f-3449-4cf6-98a6-280193bf770c.   Accessed 3 Aug 2022

25. Beiron J, Normann F, Johnsson F (2022) A techno-economic assessment of CO2 capture in biomass and waste-fired combined heat and power plants – a Swedish case study. Int J Greenh Gas Control 118:103684. https://www.sciencedirect.com/science/article/pii/S1750583 622001025. Accessed 3 Aug. 2022

26. Ember (2022) EU Carbon Tracker. https://ember-climate.org/data/data-tools/carbon-price-viewer/. Accessed 3 Aug 2022

27. EU-Parliament (2022) Revision of the EU Emissions Trading System. https://www.europarl. europa.eu/doceo/document/TA-9-2022-0246_EN.pdf. Accessed 3 Aug 2022

28. Statista (2022) Carbon dioxide emissions in the European Union from 1965 to 2021. https:// www.statista.com/statistics/450017/co2-emissions-europe-eurasia/. Accessed 3 Aug 2022

29. VCI (2019) Auf dem Weg zu einer treibhausgasneutralen chemischen Industrie in Deutschland. https://www.vci.de/vci/downloads-vci/publikation/2019-10-09-studie-roadmap-chemie-2050-treibhausgasneutralitaet-kurzfassung.pdf. Accessed 3 Aug 2022

30. Statista (2022) Umsatzanteil der EU-Chemieindustrie im weltweiten Chemiemarkt bis 2020. https://de.statista.com/statistik/daten/studie/411959/umfrage/umsatzanteil-der-eu-chemieindustrie-am-weltweiten-chemiemark/. Accessed 3 Aug 2022

31. CHEManager (2022) Erneuerbarer Kohlenstoff - Schlüssel zur Zukunft. https://www. chemanager-online.com/news/erneuerbarer-kohlenstoff-schluessel-zur-zukunft.   Accessed 3 Aug 2022

32. Zero Waste Europe, Deutsche Umwelthilfe (2021) The benefits of including municipal waste incinerators in the Emissions Trading System. https://www.duh.de/fileadmin/user_upload/ download/Pressemitteilungen/Kreislaufwirtschaft/210415_zwe_duh_policy_briefing_benefits_ MSW_in_ETS.pdf. Accessed 3 Aug 2022

33. EU-Parliament (2022) Revision of the EU Emissions Trading System. https://www.europarl. europa.eu/doceo/document/A-9-2022-0162-AM-416-425_EN.pdf. Accessed 3 Aug 2022

34. Bellona (2022) Oslo leading by example: world's first CO2 capture and storage on waste incinerator to become reality in 2026. https://bellona.org/news/ccs/2022-03-oslo-leading-by-example-worlds-first-co2-capture-and-storage-on-waste-incinerator-to-become-reality-in-202 6. Accessed 3 Aug 2022

Utilization of C1 Gases: Impact on Sustainability

22

Michael Carus

Abstract

CO/CO_2 use shows distinct environmental advantages and characteristics. Compared with other carbon sources, CO/CO_2-based fuels and chemicals generally have the lowest GHG emissions. Compared to biomass utilization, the land use efficiency of solar CCU is a factor of 50 to 150 higher than biomass. A distinctive feature of CO/CO_2 use is that this carbon source is unlimited in time and volume. Also, the potential of renewable energy is almost unlimited in terms of human demand. With CCU, access to raw materials is "democratized." In the future, anyone anywhere in the world will, in principle, have the opportunity to harvest carbon with renewable energy and CCU and use it to produce fuels or chemicals and plastics. In these products, solar energy can be stored for long periods of time without loss. CO/CO_2 use with renewables is the most sustainable pathway to fuels, chemicals, and plastics from an environmental perspective and can explicitly make a significant contribution to a sustainable economy. Nine of the 17 United Nations Sustainable Development Goals are directly addressed by CO/CO_2 use with renewable energy. CO/CO_2 use is consistent with efficiency and consistency strategies, but without requiring a strict sufficiency strategy.

Keywords

CO/CO2-based fuels · CO/CO2.based chemicals · GHG emissions · Renewable energy · United Nations Sustainable Development Goals (UN-SDG) · Carbon Capture and Utilization (CCU)

M. Carus (✉)
nova-Institut GmbH, Hürth, Germany
e-mail: michael.carus@nova-institut.de

22.1 What Does Sustainability Mean?

The term sustainability was first introduced to forestry in 1713 by Hans Carl von Carlowitz [1]. The following paragraph is considered the birth of the concept of sustainability:

> For this reason, the greatest art/science/diligence and establishment of this country will be based on how to achieve such a conservation and cultivation of wood/that there is a continuous, constant and sustainable use/because it is an indispensable thing/without which the country may not remain in its essence.

The term gained far-reaching significance with the work of the Brundtland Commission (led by the then of the then Norwegian Prime Minister Gro Harlem Brundtland) in 1983 and the Brundtland Report in 1987 [2]:

> By 'sustainable development' we mean development that meets the needs of the present generation without compromising the ability of future generations to meet their own needs and choose their own lifestyles.

In 2002, the African delegation at the RIO+10 Summit in Johannesburg reduced sustainability to a simple denominator: 'Enough—for all—forever'. And finally, in 2014, the German government used the term sustainability with temporal and spatial dimensions in its report "Sustainable Development in Germany": "Not to live today and here at the expense of people in other regions of the world and at the expense of future generations."

While there is widespread agreement on the general understanding, there are different views on the concrete application of the concept of sustainability. Frequently, the three-pillar model is used, which combines the dimensions of ecology, social affairs, and economy under the umbrella of "sustainability," as Fig. 22.1 shows.

With Rio-20, the United Nations defined 17 "sustainable development goals," as Fig. 22.2 illustrates.

Also of interest is the approach of the Wuppertal Institute from 1996, which identifies the three most important strategies for a "Sustainable Germany—A Contribution to Global Sustainable Development" [3]:

Efficiency: increasing resource productivity,

Consistency: avoiding waste through recyclable resources (today this would be called "circular economy"),

Sufficiency: frugality ("living better instead of having more"—problem: acceptance).

This should suffice as a brief insight into the complex topic of sustainability. In the following, the ecological pillar of sustainability, the strategies of efficiency and consistency as well as the corresponding sustainability goals of the United Nations will now be analyzed in more detail with regard to the use of CO and CO_2.

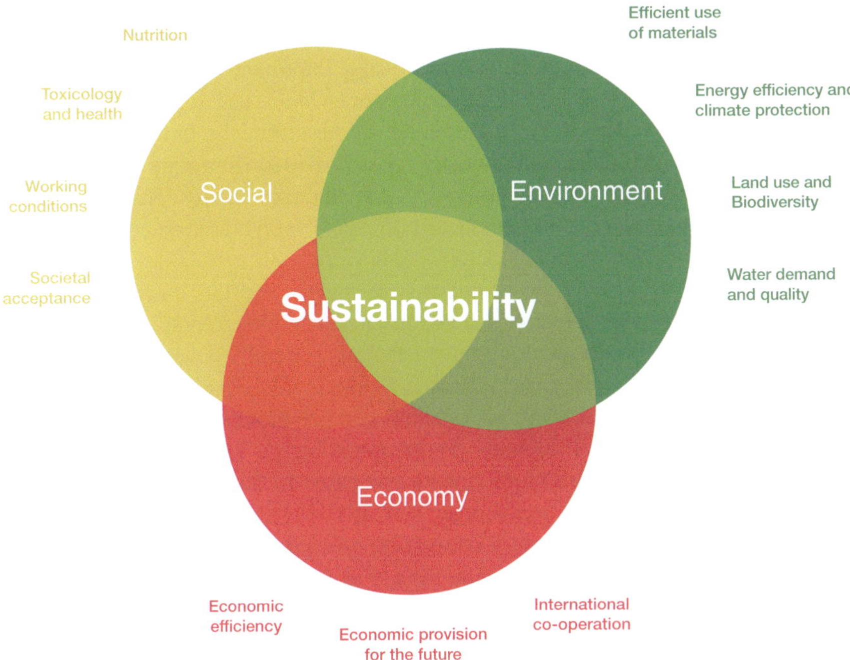

Fig. 22.1 Three pillars of sustainability. © nova-Institut

Fig. 22.2 Sustainable development goals of the United Nations

22.2 Ecological Aspects of the Conversion of C1 Gases to Chemicals and Energy Sources/Fuels

In Table 22.1, a comparative assessment of fuels and chemicals from fossil fuels, from biomass, and from CO/CO_2 is made using important ecological categories. This allows an initial assessment of the ecology of CO/CO_2 use. It is always assumed that the energy for reducing C1 gases either comes from the emission gas itself (synthesis gas, e.g. from the steel industry) or is provided from renewable sources, such as solar and wind energy, hydropower or geothermal energy. The use of fossil energy sources to reduce C1 gases leads to very unfavorable energy and CO_2 balances and is therefore not considered further here.

Table 22.1 clearly shows the ecological advantages and the special features of CO/CO_2 use, which will be presented again separately here. Compared with other carbon sources, CO/CO_2-based fuels and chemicals generally have the lowest GHG emissions. This is partly because they do not release additional fossil carbon, and partly because renewables harvest solar radiation much more efficiently than natural photosynthesis, on which all biomass production is based. The latter is due to high solar efficiency and low land requirements compared to biomass production. This point is presented in detail by Searchinger et al. 2017 [4] in a recent publication. The authors calculate that the bioenergy conversion efficiency is only 0.1 to 0.2%, i.e., only 0.1 to 0.2% of the irradiated solar energy is found in the final product (e.g., ethanol). In contrast, the net solar conversion efficiency is between 11 and 44%, which is a factor of 122 to 295 higher than for biomass utilization (compare Fig. 22.3). If solar energy is converted to liquid fuels using CCU technologies, efficiencies of over 50% can already be achieved, i.e., the conversion efficiency from solar CCU to liquid fuel is about 50 to 150 times higher than via biomass.

> The increasing practicality of solar energy tilts the land use equation further against bioenergy whenever it can be used. Even if 100 ha of good land were to become theoretically available for climate mitigation, they could generally provide at least as much energy and at least 100 times more carbon mitigation if 1 ha were used for solar and 99 to restore forests [4].

A special feature of CO/CO_2 utilization is the unlimited nature of this carbon source in terms of time and volume. On the one hand, there is enough CO_2 in the atmosphere even in the long term, and on the other hand, the carbon is recycled: It is taken from the atmosphere or industrial sources and returned to the atmosphere when (or after) it is used. This replenishes the atmosphere as a CO_2 reservoir. Also, the potential of renewable energies is almost unlimited in relation to the needs of mankind.

To meet all of the global chemical industry's growing demand for embedded carbon in 2050, only 0.7% of the area of all subtropical deserts or 1.3% of the Sahara would need to provide enough solar energy to utilise CO_2 for the chemical industry. [11]

Table 22.1 Evaluation of fuels and chemicals from different carbon sources according to environmental sustainability criteria

Sustainability criterion	Carbon source of fuels and chemicals		
	fossil-based (crude oil, natural gas, coal)	bio-based (all types of biomass)	CO/CO_2-based in connection with renewable energies (sun, wind, water, geothermal energy)
Greenhouse gas emissions (GHG)	High mainly due to the release of "fossil" carbon into the atmosphere.	Low to medium biogenic carbon is recycled; emissions due to fertilizer and pesticides	Low carbon is recycled using renewable energy
Land demand	Low	High	Low photovoltaics (PV) use solar radiation per area (at least!) 20 to 30 times more efficiently than the most efficient plants
Water demand	Low except in case of breakdowns, accidents	Low to high (depending on region)	Low
Availability	Time limited; only locally available	Limited by land requirements and problems with biodiversity; basically available globally, but in varying quantities locally	Unlimited in time and volume To meet all fuel and chemical needs, about 10% of desert land would suffice with PV; available globally anywhere on earth
Contribution to circular value creation	No	Yes	Yes
Consequences for food supply, land, water and biodiversity	Low except in case of breakdowns, accidents	Medium to high (depending on local conditions)	Low
Solar efficiency (conversion of solar radiation into end product)	–	<0.5%	>10%
Solar energy storage	Long-term storage of solar energy from millions of years	Agricultural: Seasonal Forest: Over decades	Unlimited long term storage

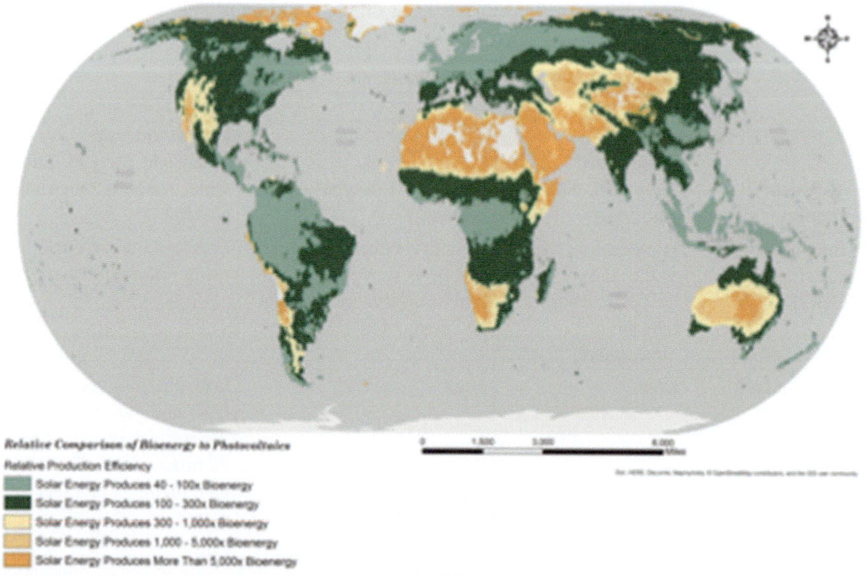

Fig. 22.3 Comparison of land use efficiency between photovoltaics (PV) today versus cellulosic bioethanol tomorrow. On 73% of the world's land, the usable yield of PV is more than a factor of 100 higher than the yield of bioethanol. On 27% of land areas, the factor is less than 100, and on average the ratio is 85: 1

There are additional important aspects: Access to raw materials is being "democratized." In the future, anyone anywhere in the world will in principle have the opportunity to harvest carbon with renewables and CCU and use it to produce fuels or chemicals and plastics. In these products, solar energy can be stored for long periods of time without loss.

The analysis so far already shows that CO/CO_2 use with renewables is the most sustainable pathway to fuels, chemicals, and plastics from an environmental perspective and can explicitly make a significant contribution to a sustainable economy. Nine of the 17 United Nations Sustainable Development Goals are directly addressed by CO/CO_2 use with renewable energy. CO/CO_2 use is consistent with efficiency and consistency strategies, but without requiring a strict sufficiency strategy. Impending raw material bottlenecks, as we know them from fossil and bio-based systems, are largely overcome in CO/CO_2 use with renewable energies.

22.3 Greenhouse Gas Emissions (GHG Emissions)

But what do the greenhouse gas balances of CO/CO_2 use with renewable energies look like in concrete terms? Are there already reliable calculations on this? Greenhouse gas balances are an essential component of comprehensive life cycle assessments, in which other impact categories, such as non-fossil energy demand, eutrophication and acidification potential, photosmog, etc., are examined in addition to GHG balances. In recent years, a number of life cycle assessments have already been carried out on CO_2-based methanol, kerosene, other fuels or even the plastic building blocks polyols and polyurethanes. The results are promising. Depending on the product and process route, the GHG savings ranged from 20 to over 90% compared to the respective fossil counterpart. Unfortunately, the results cannot be directly compared due to methodological differences between the individual LCAs.

When working with allocations of loads, one can allocate by energy (useful for fuels), mass (useful for products), or even economics—each with different results. Allocation along the value chain also leaves room for maneuver. For example, does the coal-fired power plant lose its fossil emissions (which is real if the power plant's CO_2 is passed on to a CCU process) and does the fuel produced from it get allocated its CO_2 emissions? How should it be possible to market such a fuel? If the power plant retains the CO_2, the fuel would be nearly CO_2-free, but what would the power plant operator gain from providing the CO_2? Some LCAs work with system expansion to avoid such allocation problems. Without going into this further at this point, one suspects the challenges that these new technologies bring.

All of the above methodological approaches are scientifically legitimate, but they make it difficult to compare results from different LCAs. The percentage savings shown in Table 22.2 can therefore not be directly compared.

22.4 CO and CO_2 Utilization and the Circular Economy

CO and CO_2 utilization is an integral part of an expanded circular economy, or in the words of the European Commission "sustainable carbon cycles". While the energy and transport sectors are becoming increasingly independent of carbon ("decarbonization") and are switching to electricity (or also direct solar heat use) for this purpose, a few areas of mobility will continue to be dependent on C-containing fuels, especially aviation with kerosene and also the entire organic chemical and plastics industry sector, which will remain permanently dependent on a renewable carbon source. This is not about decarbonisation, but about defossilisation: to cover the entire carbon demand by renewable carbon sources replacing all fossil resources.

The circular use of carbon, e.g. in the form of CO and CO_2, therefore plays a central role in sustainable raw material production. To this end, the carbon is to be extracted from the atmosphere or industrial waste gases, converted and used, recycled at the end of the use phase and finally released back into the atmosphere at the end of the entire life cycle. There, it can then be used again as a raw material and thus recycled.

A fundamental paper of the "Renewable Carbon Initiative (RCI) says: "Renewable carbon is the approach that will enable the chemical and derived-materials industry to become an important part of a sustainable future – it can originate from biomass, CO_2 utilisation and from recycling. As a guiding principle, renewable

Table 22.2 GHG reductions in percent for different products based on CO_2 and renewables based on different methods and sources

Product	Reduction of GHG emissions in % compared to fossil counterpart	Methodology of load sharing [source]
Methanol from industry, CO_2 and geothermal energy (Iceland)	90 (this value has never been reached by any biofuel calculated using the same method).	Allocation by energy, methodology according to EU Renewable Energy Directive (RED), CO_2 load-free, since emission/waste [5].
Kerosin via Fischer-Tropsch-Synthese oder Methanolroute	Approx. 90	Allocation by energy, methodology according to EU Renewable Energy Directive (RED), CO_2 load-free, since emission/waste [6].
Power-to-liquid with CO_2 from air capture versus fossil diesel	40 to 95 (depending on power and heat source)	Allocation according to energy. Methodology according to EU Renewable Energy Directive (RED). CO_2 load-free, as emission/waste, but: All loads for capture/provision allocated to CO_2 [7]
Power-to-liquid-biogas-CO_2 und Windenergie gegenüber fossilem Benzin, nach 200.000 km	85	CCU process gets CO_2 load-free, emissions stay with the power plant [8].
CO_2-based polyols with 20 or 30% CO_2 content compared to fossil polyol	20, resp. 30	System expansion [9]
Polypropylene carbonate (PPC) versus fossil-based polyethylene (PE) and polyethylene terephthalate (PET).	50 (against PE) and 60 (against PET)	Speaker unknown, SK Innovation [10]
Demand for embedded carbon of the global chemical industry, crude oil vs. CO_2-based methanol	67 to 100% (fully decarbonised renewable energy 2050)	System expansion [11]

carbon enables the industry to think out of the box of established boundaries and stop the influx of additional fossil carbon from the ground." [12].

Figure 22.4 shows a picture of a comprehensive circular economy in which various raw material, process, product, and recycling paths are fully integrated, as well as CO_2 as a raw material in addition to metals, minerals, fossil fuels, and biomass.

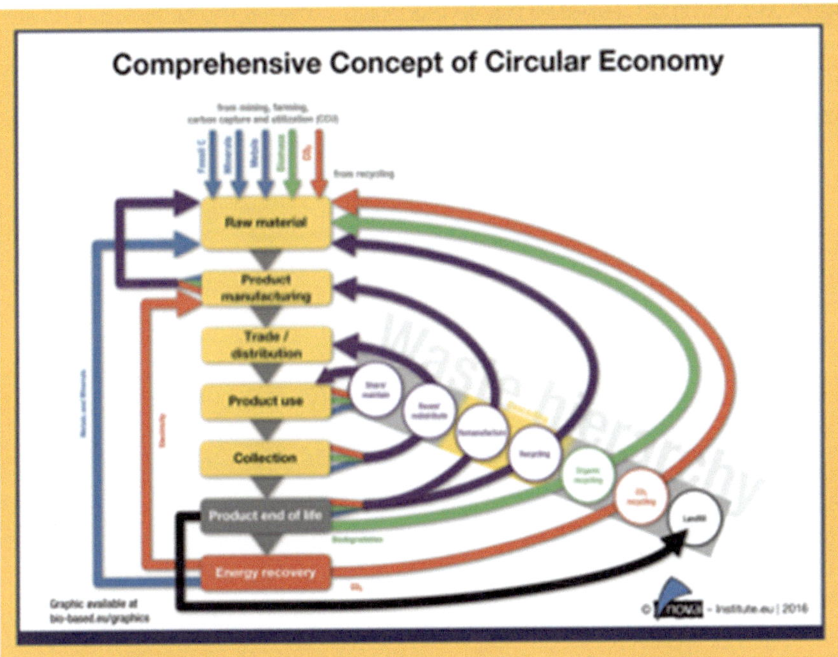

Fig. 22.4 Concept of a circular economy. © nova-Institut

References

1. von Carlewitz HC (1713) Sylvicultura oeconomica. Braun, Leipzig, p 105
2. World Commission on Environment and Development (1987) Our common future. Oxford University Press, Oxford, p 1
3. Loske R, Bleischwit R (1996) Zukunftsfähiges Deutschland – Ein Beitrag zu einer global nachhaltigen Entwicklung. Birkhäuser Verlag, Basel. ISBN 978-3-0348-5088-9
4. Searchinger TD, Beringer T, Strong A (2017) Does the world have low-carbon bioenergy potential from the dedicated use of land? Energy Policy 110:434–446
5. Schmitz N (2013) ISCC Certification schemes for CO_2 based fuels and chemicals. ISCC. Präsentation auf der "Conference of CO_2: Carbon Dioxide as Feedstock for Chemistry and Polymers 2013"
6. DBFZ (2016) Erneuerbare Kraftstoffe für Mobilität und Industrie. http://www.fvee.de/fileadmin/publikationen/Themenhefte/th2016/th2016_07_05.pdf. Accessed 6 Aug 2018
7. Universität Stuttgart (2015) Verbundprojekt sunfire – Herstellung von Kraftstoffen aus CO_2 und H_2O unter Nutzung regenerativer Energie, für das Bundesministerium für Bildung und Forschung (BMBF). Arbeitspaket D – Ökobilanz. http://ibp-gabi.de/files/sunfire_zusammenfassung.pdf. Accessed 29 Mar 2018
8. Audi (2015) Fuel of the Future: Research facility in Dresden produces first batch of Audi e-diesel. https://www.audi-mediacenter.com/en/press-releases/fuel-of-the-future-research-facility-in-dresden-produces-first-batch-of-audi-e-diesel-352/download. Accessed 29 Mar 2018
9. Müller JL (2016) LCA of CO_2-based polyols and polyurethanes. RWTH Aachen. Conference on Carbon Dioxide as Feedstock for Fuels, Chemistry and Polymers 2016

10. Ok M-A (2012) CO_2 Embedded Polyalkylene Carbonate Greenpol™. SK Innovation. Conference on CO_2: Carbon Dioxide as Feedstock for Chemistry and Polymers 2012
11. Kähler F (2022) CO2 reduction potential of the chemical industry through CCU. www.renewable-carbon-initiative.com. Accessed 29 Mar 2022
12. Vom Berg C (2022) Renewable carbon as a guiding principle for sustainable carbon cycles. www.renewable-carbon-initiative.com. Accessed 29 Mar 2022

Regional Development

Dennis Herzberg and Manfred Kircher

Abstract

North Rhine-Westphalia is Germany's leading chemical, energy and steel region and has an extensive logistics infrastructure for energy and carbon sources as well as basic chemical products. Energy and steel industries in particular emit significant volumes of CO_2. Recycling of this CO_2 by carbon-absorbing chemical industries could create a CO_2 sink and turn CO_2 emission into a locally available feedstock. C1-Chemistry can therefore make a significant contribution to regional development, especially in industrialized regions with high CO_2 emissions.

Keywords

North Rhine-Westphalia (NRW) · Regional development · Industrialized region · CO_2 emission

23.1 Current Situation

Industry, transport, commerce, trade, services and households consume raw materials, fuels, heat and energy. All sectors, including manufacturing, municipal and private waste management, communications, energy, transport, forestry and agriculture, now use carbon-based raw materials or products directly or indirectly for this purpose. For example, the smelting heat needed for steel production mostly comes from coal; agriculture needs carbon-containing agrochemicals and fuel; and

D. Herzberg (✉)
CLIB - Cluster Industrial Biotechnology, Düsseldorf, Germany
e-mail: herzberg@clib-cluster.de

M. Kircher
KADIB, Frankfurt am Main, Hessen, Germany
e-mail: kircher@kadib.de

© The Author(s), under exclusive license to Springer Nature Switzerland AG 2023
M. Kircher, T. Schwarz (eds.), *CO2 and CO as Feedstock*, Circular Economy and
Sustainability, https://doi.org/10.1007/978-3-031-27811-2_23

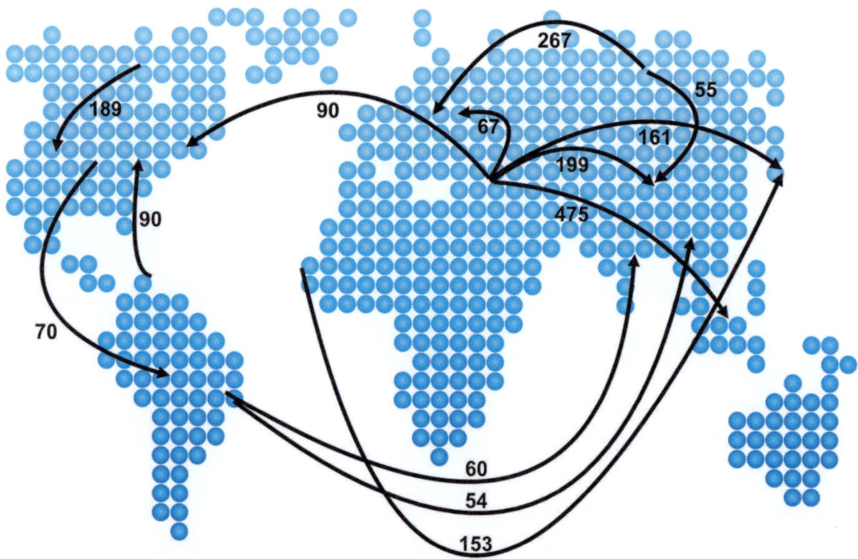

Fig. 23.1 Routes and volumes of global trade in crude oil (2015; in MT; data from [1])

even seemingly distant sectors such as communications are linked to the chemical industry, e.g., via polymers for cable insulation, and thus depend on carbon. The same is true for the real estate industry, where carbon-containing chemical products are used in construction and heating is derived from carbon-based energy sources.

Today, the carbon sources for the generation of heat, energy, fuel and organic chemicals in Germany are predominantly coal, mineral oil and natural gas and thus of fossil origin. Natural gas and mineral oil are almost entirely imported, and only domestic coal production was relevant until 2018 (176 MT; 2016). Thus, the German carbon economy is an integral part of global trade, especially of mineral oil (Fig. 23.1) and natural gas (Fig. 23.2). Both illustrate Germany's and Europe's dependence on distant carbon and energy sources.

Oil and gas can be transported cost-effectively over very long distances because they offer high carbon and energy density, are largely homogeneous in composition and are biologically/chemically stable to store. A large-volume logistics network over sea with tankers and over land via pipelines ensures worldwide supply. Based on this global infrastructure for the supply of carbon and energy carriers, industrial regions with heavy and chemical industries as well as processing plants have established themselves. One example is the state of North Rhine-Westphalia (NRW) in Germany, where, starting in the nineteenth century, the steel industry initially created jobs and prosperity on the basis of domestic coal, and later, in the twentieth century, the chemical industry, supplied with oil via the Rhine and pipelines, gained in importance. Today, NRW is Germany's leading chemical region, which together with the neighboring southern Netherlands and Flanders accounts for 30% of European chemical production. Large capacity refineries have

Fig. 23.2 Routes and volumes of global trade in natural gas (2015; in million cubic meters; data from [1])

been established at locations with direct access to global oil logistics (seaports, pipelines), producing fuel (gasoline, diesel, kerosene) and basic chemical products (ethylene, propylene, etc.). The most important chemical sites, in turn, are connected to basic chemicals by pipelines for crude oil and products (Fig. 23.3). In Feburary 2024, Russia's war of aggression on Ukraine has fundamentally changed the situation described above regarding the supply of Russian oil and natural gas to Europe, but nevertheless the basic message of low-cost transportation of fossil carbon and energy sources over long distances remains true.

The multi stage processing of basic chemicals, on the other hand, is becoming smaller in volume as the value chain progresses through to the consumer product and is therefore not tied to large sites. This more medium-sized industry creates jobs in the countryside and is therefore, together with large-scale industry, an indispensable part of Germany's prosperity. Together, both large-scale industry and SMEs form regional clusters and shape entire regions. NRW, with its focus on the chemical and processing industry, is an impressive example of a chemical region.

23.2 The Challenge

The undeniable climate change and the Paris climate protection agreement [3] agreed in 2015 and entered into force in 2016 in this context call for an extensive shift away from fossil raw materials by 2050 (Chap. 1). This means that the energy and chemical industries, which have so far been largely based on fossil raw materials, and the corresponding regions will have to undergo an enormous transition. While

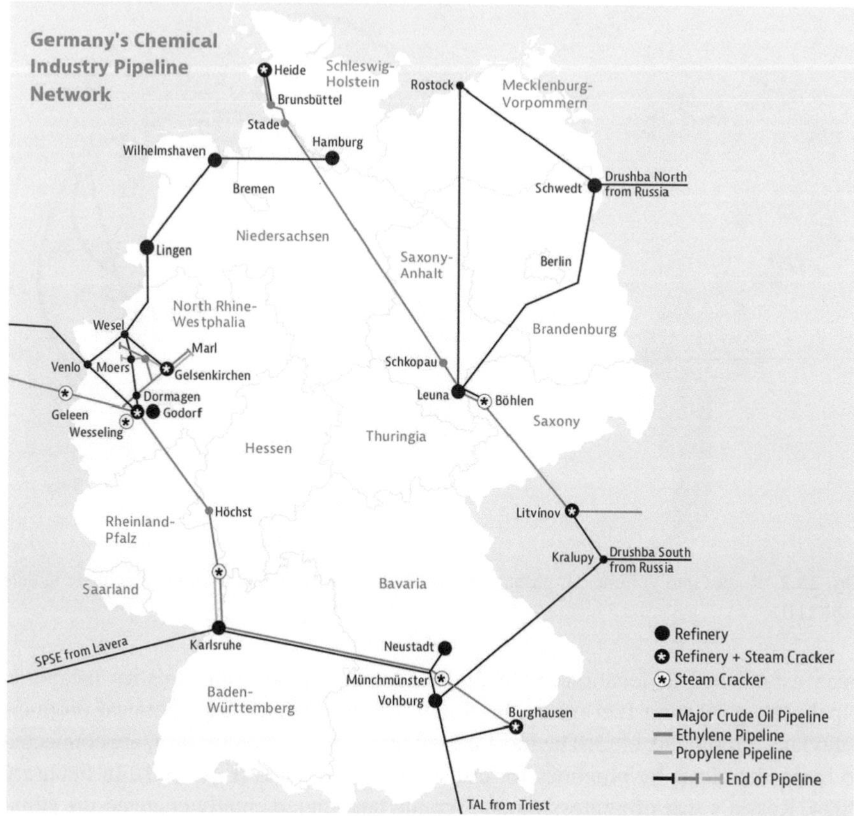

Fig. 23.3 Chemical pipelines in Germany [2]. © Germany Trade and Invest

carbon-free alternatives are available with alternative energies for energy generation, (organic) chemistry remains dependent on carbon. This industry needs to find sustainable alternatives to fossil carbon sources. Plant-based ingredients and biomass are one option. Starch, sugar, vegetable oils, animal fats and cellulose already account for 13% of the chemical raw materials processed in Germany [4]. They are mainly used in processes for which they bring technical advantages. They cannot (yet) compete with fossil carbon sources in terms of cost, because production, logistics and processing entail cost disadvantages. Compared to mineral oil, bio-based raw materials have a significantly lower carbon and energy density, which is why transport is limited to small-volume units by ship, rail and truck. Large-scale supply to major industrial centers from distant biomass regions will therefore be problematic, and it is foreseeable that at least the initial processing stages in the biomass regions will be decentralized and in relatively small facilities. The production of basic chemicals, which is still located in Europe today, could therefore migrate to global biomass regions. For example, ethylene based on bioethanol is technically easily available. In Brazil, ethylene is already produced from sugar cane, and the question is whether and when downstream industries will follow. This means that Europe's chemical regions may face a fundamental change if other sustainable raw materials are not developed in addition to biomass. In fact, C1 gases play a key role here.

23.3 The Solution

Today's production and consumption processes are largely linear in nature. Raw materials are processed into products and these are disposed of after use or recycled for energy. Although the material flows of the various processing chains are very efficiently interlinked according to the principle of the production network, the carbon emissions generated in production and, at the latest, in the disposal of consumer products (in Germany, through energy recovery) ultimately release all the carbon fed into the chemical industry into the atmosphere in gaseous form. This not only damages the climate, it is also a major loss. Especially for established industrial regions with high emissions, the use of this carbon offers enormous potential in the medium and long term (Chap. 12–17). This applies both to the transition phase to the bioeconomy and thereafter, because the raw material change that has already begun is to be designed as a decades-long process up to 2050, in which fossil carbon carriers are processed in parallel with biobased ones, albeit with a decreasing trend, for many years. The material utilization of CO and CO_2 therefore offers an early opportunity to reduce emissions and at the same time the consumption of fossil raw materials. The decisive potential for the economic region lies in the fact that new cooperation opportunities arise for the entire carbon processing industry, from the energy to the steel to the chemical sector. Carbon emitters can transform themselves into carbon suppliers, and CO_2 emission becomes a regionally available raw material. That this is not a distant vision is demonstrated by projects such as "Dream Production"(Covestro in Dormagen [5], Chap. 26), "CO-based ethanol"(ArcelorMittal in Ghent [6]), "Carbon2Chem"(thyssenkrupp Steel Europe in Duisburg [7]) and "CO_2 to methanol"(BASF in Ludwigshafen [8]).

The decisive keyword in these processes is raw material efficiency, because the primary carbon source is used far more efficiently with emissions recycling than before. Such processes thus make a lasting contribution to sustainability. While today emission gases of fossil origin are recycled, future bio-based processes will also emit CO_2. For example, biotechnical fermentation plants for the production of bioethanol, bio-based basic chemicals such as lactic and succinic acid, or pharmaceuticals based on sugar emit considerable amounts of CO_2. Biogas plants, which are widely established in Germany, also produce a mixture of methane (50–60%) and CO_2 (40–50%). CO_2 emission is therefore not only a problem of fossil raw materials, but also an issue for the industrial raw material biomass. Funds invested in emissions recycling today will therefore pay off in the long term after the raw material transformation has taken place. In addition, the aspect of resource efficiency is also relevant with regard to biobased raw materials, because they are limited. Primarily, they have to secure the food supply for a growing world

Table 23.1 Reuse rates of materials containing carbon

	So far	From 2019	From 2022
Material	75%	80%	90%
Cardboard, paper, carton	70%	85%	90%
Beverage carton packaging	60%	75%	80%
Other composite packaging	60%	55%	70%
Plastics (material recycling)	36%	58.5%	63%

population, and there is only limited land available worldwide. In addition, the greenhouse gas emissions caused by agriculture (around 30% of total global emissions) and the threatened biodiversity demand that land be conserved. Consequently, agricultural raw materials must be used as efficiently as possible. This also applies to forestry and marine biomass, which also have industrial potential.

Whether fossil-based or bio-based, the use of previously neglected carbon-containing gases offers industrial sites in particular, which are usually strong sources of emissions, enormous leverage for increasing resource efficiency. However, it is to be expected that the carbon volume still coming from imported fossil raw materials today will increasingly decrease as the raw material change progresses and can only be replaced by bio-based carbon at great expense. For this reason, the EU [9] and the German government [10] are calling for a further increase in raw material efficiency. In order to convert waste into valuable materials, the slogan "reduce, reuse, recycle" calls for the economical use of products (reduce), reuse (reuse) and recycling (recycle). Consequently, the German Packaging Act prescribes the continuous increase of the partly mechanical reuse rate of certain materials, including those containing carbon (Table 23.1) [11].

"Reduce, reuse, recycle" unquestionably increases the raw material efficiency of the process chains concerned, but reuse and mechanical recycling are also finite. If reuse is no longer possible due to contamination and wear, the only remaining option is energy recovery. Combustion occurs and thus, depending on the process, CO or CO_2 is generated, which, as mentioned above, can be recycled. Non-recyclable solid industrial and municipal wastes can also serve as a carbon source (Chapters 18, 19 and 21). They are gasified to synthesis gas, as Concord Blue (Herne; Germany) [12], for example, operates worldwide (Chap. 5). The same process that processes CO from metallurgical gases can be used to recycle this CO-containing gas (see ArcelorMittal; thyssenkrupp above). This brings other solid carbon sources within reach in addition to gaseous ones. It is obvious that waste can be recycled quite independently of the original carbon source (fossil- or biobased). This opens up further carbon potential for densely populated (industrial) regions in particular, both for the transition phase of the raw material change and afterwards.

Of course, it should not go unmentioned that the utilization of CO and CO_2 is dependent on the availability of hydrogen (Chap. 3). Limited volumes are available at individual chemical sites as a by-product of chlor-alkali electrolysis, but for widespread application, hydrogen must be produced at high energy cost by water electrolysis. What initially appears to be an obstacle, however, turns out to be an opportunity on closer inspection. The energy transition associated with the raw material change generates volatile electricity (solar, wind energy), some of whose peak supply cannot be consumed. These peaks can be used in water electrolysis and produce hydrogen. Referred to as power2gas, the first such plants are already in operation [13]. This hydrogen can serve directly as an energy carrier or as a reducing agent for CO_2 (to methane). This "methanation" has also already reached pilot scale [14]. It shows the potential that lies in linking the energy sector with CO_2-producing industries.

If so far the focus has been on technical possibilities of recycling, now the possibilities of politics and legislation to accelerate the transformation to circular value creation come into play. The example of the Packaging Act has already been mentioned above. Emissions allowance trading is also one of the steps forcing the recycling of gaseous carbon and thus improving raw material efficiency, because recycling carbon reduces emissions and creates the carbon sinks called for by the Paris Climate Agreement (Chap. 1).

As noted above, industrial regions of high emission and population density, in particular, have the potential for gas streams with a total volume and concentration of CO and CO_2 high enough for economic use. Another prerequisite is CO- and CO_2-converting plants, i.e. chemical industry, close to the gas generation sites, because further conversion preferably takes place in the same region. In this way, established regions can use their existing infrastructure (diverse industry, integration along the value chain, skilled personnel, research facilities, authorities, logistics infrastructure, etc.) as an advantage. Use of bio-based raw materials, raising raw material efficiency, waste recycling and the use of CO and CO_2 by integrating heavy industry, chemical and energy industries are among the required measures.

This scenario shows that the corresponding European industrial regions are not helpless in the face of raw material change, but have a long-term future. A basic prerequisite for their implementation, however, is social acceptance. The importance of this factor is illustrated by the example of a 67 km CO pipeline in NRW. Although already completed in 2009, it has not been put into operation to date due to protests by residents and legal proceedings [15, 16].

23.4 The Regional Innovation Network (RIN) Material Flows

The factors that are decisive for the establishment of such regional material flow concepts are being investigated in the Regional Innovation Network (RIN) Material Flows, which has been funded by the North Rhine-Westphalian Ministry of Science since 2014 and is coordinated by the Industrial Biotechnology Cluster (CLIB-Cluster) in cooperation with the two core partners Deutsche Gesellschaft für Abfallwirtschaft (DGAW) and EnergyAgency.NRW (now NRW.Energy4Climate). The spatial focus is on the Rhineland (NRW). It includes cities and municipalities in the Middle and Lower Rhine regions and combines in a small area both densely populated areas with a high industrial density and rural areas with intensive agricultural use. This combination is of decisive importance for the goal of recycling previously unused or hardly used side and residual material flows of different origins. Material flows available in the region are to be used, regardless of whether they originate from the region itself (e.g. biogenic residual materials) or are only fed into secondary utilization in the region in the sense of cascade utilization. This includes, for example, municipal waste or CO and CO_2 emissions from the process industry, waste recycling and agriculture. In this way, technical and biological carbon cycles are to be closed locally in order to increase resource efficiency in agricultural and industrial production and to develop contributions to saving raw materials and protecting the climate. These concepts should be economically, ecologically and socially sustainable.

This requires proof of economic viability, at least with a medium-term perspective. New, alternative processes are always in competition with existing processes. Due to the technical advantage of established processes, which—e.g. in the case of the chemical industry—have been optimized over decades, innovative processes cannot usually be competitive from the outset. However, by involving practitioners from industry as early as possible, it is possible to ensure the selection and development of processes on the basis of relevant criteria and the focus on promising products. This can minimize risks in development and technology transfer. The environmental compatibility of new approaches plays an equally important role. Here, too, it is important to demonstrate that new processes have a lower environmental footprint than existing ones. This can be achieved, on the one hand, through the use of secondary raw materials (side streams and exhaust gases) and, on the other, through gentler process parameters. The first aspect in particular is being pursued in the RIN Material Flows, since great potential is seen here with regard to the structures in NRW described above. Ultimately, quantitative analysis methods such as Life-Cycle Assessment (LCA) are then required to demonstrate better environmental compatibility compared to fossil-based processes. Finally, social sustainability can only be ensured by social acceptance of the developed processes. This requires an early discourse with citizens and institutionalized actors, already during the development of the approaches. Here, the RIN Material Flows aims to develop new ways of involving citizens in such discourses. Building on existing concepts, this is done by adapting them to the specific topics of the RIN Material Flows.

To develop these regional concepts, the RIN Material Flows focuses on the three thematic pillars "Raw material potentials", "Technical implementation" and "Networking and acceptance". They build on the actual actor network that forms the foundation of these three pillars. In all three thematic pillars, the network has initiated concrete implementation projects that examine the various aspects of establishing local closed-loop concepts. Examples include studies on specific residual material potentials from agriculture and the food and animal feed industry in the Rhineland. Research projects on the use of different agricultural residual material streams, on the recovery of phosphates from press cakes of the food industry or on the development of participatory instruments for the involvement of a broad public in the development of concepts for carbon recycling were also initiated. Furthermore, the involvement of public bodies such as cities and municipalities was worked on intensively, especially using the example of the city of Krefeld.

The experiences from the network and the implementation projects show that regional location factors play a decisive role for the concepts for carbon recycling discussed here. This does not only refer to the spatial distribution of carbon sources and sinks or the existing infrastructure. The combination of different economic sectors in the given region, direct access to scientific facilities and, last but not least, established communication channels to involve the local population are further decisive factors.

References

1. BP (2017) BP Statistical Review of World Energy June 2017. https://www.bp.com/content/dam/bp/en/corporate/pdf/energy-economics/statistical-review-2017/bp-statistical-review-of-world-energy-2017-full-report.pdf
2. Germany Trade and Invest (2012) Germany's Chemical Industry Pipeline Network. https://www.gtai.de/GTAI/Content/EN/Invest/_SharedDocs/Downloads/GTAI/Maps/C-H/map-pipelines-for-germany-chemistry.pdf?v=4
3. Europäische Kommission. Pariser Übereinkommen. https://ec.europa.eu/clima/policies/international/negotiations/paris_de#tab-0-0
4. VCI (2019) Daten und Fakten – Rohstoffbasis der chemischen Industrie. https://www.vci.de/vci/downloads-vci/top-thema/daten-fakten-rohstoffbasis-der-chemischen-industrie-de.pdf
5. Covestro (2016) Kohlendioxid – ein neuer Rohstoff. https://www.covestro.de/de/projects-and-cooperations/co2-project
6. AcelorMittal (2015) AcelorMIttal, LanzaTech und Primetals Technologies investieren 87 Millionen Euro zum Bau einer bahnbrechenden Produktionsanlage für Biokraftstoffe. https://germany.arcelormittal.com/icc/arcelor/broker.jsp?uMen=d1a7069a-d644-9831-84a6-8e5207d7b2f2&uCon=39776b25-1396-e417-6d28-46407d7b2f25&uTem=aaaaaaaa-aaaa-aaaa-aaaa-000000000011
7. ThyssenKrupp. https://www.thyssenkrupp.com/media/c2c/presse/pm_carbon2chem_technikum.pdf
8. BASF (2017) Gemeinsame Presseinformation. https://www.basf.com/de/company/news-and-media/news-releases/2017/08/p-17-293.html
9. Europäische Kommission. Auf dem Weg zu einer Kreislaufwirtschaft. https://ec.europa.eu/commission/priorities/jobs-growth-and-investment/towards-circular-economy_de
10. BMU. Gesetzesübersicht Kreislaufwirtschaft. http://www.bmub.bund.de/themen/wasser-abfall-boden/abfallwirtschaft/abfallpolitik/kreislaufwirtschaft/
11. VERPACKG/Landbell AG. http://verpackungsgesetz-info.de
12. Concord Blue (2016) Worldwide facilities. http://www.concordblueenergy.com/worldwide-facilities.aspx
13. Energiepark Mainz/Mainzer Stadtwerke AG. http://www.energiepark-mainz.de
14. BioPower2Gas. Microbenergy GmbH – Viessmann Group. http://www.biopower2gas.de/partner/microbenergy-gmbh-viessmann-group
15. BUND. Widerstand gegen CO-Pipeline. https://www.bund-nrw.de/themen/technischer-umweltschutz/im-fokus/co-pipeline/
16. Kölnische Rundschau (2017) Pläne der Bayer-Tochter Covestro - Co-Pipeline endgültig vor dem Aus? https://www.ig-erkrath.de/cms/upload/pdf/presseberichte/co-pipeline/pb-2017-04-11-koelnische-rundschau-plaene-der-bayertochter-covestro-co-pipeline-endgueltig-vor-dem-aus.pdf

Utilization of C1 Gases: The Regulatory Framework

24

Jörg Rothermel, Dennis Krämer, Tilman Benzing, and Tina Buchholz

Abstract

Climate change and the associated energy transition are challenges of our time that can be addressed and even better linked by a sustainable material use of CO_2. Existing European and national legal regulations and programs can support or possibly hinder the use of CO_2 as a raw material if designed accordingly. Examples would be European emissions trading or relevant directives for raw materials, fuels and waste. The Circular Economy Package, the European Emissions Trading Scheme (EU ETS) and the new Renewable Fuel Directive (RED II) are three policy instruments that can have a particular impact on CO_2 utilization; these policy frameworks have the greatest potential to accelerate the realization of CO_2 utilization processes.

Keywords

North Rhine-Westphalia (NRW) · Regional development · Industrialized region · CO_2 emission

J. Rothermel · T. Benzing · T. Buchholz (✉)
VCI - Verband der chemischen Industrie e.V., Frankfurt am Main, Germany
e-mail: rothermel@vci.de; tbenzing@vci.de; buchholz@vci.de

D. Krämer
DECHEMA - Gesellschaft für chemische Technik und Biotechnologie e.V., Frankfurt am Main, Germany
e-mail: kraemer@dechema.de

24.1   Introduction

Climate change and the associated energy transition are challenges of our time that can be addressed and even better linked by a sustainable material use of CO_2. Furthermore, CO_2 can serve as a carbon source for the chemical industry and thus represents an expansion of the raw material base. Some pathways to material use of CO_2, particularly pathways with the greatest CO_2 utilization potential, require a significant amount of renewable electricity due to the hydrogen required as a cofactor, which must be generated CO_2-free via water electrolysis. In addition, the opportunities to create value with CO_2 as a carbon source for products based on carbon as an important element are complex. The motivating reasons for using CO_2 as a raw material can vary between climate protection (reducing the greenhouse gas CO_2 by recycling it) or using an alternative carbon source to conserve fossil resources.

Chemists, engineers, biotechnologists, as well as life cycle assessors, economists and politicians must work together to identify and set a sensible framework for the material use of CO_2. Processes can be made more efficient through innovation, but in many cases technical progress is not the only decisive factor in terms of economic efficiency. For example, there are hurdles to CO_2 utilization that exist because of the current market situation: Production of carbon-based products (especially in chemicals and fuels) will be and remain more expensive than production based on fossil raw materials. Therefore, for competitive reasons, only a few products will remain for the time being.

Existing European and national legal regulations and programs can support, but also hinder, the use of CO_2 if they are designed accordingly. Examples include European emissions trading or the relevant directives for raw materials, fuels and waste.

24.2   Circular Economy Package of the European Union

The circular economy is a system in which resource use and waste, emissions, and energy losses are minimized by closing, restricting, and slowing down material and energy cycles, which can be achieved through durable design, maintenance, repair, reuse, remanufacturing, refurbishment, and recycling. This is in contrast to a linear economy, which is a "take, make, dispose" production model.

In 2015, the European Commission (EC) presented what it called the Circular Economy Package - a program to accelerate progress in the areas of production, consumption, waste management and use of secondary raw materials, all of which play an important role in a circular economy. The package can be seen as stimulating the application of new horizontal working methods, addressing the entire economic cycle and not just waste prevention goals. In addition to changes in waste legislation, the package includes a comprehensive action plan.

The material use of CO_2 is mentioned in a footnote in the Circular Economy Package. The rather low positioning for CO_2 recycling is caused by the lack of a definition of what exactly CO_2 is: waste/emission, a by-product or actually a raw material. As part of the action plan, the EU Commission is analyzing the legal regulations that exist in chemicals law and waste law regarding the ingredients of products to be recycled. The aim is to strengthen recycling.

The circular economy package itself does not integrate any activities to support research and implementation of CO_2 utilization pathways, but the mere naming of the topic in the text paves the way for follow-up activities. The impact will be reflected in the research funding landscape and in the design of EU regulations. Furthermore, decision-makers will be encouraged to pursue activities in the field of CO_2 use, to see CO_2 use as an opportunity to close the carbon cycle and thus contribute to the circular economy.

24.3   The EU Renewable Energy Directive

The 2009 EU Renewable Energy Directive (2009/28/EC) states that the EU as a whole must achieve a 20% share of energy from renewable sources in gross final energy consumption by 2020. Gross final energy consumption is defined as "energy products supplied to industry, transport, households, services [...] and agriculture, forestry and fisheries for energy purposes" (Art. 2f).

Member states are given binding national targets ranging from 10% (Malta) to 49% (Sweden). In addition, each member state must guarantee a share of 10% from renewable sources in the transport sector. In order to count biofuels and bioliquids towards the targets, there are also requirements for sustainable cultivation of biomass and, since 2015, for the reduction of greenhouse gas emissions (Art. 17ff). In addition, the directive allows for cooperation between member states and between member states and third countries to achieve the targets.

The member states themselves regulate how they achieve their targets. In Germany, for example, electricity generation from renewable energies is subsidized by a feed-in tariff under the Renewable Energy Sources Act (EEG) and financed by apportioning the costs to electricity consumption (EEG apportionment). In the case of fuels, on the other hand, the Federal Immission Control Act (BImSchG) regulates the implementation of EU requirements: until 2014, through the obligation to blend gradually increasing minimum proportions of biofuels, and since 2015, through a gradually increasing percentage reduction in greenhouse gas emissions from fuels placed on the market.

24.3.1 Redesign of the Renewable Energy Directive

With the 2020 reference year for the EU's Renewable Energy Directive (RED) targets approaching, the Commission presented an energy package at the end of November 2016 that includes a draft for a new Renewable Energy Directive. This sets an EU-wide target of 27% for the year 2030; national targets are no longer envisaged. Against the backdrop of the "tank/plate" criticism, the targets for conventional biofuels are to be lowered in the transport sector and instead renewable alternatives that do not compete with food and feed are to contribute increasingly to meeting the targets.

Since RED II was still in the legislative process when this paper was written and will be implemented through national regulations, the following question can only be analyzed in a general way.

24.3.2 What Regulatory Content Interacts with the Material Use of CO_2?

Four levels can be distinguished along the process chain for the technologies referred to as "power-to-X" (Table 24.1).

Levels 1–3 affect all power-to-X processes. Cumulatively, the effects are negative, since the costs for subsidizing electricity generation from renewable energies are passed on to the use of the electricity. Due to this burden on the electricity price, power-to-X is currently hardly feasible in Germany at competitive conditions. This would require a fundamental reform of the EEG to reduce costs or at least an alternative to the current pay-as-you-go financing.

A similar conflict of objectives also occurs at level 4 between the different uses "horizontally" if they are promoted in different ways.

Finally, the immense complexity of the regulations (EEG/BImSchG/BImSchVen), which increases with each amendment, should be pointed out, making an impact assessment of the effects and their interactions and conflicting goals increasingly difficult.

Table 24.1 Levels of power-to-X technologies with the corresponding effect

Level	Effect	Explanation
1. Generation of the RES-E	Positive	Promotion through EEG feed-in tariff
2. Production of renewable hydrogen from renewable electricity and water via electrolysis	Negative	Burden on electricity consumption due to the EEG levy (allocation of costs from level 1 to the electricity price).
3. Synthesis of hydrocarbons from CO_2 and hydrogen	None	CO_2 is not considered as "renewable energy"
4. Utilization of hydrocarbons:		Varies according to use
(a) In the transport sector	Positive	Can be counted toward quota as alternative renewable fuels
(b) Energetic (use after storage)	None	Promotion of renewable energies takes place at the level of generation (e.g. biogas)
(c) Material for chemical products	None or negative	No direct impact, as there are sensibly no targets/quotas Possible competition for use if the other uses are promoted and hydrocarbons become expensive/scarce as a result

24.4 EU Emissions Trading

The European Emissions Trading Scheme (EU-ETS) is the first multi-state and world's largest trading system for greenhouse gas emission allowances. It regulates the contribution of the emissions trading sector to the EU's climate protection targets and is thus the EU's central climate protection instrument. The emissions trading sector includes the energy industry and large parts of energy-intensive industry. EU emissions trading was introduced in 2005 and currently covers around 12,000 plants across Europe, which together account for almost half of all greenhouse gas emissions in Europe. The basic principle of emissions trading forms a "cap and trade" system. This means that a cap determines by law how much greenhouse gas emissions may still be emitted by the plants subject to emissions trading. A corresponding quantity of emission allowances (= certificates) is auctioned by the legislator or issued free of charge to eligible plant operators. The emission allowances can be freely traded on the market. The EU emissions trading system is therefore a quantity system that guarantees target achievement regardless of the price of a certificate.

24.4.1 What Regulatory Content Interacts with the Material Use of CO_2?

The current Emissions Trading Directive, which regulates emissions trading until the end of 2020, does not provide for special treatment for plant operators that capture CO_2 from point sources and use it as a material. The CO_2 that escapes from the plant is considered an emission from the generating plant, despite the capture. And only if the captured CO_2 is also geologically stored in a CCS project is special consideration given. This means that CO_2 that is captured, transferred and recycled from a plant subject to emissions trading is subject to emissions trading costs. The only exception to this so far is the possibility, only opened up in 2017 by a ruling of the European Court of Justice, of using carbon dioxide emissions to produce precipitated calcium carbonate in the lime industry. In doing so, the producer of the carbon dioxide emissions now has the option of deducting the forwarded amount of CO_2 from its total amount of greenhouse gas emissions for which it must surrender allowances. The entitlement to free allocation remains unaffected by this regulation.

For the future regulations in EU emissions trading, it is being discussed in the current legislative process for the Emissions Trading Directive from 2021 onwards to take CCU applications more into account, e.g. via research funding within the framework of a new innovation fund, which is fed by EU emissions trading certificates, as well as through possibilities to take CCU into account in the regulations for the Monitoring Regulation. The Monitoring Ordinance regulates the monitoring obligations of plant operators subject to emissions trading. The mandatory monitoring of emissions determines the total quantity of emissions for which the facility must surrender allowances. Here, regulations could be implemented that could differently consider or even promote the capture, transfer and material use of CO_2. Further consideration of CCUs in the new emissions trading directive can be done through the calculation of benchmarks. Based on the top 10% of the most efficient plants in Europe, the benchmarks determine the maximum number of allowances per ton of product a plant can receive. CCU applications should also be taken into account here. However, it remains to be noted that CCU applications can often only be realized competitively in a special environment (e.g., spatial or economic) (sales markets). Therefore, the consideration of CCU applications in the determination of benchmarks must not lead to an unjustified allocation reduction for plants to which these reduction potentials are not open. In principle, CO_2 captured and transferred at the point sources of plants subject to emissions trading can be used for the entire C1 rail.

R&D&I and Industry Examples: Challenges and Opportunities in Scaling Up

Thomas Schwarz

Abstract

Currently, several technologies for the chemical or biotechnological conversion of C1 gases are on the threshold of industrialization. Several process developments have now successfully passed the laboratory and pilot scale and are now facing the next phase on the way to industrialization: the transfer to demonstration or production scale. This chapter describes some such projects by way of example.

Keywords

Laboratory scale · Pilot scale · Demonstration scale · Production scale

A distinction can be made between demonstration systems that are only designed for the production of a specific chemical substance (dedicated systems) or that can be used variably for the production of different products (multi-purpose systems). Especially in the case of multi-product systems, it is advisable to rely on modular and flexibly adaptable process units. The biorefinery plant of Global Bioenergies in Leuna for the production of isobutene, for example, is a monoproduct plant, while the plant of BioBaseEurope in Ghent can demonstrate production on an industrial scale.

The scale-up of the processes and the integration of the different process steps into an overall process to demonstrate technical, economic and sustainable feasibility are essential for commercialization and fulfill several important tasks. Thus, both the suitability of the developments under industrial and at least near-production conditions is demonstrated, and the efficiency of the processes is further increased

---

Author Thomas Schwarz has died before the publication of this book.

by further process optimization (economy of scale) in continuous operation. In any case, demonstration plants serve to validate the newly developed processes in an industrial environment and to be able to supply several batches of chemical products to the chemical industry in sufficient quantity and quality in conformity with specifications for testing and further application.

Managing the technical risks in a demonstration or industrial scale and operating the plants with consistent product quality and yield over a longer period of time is the basic prerequisite for a sustainable willingness to invest in further development work as well as production facilities. This important key task is fulfilled by the first large-scale reference projects with industrial participation described in this chapter. Some of them are pure demonstration projects, but they are also the first industrial production facilities. In addition to industrial validation and demonstration of technical feasibility, it is important to use these lighthouse projects to test and improve the interaction between the various industrial partners (raw material suppliers, technology developers, chemical industry, plant manufacturers, etc.) in order to gain attention and acceptance among politicians and the general public. For the initiating companies, the successful commissioning of demonstration plants often leads to the initiation of new cooperation agreements with industrial companies and to the acquisition of further capital via investors (venture capital, strategic investors) as well as on the part of public funding agencies. Ultimately, this important milestone of successful "demonstration at industrial scale" means a significant and sudden increase in the value of the technology and thus of the developing company.

Ultimately, the operation of the demonstration plant must demonstrate, in addition to the ecological meaningfulness through e.g. the closing of carbon cycles and the improved carbon footprint of the products, that the processes have sufficient value creation potential and production efficiency to be able to produce competitively in the long term. If pure drop-in chemicals (identical chemical raw material for further processing via a new synthesis route, e.g. acetone or isobutene) are produced, it must be possible to produce them at least at comparable costs or even more cheaply. This is the only way to justify the high plant investments and the enormous costs incurred for the development work that has been carried out and is still outstanding in the companies. Although the increased raw material flexibility and unlimited availability of the "new" raw material C1 gas represent strategic added value for the chemical industry in the medium to long term, there is currently only a very limited willingness on the part of the industry to invest in large-scale plants alone. This is due not least to the current price situation for fossil raw materials. The provision of public funding for the construction of demonstration plants and the mobilization of risk capital is therefore enormously important. However, this often only succeeds with the necessary momentum when the industrial feasibility can be demonstrated by initial demonstration plants in the respective chemical region or on site. Thus, pioneering companies and strategic investors who have the courage and vision to build an industrial demonstration plant for the first time are of enormous importance in bringing these new technologies into industrial practice. The technologies presented for the conversion of C1 gases are currently in this sensitive phase of industrialization.

The biotechnological production of highly functionalized molecules, which have not yet been available on the market at all or only in small quantities, from C1 gases offers an opportunity to significantly increase the value creation potential and the unique position of these technologies. However, in this case, the time required for a successful market launch is significantly longer and the associated start-up costs are much higher.

R&D&I and Industry Examples: Covestro's Dream Production

Christoph Gürtler

Abstract

In the "Dream Production" project, we exemplify the use of CO_2 as a raw material for high-quality materials—a highly interesting and visionary technology for the future. Fossil resources are finite, thus the use of CO_2 as a chemical raw material is an important resource-saving approach. With the cardyon® technology, we were able to start up an industrial process for the production of polyols for sustainable polyurethane applications, in particular foams and thermoplastic polyurethanes, for the first time using CO_2 as a raw material. LCA studies show that the new products based on the CO_2-based polyols are ecologically more advantageous than conventionally manufactured products.

Keywords

Dream Production · cardyon® technology · Polyols · Polyurethane · Life-cycle assessment (LCA)

26.1 Introduction

It is actually an obvious thought: If mankind is already producing too much carbon dioxide (CO_2) and thus creating an imbalance in the natural carbon dioxide cycle, why not try to use at least a small part of this intrinsically harmless gas for something useful? This would make it possible to conserve other carbon sources and use the "waste product" CO_2 as a raw material for the chemical value chain, for example.

With this dream in mind, Covestro (formerly Bayer MaterialScience, BMS), Bayer Technology Services (BTS), RWTH Aachen University (represented by the

C. Gürtler (✉)
Covestro Deutschland AG, Leverkusen, Germany
e-mail: christoph.guertler@covestro.com

© The Author(s), under exclusive license to Springer Nature Switzerland AG 2023
M. Kircher, T. Schwarz (eds.), *CO2 and CO as Feedstock*, Circular Economy and
Sustainability, https://doi.org/10.1007/978-3-031-27811-2_26

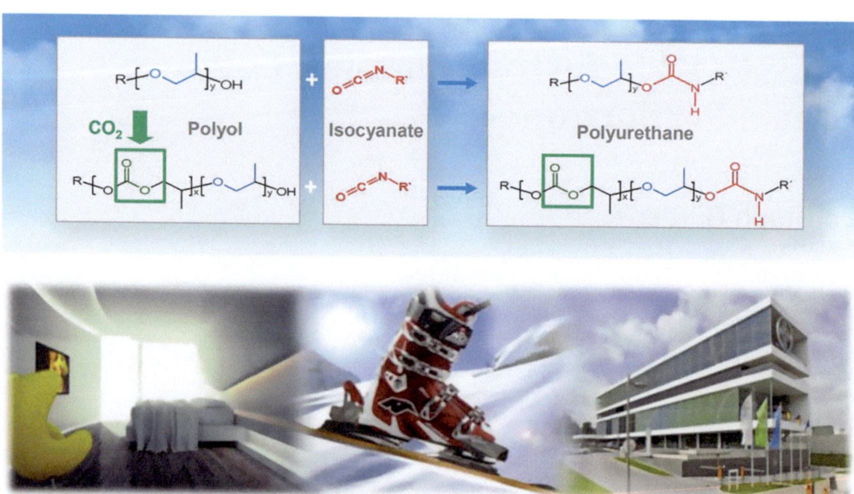

Fig. 26.1 Conversion of CO_2 with propylene oxide to polycarbonate polyols and further to polyurethanes for various applications. © Covestro

CAT Catalytic Center, AK Leitner, AK Liauw and AK Bardow) and RWE Power AG launched the "Dream Production" project in 2010, which is funded by the German Federal Ministry of Education and Research (BMBF). The common goal was to provide CO_2 from the exhaust gas stream of a coal-fired power plant in a purity sufficient for synthesis chemistry (per synthesis) and to use CO_2 as a C1 building block for the production of polyether carbonates. The basis for this were to be catalysts that had already been made available among other things, as part of the precursor project "Dream Reactions". In addition, the novel polyether carbonates and their use in the production of polyurethane samples were to be tested.

Using CO_2 as a chemical building block is not a new idea. Chemists were already conducting research in this field in the 1960s. The challenge was to find an economically and ecologically viable process for using the inert CO_2 molecule.

Before the Dream Production project began, the separation of CO_2 from power plant flue gases was in the development stage (pilot plant stage). Here, the priority was to work on energy-efficient scrubbing processes for CO_2 and on stable scrubbing agents in order to make power plants low in CO_2 as cost-effectively as possible. At the same time, the quality of the separated CO_2 had to be sufficient for the corresponding use.

The targeted CO_2-containing polyether carbonates were to be an alternative to conventional polyether polyols, which are based entirely on fossil building blocks (Fig. 26.1). First laboratory quantities could be prepared in the predecessor project "Dream Reactions". Due to the high pressures involved, novel reaction engineering solutions were required to scale up the reaction. In addition, it was not clear how the impurities in the CO_2 from the flue gas system would affect the catalyst behavior.

26.2 Project Description

The planned "Dream Production" research project was intended to provide incentives and impetus for subsequent work by preparing innovations and enabling further developments. Germany as a technology location and, in particular, the competitiveness of German companies in the field of CO_2 capture and recycling were to be considerably strengthened by the project.

The participation of an internationally recognized and renowned university in the project consortium also offered the opportunity to further advance and establish the technical development of CO_2 as a synthesis building block for polymers in the scientific environment.

The project was divided into seven work packages, which were handled by the respective project partners. Work package 1 (WP1) described the separation of CO_2 from power plant flue gases, the liquefaction, filling and supply of this CO_2, the supply to the partners and the quality monitoring. The separated CO_2 was tested in WP2 for compatibility with chemical catalysts, and reaction and microkinetic data were obtained. Spectroscopic measurements covered WP5 as well as mathematical modeling. In WP3, a miniplant system for the reaction of propylene oxide with CO_2 was designed, built and operated.

The material samples produced were tested in WP4 for their product properties and technical applicability. WP6 comprised basic research on the activation of CO_2. Accompanying the project, an ecological evaluation (eco-efficiency analysis) was carried out in WP7 by means of life cycle assessment.

26.3 Results

Within the framework of "Dream Production", the pilot plant for CO_2 scrubbing at RWE's power plant in Niederaussem (Germany) was successfully supplemented by a liquefaction and filling plant and put into operation. The separated CO_2 was made available to all project partners after CO_2 scrubbing. Regardless of the various purification methods tested, the quality of the CO_2 recovered was always high enough to be used for the planned syntheses.

For the production of sufficient sample quantities of polyether carbonate polyols, a scale-up of the existing laboratory-scale plant was carried out based on the results from the previous project (Fig. 26.2). This miniplant facility was used to advance the process and formulation development for the production of the polyether carbonate polyols. According to the increasing complexity, first a "semi-batch", then a "Continuous Addition of Starter (CAOS) semi-batch" and finally a "COAS conti" process were successfully implemented.

With the help of the miniplant, sample quantities of different polyether carbonate polyol grades could be produced and tested in various polyurethane applications. The accessible material quantities made it possible to test polyurethane materials not only on a laboratory scale but also on a larger scale for the main target application, flexible foam. It was shown that in standard formulations the conventional, purely

Fig. 26.2 Completed miniplant plant for the production of CO_2-containing polyols. © Covestro

petroleum-based polyol can be replaced by the CO_2 polyether both in parts (as a blend with conventional polyol) and 100%. The polyurethane materials obtained in this way showed similar property profiles to those produced from purely petroleum-based polyols. In addition, a reduced fire load of the foams based on the CO_2 polyols was observed.

The accompanying eco-efficiency analysis [1] clearly showed that the polyols produced in the "Dream Production" project with about 20% CO_2 content generate a lower amount of kilograms of CO_2 equivalents per kilogram of polyol produced compared to the conventional petroleum-based polyol. In addition, the consumption of fossil raw materials per kilogram of product can be significantly reduced by incorporating CO_2.

26.4 Utilization

In numerous publications, lectures, master's and diploma theses, patents and trade fair presentations, the results achieved were recorded and made accessible to the specialist scientific public as well as the general public. The focus was not only on seeking scientific discussions with experts, but also on promoting public acceptance of CO_2 as a useful building block (e.g. by participating in roundtable discussions of the IASS, framework activities of DECHEMA). The detailed life cycle assessment of industrial processes for the production of CO_2-containing products, such as the polyether carbonate polyols, was essential for the discussions addressed and forms a

Fig. 26.3 Marketing of CO_2 polyols under the name cardyon™. © Covestro

basic understanding for many further projects (such as "Dream Polymers", "Dream Polyols" or "Production Dreams").

In addition, the knowledge gained in "Dream Production" in CO_2 separation, especially with regard to CO_2 quality, is already being used in the implementation of other CO_2 utilization projects. Since 2013, for example, a catalyst test rig has been in operation in Niederaussem to investigate the conversion of power plant CO_2 with hydrogen to methane or methanol. This process is known as power-to-gas, among other things. To represent real operating conditions, carbon dioxide from the CO_2 liquefaction and filling plant is also used here. The evaluation of this process can be based on the results obtained in "Dream Production".

Overall, the chemical engineering, reaction engineering and process engineering results on polyether carbonate polyol technology achieved in the project provided the basis for a successful scale-up to a demonstration plant. Covestro built a demonstration plant with a production volume of around 5000 metric tons per year at the Dormagen site with an investment of approximately €15 million and started up the plant in 2016. The CO_2-containing polyol is marketed under the name cardyon™ (beYONd CARbon Dioxide) (Fig. 26.3). Among the first customers are mattress manufacturers such as Recticel. Germany as a technology location and, in particular, the competitiveness of German companies in the field of CO_2 capture and recycling have been considerably strengthened by the project.

Recognition for the innovative performance of this project was given by various national and international awards: Top 3 German Sustainability Award/Initiatives (2011); Land of Ideas (2012), KlimaExpo.NRW (2015), EUROPUR Sustainability Award (2016), ICIS Innovation Award (2016), Sustainia100 (2016), CEFIC

European Responsible Care Award, Category: Environment Award (2016), Nomination for the German Future Award (2019).

Reference

1. Van der Assen N, Bardow A (2014) Life cycle assessment of polyols for polyurethane production using CO_2 as feedstock: insight from an industrial case study. Green Chem 14:3272

R&D&I and Industry Examples: LanzaTech's Gas Fermentation

27

Christophe Mihalcea, Freya Burton, Robert Conrado, and Sean Simpson

Abstract

LanzaTech is the first company to commercialize a CO/H_2 gas fermentation process for the production of ethanol. Beyond ethanol, LanzaTech is developing fermentation processes for other chemicals from industrial waste gases containing CO, CO_2 and H_2. Prior to commercialization, the technology was tested in demonstration plants converting gases from steel mills or from municipal solid waste gasification. Depending on the gas source, the gas can range from CO-rich gas mixtures to those containing essentially H_2. The entire gas spectrum is suitable for the production of chemicals. By producing chemicals, LanzaTech is enabling entry into the carbon cycle economy. Waste carbon is captured in new products such as plastics and synthetic fibers that can be further recycled after use. LanzaTech thus transforms waste carbon from a pollutant into a raw material.

Keywords

LanzaTech · Gas fermentation · Ethanol · CO · CO_2 · H_2 · Steel mill · Municipal solid waste

27.1 LanzaTech

Founded in 2005, LanzaTech is commercializing a breakthrough technology for capturing and utilizing carbon from energy-rich exhaust gases containing carbon monoxide, carbon dioxide and/or hydrogen. Products range from everyday products, such as ethanol and jet fuel, to chemicals used in the manufacture of nylon and other

C. Mihalcea · F. Burton (✉) · R. Conrado · S. Simpson
LanzaTech, Skokie, IL, USA
e-mail: christophe.mihalcea@lanzatech.com; Freya@lanzatech.com;
robert.conrado@lanzatech.com; sean.simpson@lanzatech.com

© The Author(s), under exclusive license to Springer Nature Switzerland AG 2023
M. Kircher, T. Schwarz (eds.), *CO2 and CO as Feedstock*, Circular Economy and
Sustainability, https://doi.org/10.1007/978-3-031-27811-2_27

333

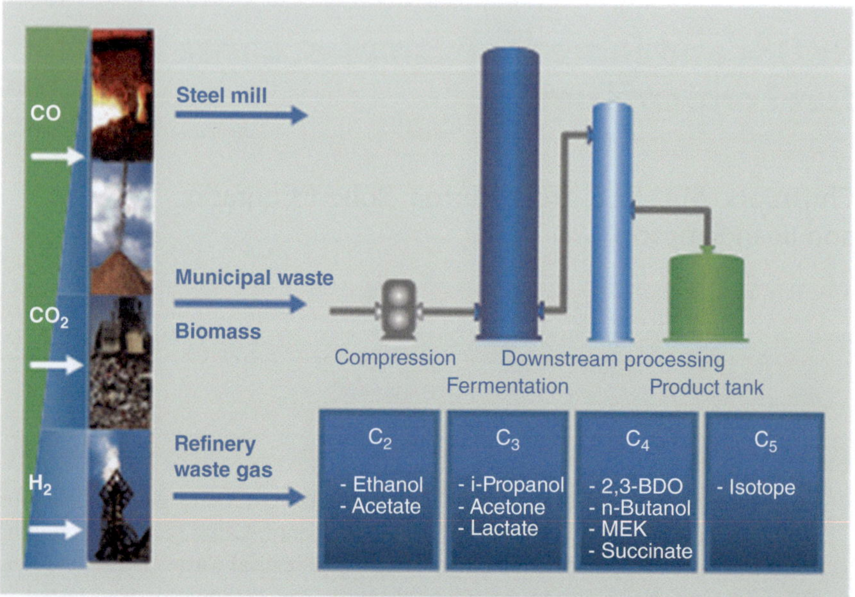

Fig. 27.1 LanzaTech process: The gaseous feedstocks can come from different industrial sectors and processes and therefore generally have different compositions. Regardless, these gas mixtures can always be used for fermentation into chemicals with a carbon chain length of 2–5. © LanzaTech

plastics. By using waste gases, LanzaTech is able to offer environmentally sustainable fuels and chemicals that do not compete with food production value chains and have no impact on land and water use.

The LanzaTech process (Fig. 27.1) continuously feeds energy-rich waste gas or residual gas streams into a fermentation reactor containing LanzaTech's proprietary microorganisms. The microorganisms grow and multiply consuming the gas contents, producing ethanol and/or other chemicals and fuels. These products are separated from the fermentation broth and can be immediately recycled or further processed into products such as plastics, nylon, rubber, and drop-in fuels, including jet fuel.

27.2 Metabolic Pathways of C1 Conversion to Products

LanzaTech's microorganisms for industrial processes belong to the acetogens. Acetogenic bacteria are able to convert CO and/or CO_2 with H_2. In contrast to the energy source sugar, in whose utilization pyruvate is the key intermediate, the utilization of gaseous carbon first leads to acetyl-CoA, which is converted via the reductive acetyl-CoA metabolic pathway (Wood-Ljungdahl biosynthesis) to acetic acid, ethanol and other end products. The biochemistry of this metabolic pathway has been extensively documented in numerous review articles, including Ragsdale and Pierce [1] and Fackler et al. [2]. Figure 27.2 shows the metabolic pathway including reduction equivalents and ATP consumption for each step. In addition to acetic acid and ethanol, the LanzaTech organism naturally produces 2,3-butanediol, a compound otherwise derived from pyruvate rather than acetyl-CoA. Under certain conditions, lactic acid is also formed, but only in small amounts. An overview of the reaction steps leading to the major fermentation products is given Fig. 27.3.

27.3 Feedstock Flexibility and Scale-Up Experience

LanzaTech is a pioneer in the development and deployment of gas fermentation technology. The company has gained experience with a spectrum of gas streams of different carbon monoxide, carbon dioxide, and hydrogen compositions. Figure 27.4 shows examples of gas streams from different industries and processes. Here, LanzaTech prefers waste gases that have a high concentration with respect to a) carbon monoxide, b) carbon monoxide and hydrogen, and c) hydrogen. Figure 27.4 shows the time course of the development of gas fermentation from different sources.

27.3.1 Exhaust Gas with High Carbon Monoxide Content

Since 2008, LanzaTech has successfully demonstrated carbon monoxide-based gas fermentation of ethanol with four pilot and scale-up plants installed at start-up industrial sites around the world. In more than 40,000 h of operation, including several fermentations with more than 2000 h of operation, all targeted performance parameters have been met or exceeded.

Currently, LanzaTech operates four two commercial steel-based and ferro alloy emission based projects, the first of which was commissioned in May 2018 and is now continuously used for ethanol production in China. Figure 27.5 shows a recent photo of the plant in Caofedian, China.

Figure 27.6 shows data obtained over a long period of time from the Shougang demonstration plant in China. The continuation of CO conversion and ethanol production at a high level despite unplanned production interruptions should be highlighted. The continuous production of chemicals using the LanzaTech gas fermentation process under real conditions of a steel mill demonstrates the robustness of the process and thus its basic suitability for commercial operation.

27.3.2 Synthesis Gas with High Carbon Monoxide and Hydrogen Content

Gas fermentation based on syngas has been developed by LanzaTech on a laboratory scale for 10 years. The process has been tested in a pilot plant built by Sekisui

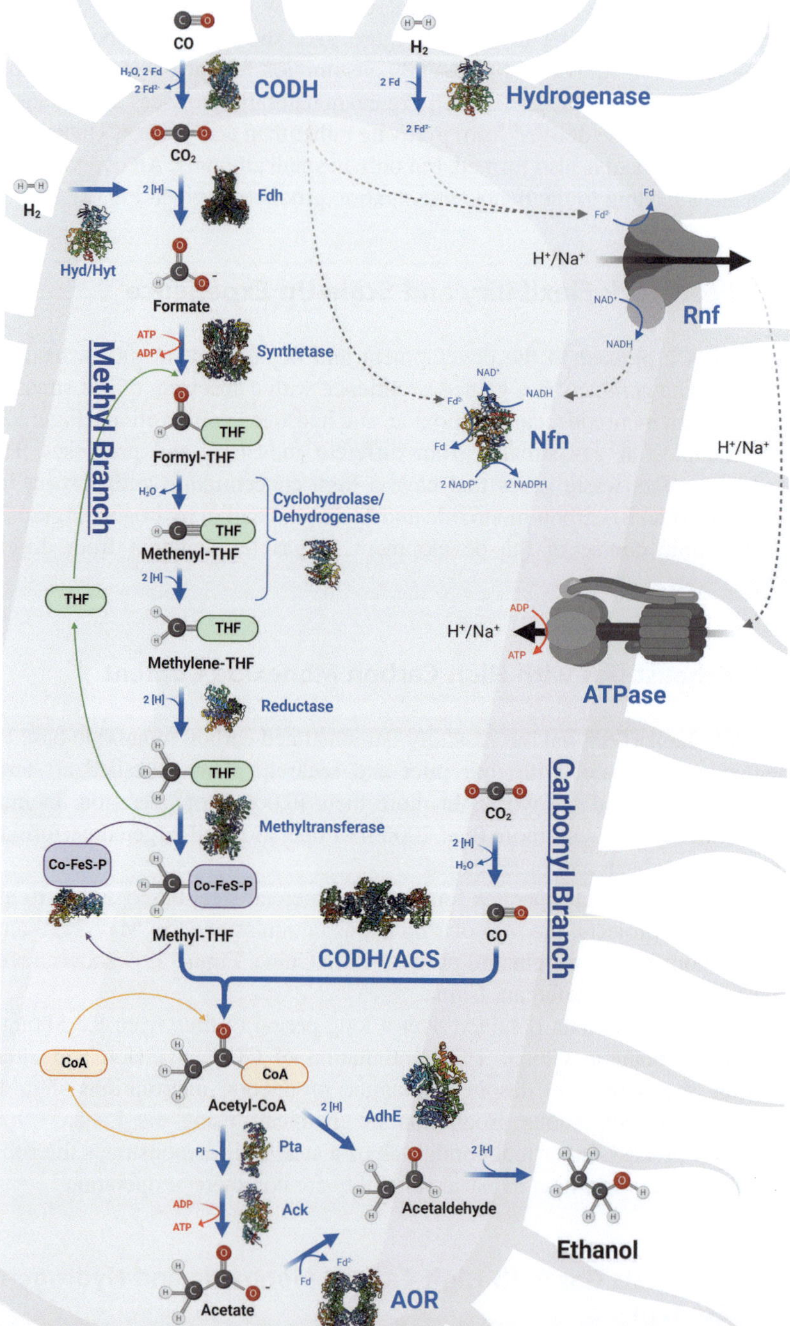

Fig. 27.2 Sequence of the Wood-Ljungdahl or the reductive acetyl-CoA pathway. In the methyl pathway, CO_2 is reduced by integration into tetrahydrofolate. The formyl group is reduced in several steps to a methyl group, which is transferred by a methyl transfer protein. Another CO_2 molecule is reduced to CO by the enzyme CO dehydrogenase/acetyl-CoA synthase (CODH/ACS). By combining with the methyl group and coenzyme A, acetyl-CoA is formed. Abbreviations: *Ack*

(Japan) in Japan since 2014. The syngas used there is produced by gasification of municipal solid waste (household waste). This pilot plant also met or exceeded the target parameters, including a long-term test over 60 days. In addition, syngas was tested and successfully fermented at sites operated by other suppliers. The different feedstocks used for gasification each result in a different gas composition and specific contamination profile. The microorganisms used in gas fermentation have successfully fermented syngas from flooring, rubber and other waste materials, demonstrating the robustness of the bacteria and the entire process.

LanzaTech is also cooperating with partners in India to build a commercial-scale gas fermentation plant there starting from agricultural residues (agricultural waste, biomass).

27.3.3 Waste Gas with High Hydrogen Content

Many industrial off-gases contain significant levels of hydrogen and carbon dioxide with low levels of carbon monoxide. LanzaTech has been optimizing the fermentation of such hydrogen-containing gases since 2015. It has been shown that more than 80% of the carbon bound in the products comes from the CO_2 content. At laboratory scale, all performance parameters could be achieved with gas mixtures with a high hydrogen content, and in the meantime, suitability for commercial operation has also been demonstrated at pilot scale in continuous operation of almost 1000 h. Technical planning for such a gas fermentation plant with Indian Oil Company (IOC) was started at the end of 2017, and commissioning is scheduled for 2023.

27.4   Innovation in Gas Fermentation Reactor Technology

Reactor design is at the heart of the gas fermentation process. Ongoing development is expected to improve energy and overall process efficiency to further reduce production costs, making gas fermentation an economical alternative to burning feedstock for power generation or simply flaring. LanzaTech is the only gas fermentation company working on the reactor design itself and has a vigorous patent strategy in this regard. LanzaTech's CO-based demonstration plant in China (370,000 liters of ethanol per year) and commercial production in Caofedian

Fig. 27.2 (continued) acetate kinase, *ACS* acetyl-CoA synthase, *AdhE* bifunctional aldehyde/alcohol dehydrogenase, *AOR* aldehyde:ferredoxin oxidoreductase, *CODH* CO dehydrogenase, *Co-FeS-P* corrinoid iron sulfur protein, *Hyd/Hyt* NAD/NADP-specific electron-bifurcating hydrogenase, *Fdh* formate dehydrogenase, *MTHFR* methylene-THF reductase, *Nfn* transhydrogenase, *Pta* phosphotransacetylase, *Rnf* ferredoxin:NAD+-oxidoreductase, *THF* tetrahydrofolate. Image used with permission from the Annual Review of Chemical and Biomolecular Engineering, Volume 12 © 2021 by Annual Reviews, http://www.annualreviews.org/

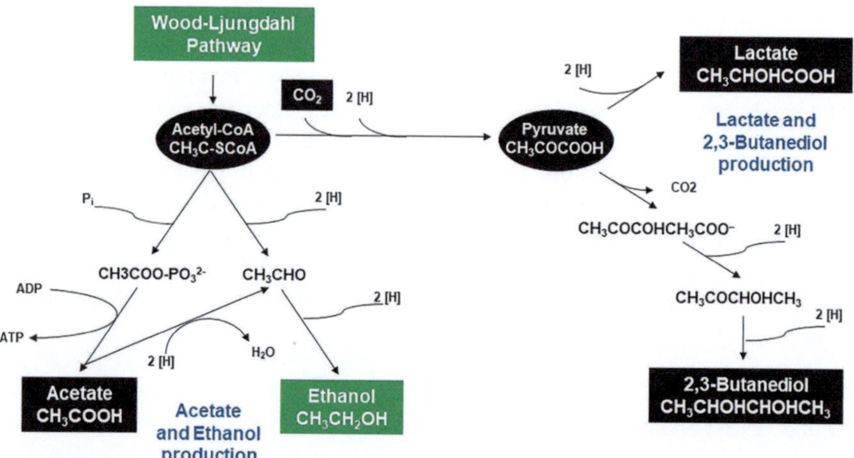

Fig. 27.3 Biosynthetic pathways and reduction equivalents originating from acetyl-CoA in Clostridium autoethanogenum. Acetate and ethanol are derived directly from acetyl-CoA. The formation of biomass and of 2,3-butanediol requires the synthesis of pyruvate as an intermediate. 2,3-Butanediol is subsequently formed from α-acetolactate and acetoin. © LanzaTech

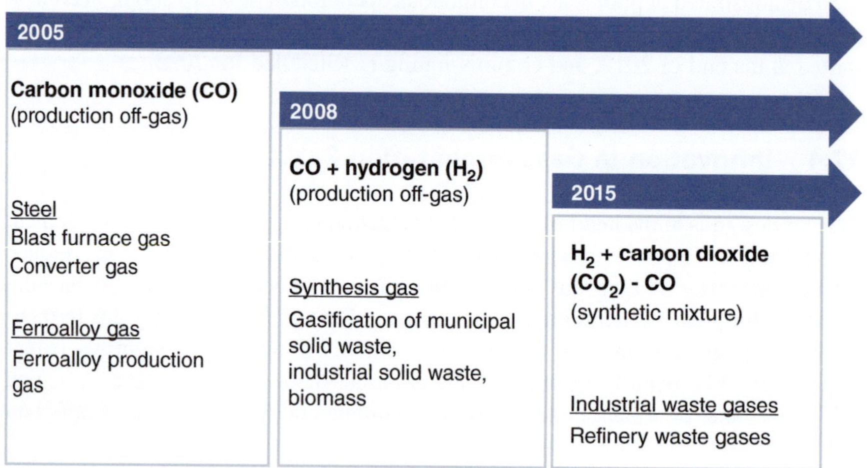

Fig. 27.4 LanzaTech's development of fermentation processes starting from steel mill waste gases to other waste gases of different compositions. © LanzaTech

(China) (50,000 liters of ethanol per year) now represent the state of the art for the gas fermentation reactor.

Fig. 27.5 LanzaTech commercial shougang plant commissioned in Caofedian, China, in May 2018. © Beijing Shougang LanzaTech New Energy Technology Co, Ltd.

The biggest challenge in gas fermentation is the limited gas-liquid mass transfer rate. LanzaTech has developed a series of forced circulation loop reactors that have been successfully used in CO gas fermentation. To better understand the fundamentals of this reactor concept, mass transfer and fluid dynamics are mathematically modeled. On this basis, the reactor was developed for production scale. Figure 27.7 shows the modeling and measured data confirming the theoretical expectation. The experimentally determined CO conversion could be increased to about 90% while maintaining the desired product titers, which ultimately improved the productivity of the bioreactors.

27.5 Robustness of the Bacteria

As living organisms, the bacteria used in gas fermentation are self-replicating biocatalysts. They are resistant to numerous potentially toxic compounds found in industrial waste gases. These include substances that damage traditional metal catalysts used in gas conversion. In addition, the bacteria were exposed to unstable operating conditions, such as interruption of gas, nutrient, or power supplies. In each

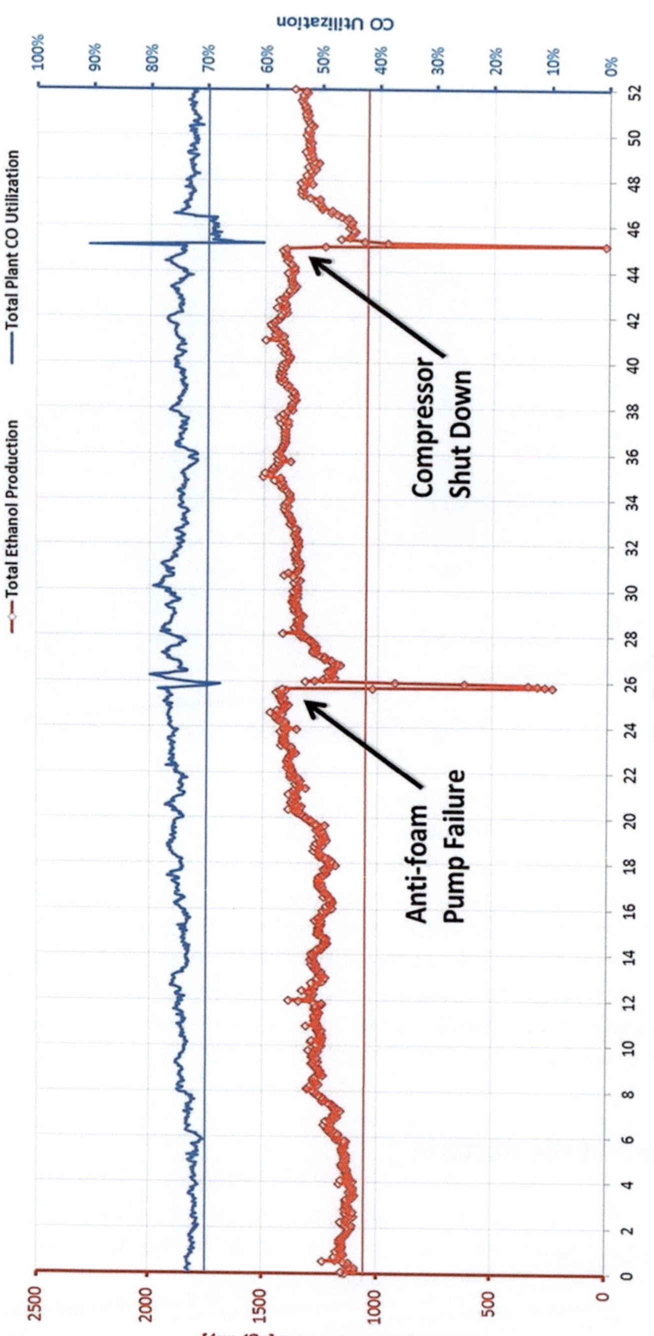

Fig. 27.6 Ethanol production and gas conversion data from a fermentation run at the Shougang demonstration plant in China. The horizontal lines indicate the company's target parameters. Both the ethanol productivity and gas conversion targets were achieved over a period longer than 6 weeks. © LanzaTech

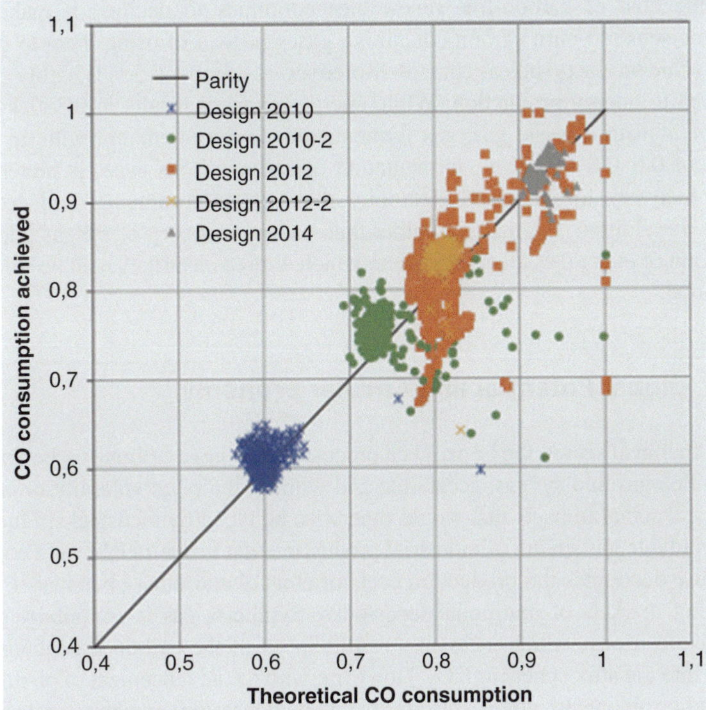

Fig. 27.7 Shown are results from LanzaTech's scalable bioreactor and corresponding data from the successively commissioned fermentation plants. kL*(1-Sat%) is a dimensionless parameter that combines all variables affecting mass transfer. © LanzaTech

case, standard performance parameters were regained after a short interruption. This demonstrates the robustness of the bacteria under production conditions.

27.6   Life Cycle Analysis

Based on the (fossil-based) steel mill off-gas from a European plant, LanzaTech's ethanol achieves a 70% reduction in greenhouse gases compared to the fossil-based alternative. This average value can vary with the gas source, with the process variant, and with the emission factors specific to the site. The most significant emission factor from gas fermentation today is electricity consumption, which will be reduced as the supply of renewable energy increases.

A further improved greenhouse gas balance is achieved when syngas from gasified biomass or municipal solid waste is used. Ethanol produced in this way has emission values of 1.5–11.7 g CO_2eq/MJ. Compared to gasoline, this corresponds to an emission reduction of 87–98%.

As the cost of carbon-free renewables continues to decline, it makes little economic sense to burn carbon-containing gases instead of using them to produce higher-value products such as ethanol. Moreover, gas fermentation is highly efficient compared to energy production. While energy recovery results in 0.3–0.4 MJ per 1 MJ of industrial waste gas, gas fermentation produces ethanol with an energy content of 0.6–0.8 MJ. Thus, fermentation of carbonaceous gases is preferable to energy recovery for heat and electricity for both environmental and economic reasons. In addition, the material rather than energy recovery of syngas allows the production of ethanol as an energy store, which, unlike electricity, can be stored and transported.

27.7 Global Potential and Circular Economy

For the chemical sector, the LanzaTech process makes large-volume carbon sources, such as flue gas and syngas, accessible and without the price volatility or environmental and social impacts that would otherwise be faced by feedstocks. Flue gas is an unavoidable gas stream of industrial production that is not tradable as a commodity. Its use decouples the production costs of ethanol and other chemicals from the fluctuating markets of traditional feedstocks. Synthesis gas is commonly used to generate electricity, which inevitably results in all of the carbon it contains being emitted into the atmosphere as CO_2. However, with the advancement of photovoltaic and wind turbine technologies, electricity can be generated on a large scale without the release of CO_2. From an ecological perspective, it therefore makes sense to use this resource for the production of fuels and chemicals. Table 27.1 gives an idea of the enormous potential. Steel mill gases alone could provide the carbon source for 90 MT or for 114 billion liters of ethanol. This corresponds to today's world production of bioethanol.

Technologies like LanzaTech's enable a circular economy for carbon by converting waste gases into fuels and chemicals for our daily needs. Carbon, which would otherwise be released into the atmosphere as a greenhouse gas, thus becomes a raw material and is bound in products.

Table 27.1 Ethanol production potential in Europe, the Middle East, Africa, Asia, and South and North America from various gases (1000 tons per year)

Region	Blown steel converter gas (steel mill)	Blast furnace gas (steel mill)	Exhaust gas (ferro-alloy production)	Synthesis gas (from municipal solid waste)
Europe, Middle East, Africa	2844	11,796	1593	25,947
Asia	12,995	54,490	1040	28,759
South and North America	1729	6828	239	23,813
Total	17,568	73,114	872	78,519

References

1. Ragsdale SW, Pierce E (2008) Acetogenesis and the Wood-Jlungdahl pathway of CO_2 fixation. Biochim Biophys Acta, Proteins Proteomics 1784(12):1873–1898
2. Fackler N et al (2021) Stepping on the gas to a circular economy: accelerating development of carbon-negative chemical production from gas fermentation. Annu Rev Chem Biomol Eng 12 (1):439–470. https://doi.org/10.1146/annurev-chembioeng-120120-021122

Torsten Müller

Abstract

In the reduction of CO_2 emissions from industrial plants, an economic or technical limit has been reached in many places. A further noticeable reduction of emissions can only be achieved by cross-industrial cooperation between different industrial sectors. The coupling of processes allows the material use of process gases and residual gases containing CO and CO_2. In addition to the reduction of CO_2 emissions, a reduction in the use of fossil raw materials can thus also be achieved. In the Carbon2Chem® joint project, companies from various industries as well as scientists in the field of basic and applied research are working to realize this coupling. The central aim is for the targeted integrated system comprising chemicals, steel production and energy generation to ultimately be more economical and sustainable than the sum of the individual systems involved. In addition to the preparation of a synthesis gas suitable for material use from metallurgical gases, the integration of renewable energies and the dynamic accumulation of metallurgical gases represent a major challenge. The flexibilization of the overall system required for this purpose calls for new technical and organizational solutions, which are to be found on an industrial scale in the Carbon2Chem® project funded by the BMBF.

Keywords

Reduction of emissions · Cross-industrial cooperation · CO · CO_2 ·
Carbon2Chem® · Renewable energies · Metallurgical gases

T. Müller (✉)
Fraunhofer UMSICHT - Fraunhofer Institute for Environmental, Safety and Energy Technology,
Oberhausen, Germany
e-mail: torsten.mueller@umsicht.fraunhofer.de

28.1   Introduction

The success of the energy transition and climate protection cannot be achieved by optimizing individual industrial systems alone, but requires networking between industries and sectors in order to be able to leverage existing potential. As part of the Carbon2Chem® joint project funded by the BMBF, companies from the chemical, steel, energy and plant engineering sectors are working to make a tangible contribution. The central objective is to reduce CO_2 emissions, taking the steel industry as an example, by economically utilizing the co-product gases generated during smelting processes.

In today's industrial plants, it can be assumed in chemical production, energy generation and steel production that the plants are each operated close to the technically possible optimum. In view of the far advanced optimization, a further reduction of CO_2 emissions in the individual processes is technically hardly feasible, since in steel production in the blast furnace process, for example, a minimum quantity of coal cannot be undercut for thermodynamic reasons. However, a significant reduction in CO_2 emissions could be achieved through the material utilization of the CO- and CO_2-containing co-product gases generated in this process.

Knowledge of coal-based chemistry forms the basis for the production of chemical products from CO- and CO_2-containing gases. In some cases, this involves drawing on knowledge that has been known for 100 years. In principle, many chemical processes can be operated on the basis of CO_2 as a starting material. CO_2 can thus become an important raw material. The production of basic chemicals by means of CO_2 recycling offers the opportunity to reduce overall CO_2 emissions, since closing loops displaces fossil raw materials such as crude oil. The alternative use of CO and CO_2 as carbon carriers in chemical production creates a high demand for hydrogen, which can be provided by electrolysis processes, among other methods. By using renewable energies, this hydrogen can be produced sustainably; due to the volatility of renewable energies, this introduces an additional dynamic into the interconnected system.

In the case of previous energetic use of CO-containing residual gases, as in the example of integrated smelters that operate largely self-sufficiently in terms of energy, the switch to material use of the CO- and CO_2-containing co-produced gases requires at the same time sufficient energy supply from renewable sources to cover the smelter's energy requirements. Suitable concepts must also be developed for this purpose.

28.2   The Project

As part of the Carbon2Chem® joint project, work is being carried out on providing the necessary technologies for the material utilization of the dome gases produced in smelter operations. In several subprojects, the necessary technologies are being developed for a promising process concept, even under dynamic boundary conditions. A key challenge is the purification and treatment of the dome gases to enable their catalytic conversion. Individual trace components can significantly disrupt the processes at this point.

In addition, it must be possible to operate the process network flexibly, since both the CO- and CO_2-containing metallurgical gas streams and the sustainably generated hydrogen required for chemical production can fluctuate greatly in their timing. Suitable dynamic concepts must be developed for this purpose through cooperation between science and industry at both the technical and organizational levels.

Technical development has a major focus on catalysis research (catalysts, processes, regeneration). Here, the development must partly start at the laboratory level and should be completed at the production level at the end of the joint project. For this reason, the project is divided into three project phases, starting with basic research, continuing with testing on a pilot plant scale and ending with large-scale implementation. To this end, joint structures are being created within the project to enable close cooperation between the project partners in all phases.

To this end, a superordinate subproject will take on the task of bringing together the technologies from the individual subprojects under process engineering and energy aspects as well as sustainability criteria. The aim of Carbon2Chem$^{®}$ is not to find a singular solution, but to develop a modular concept that can be used worldwide in the area of large CO_2 sources. In all areas, the aim is to reduce the carbon footprint of the processes compared with fossil-based reference processes. Industry and science are jointly preparing the tools required for this as part of the project.

The main issues of the project relate to the coupling of electricity, heat, steel and chemical processes under dynamic process control. From an economic point of view, it is necessary to examine how interactions of a plant network consisting of, for example, an integrated steel mill and chemical plants with a higher-level energy network can be used synergistically in order to be able to operate such a network economically. This includes the question of the economically relevant products that can be produced by such a plant network. The question of the dimension of the processes and the product quality is essential here, since such a plant network can have a considerable influence on existing markets, especially in the area of basic chemicals. Due to the coupling of different production plants, the dimensioning is an important aspect to enable an efficient turnover in combination with a flexible process control.

R&D&I and Industry Examples: The CO$_2$ Electrorefinery: A New Concept for Carbon Dioxide (CO$_2$) Capture and Utilization (CCU)

Deepak Pant, Metin Bulut, and Heleen De Wever

Abstract

Among the available CO$_2$ conversion processes, electrochemical technologies have proven to be efficient approaches in which electrodes play a crucial role. Currently, electrochemical CO$_2$ reduction processes are used for a variety of products such as fuels, basic chemicals, and fine chemicals. Proof-of-concept studies are underway for bio/electrochemical technologies and suitable products are being selected. Further studies on the commercial aspects of this technology are being conducted at VITO.

Keywords

Electroynthesis · Formic acid · Formatotrophic microorganisms

29.1 Electricity-Driven CO$_2$ Conversion

Renewable-fed electrobioproduction is a promising technology that addresses many drawbacks of power generation from agricultural side streams. Electricity is a versatile energy source that should be abundant from renewable sources in the medium term. However, it cannot be produced continuously from most renewable energy sources, but accrues intermittently. Thus, electricity must be stored when there is excess production, and fed back from this supply during periods of insufficient power generation. Electricity also cannot be integrated directly into the current fuel- or chemical-based system. Therefore, a technology is needed that can convert electricity directly into fuels and chemicals. This can be achieved via photoelectric, biological, bioelectrochemical, or electrocatalytic processes using either

D. Pant (✉) · M. Bulut · H. De Wever
VITO - Vlaamse Instelling voor Technologisch Onderzoek, Mol, Belgium
e-mail: deepak.pant@vito.be; metin.bulut@vito.be; heleen.dewever@vito.be

© The Author(s), under exclusive license to Springer Nature Switzerland AG 2023
M. Kircher, T. Schwarz (eds.), *CO2 and CO as Feedstock*, Circular Economy and Sustainability, https://doi.org/10.1007/978-3-031-27811-2_29

organometallic, organic, or bioorganic catalysts [1]. A CO_2 electrorefinery, as described below, includes the last three options.

29.2 Bio/Electrochemical Systems and the Role of Gas Diffusion Electrodes (GDE)

The reduction of CO_2 through bio/electrochemical technologies is expected to have tremendous potential for the renewable chemicals and fuels sector. In this context, the performance of the electrodes is not only a key parameter for the technical/operational application, but also for the economic viability of this technology. The electrodes used should be available at low cost, have high performance (i.e., high electron transfer capacity), and be compatible and flexible with different processes (i.e., with a wide range of electrolytes and with bacterial colonization or biofilm formation). From an operational standpoint, it requires long-term use with low maintenance, as well as good resistance to variations in temperature or electrical potential.

The electrodes developed by VITO under the brand names VITO CORE® and VITO CASE® are efficient compared to conventional platinum group metal (PGM)-based electrons, are inexpensive and have sufficient volume for the rapidly growing electrosynthesis market. The starting materials for these electrodes are activated carbon, polymer binder and a stainless steel mesh of minimal thickness. Recently, carbon-free metal based gas diffusion electrodes have also been developed at VITO. The manufacturing process of VITO electrodes follows a dry synthesis method that does not use solvents. Low energy use due to milder temperature conditions during manufacturing means equally low CO_2 emissions. VITO electrodes have a lower CO_2 footprint compared to other commercial electrodes (PGM-based materials) in both raw material synthesis and electrode manufacturing [2].

The hybrid structure of VITO electrodes, which achieves efficient homogeneous current distribution in the electrode material, ensures high electron transfer rates and minimizes the energy losses of the process. Carbon-supported electrodes maintain their high efficiency in different electrolyte conditions and exhibit a high electron transfer rate. Electrodes made of graphite, activated carbon or other carbon-based materials have the disadvantage of not being effective current collectors in their pure form. Therefore, the electrode technology developed by VITO has an integrated current collector that ensures efficient electron transfer and lower losses due to electrical resistance in the system. Adaptability, wide range of applications, cost efficiency, and their gas diffusion properties are important advantages of VITO electrodes, which are summarized in Table 29.1. Potential application fields of VITO electrodes are bioelectrochemical systems (BES) as well as electrochemical systems for CO_2 electroreduction using metal-based electrodes.

Table 29.1 Special characteristics of the VITO electrodes, essential components of the CO_2 electro-refinery

USP VITO-electrodes	CAPEX	OPEX
Low cost	Low investment costs	–
Long life	–	Low exchange rates
Large surface area	Low investment costs	High CO_2 turnover rate and product yields
Batch to continuous process	Smaller installation area (with the same productivity as in batch)	Fewer person hours/higher productivity
Compatibility	–	–

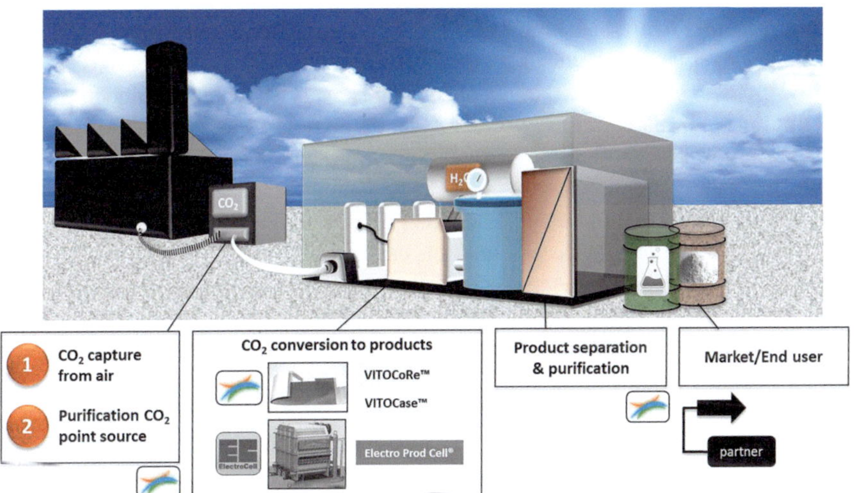

Fig. 29.1 Schematic diagram of a CO_2 electrorefinery converting industrial CO_2 into chemicals

29.3 CO_2 Electorefinery

An electorefinery is defined as a process of electrolytic refining in which a metal such as copper is deposited by electrolysis. In this process, the raw metal is used as the anode, where it goes into solution, while the ultrapure metal is deposited on the cathode. Based on this principle, VITO proposed the concept of the CO_2 electric refinery in 2015. This is intended to provide a production platform for converting CO_2 into basic chemicals (so-called building blocks) as well as higher-value organic compounds (Fig. 29.1). The core of the CO_2 electrorefinery is an industrially applicable bio/electrochemical system, as described in Fig. 16.13, that can convert CO_2 side or waste streams from industry into hydrocarbons and/or more complex organic compounds. The approach described covers the entire process chain from

CO_2 capture and purification, conversion to products, and product separation and purification. Here, the conversion can be carried out either by microbial electrosynthesis or by electrocatalytic CO_2 reduction. In both cases, the previously described VITO gas diffusion electrodes (GDE) can be used, which can be flexibly applied in different reactor designs [3].

BES refers to a bioreactor technology that exploits the ability of electrochemically active microorganisms to transport electrons to or from a solid electrode [4, 5]. Research in recent years has expanded the utility of BES beyond power generation in microbial fuel cells (MFCs) to more complex processes such as for chemical synthesis, bioremediation, and resource recovery. Uniquely, BESs can also produce valuable fuels and chemicals from low-value waste or CO_2 with a small input of electrical energy. Such a production system is known as microbial electrosynthesis (MES), in which cathodic biocatalysts reduce the available terminal electron acceptor to produce high-value products [6]. Bioelectrosynthesis has several advantages over electrochemical synthesis methods because it requires only a lower energy input [7], does not require specialized catalysts, and has higher product specificity [8]. To date, most research efforts in the field of MES have been limited to laboratory scale, with acetate being the main product [9].

Electrocatalytic reduction of CO_2 has been demonstrated and extensively studied on a variety of different electrode materials [10]. Depending on the cathode material and the electrolyte used, different products can be obtained, mainly formic acid and CO, but also hydrocarbons with relatively low molecular masses, such as methane or ethylene. The use of a gas diffusion electrode in combination with optimized catalyst layer deposition processes has often led to a significant improvement in electrode performance for CO_2 reduction [11]. With the use of GDEs, current densities in CO_2 electroreduction could be increased by one to two orders of magnitude [12]. In CO_2 electrochemical reduction, GDEs with a Sn electrocatalyst produced formate with up to 2 kA/m^2 and a Faraday efficiency of 90% [13].

29.4 Value Creation Perspective

Several companies, research institutes and universities are currently focusing on the commercialization of CO_2-based technologies. Companies such as Audi, Siemens, Arcelor, Suez Environment, etc. in Europe and Indian Oil Corporation Limited (IOCL) in India, Sabic from the Middle East and Singapore are some examples of companies that have invested in the research and development of CO_2-based electrochemical technologies. Examples that provide an assessment of technology maturity include MANTRA Energy's (Canada) pilot project on an electro-reduction of carbon dioxide (ERC) technology based on a 3D cathode for the reduction of CO_2 to formic acid [14]. Another example is an electroreduction plant from DNV that can produce 350 kg of formic acid per year [15]. The above examples highlight the importance of suitable electrodes in these processes. A recent techno-economic analysis of the CO_2 conversion electroreduction plant by VITO found that the success of a business case depends on the chosen product and its market size

Table 29.2 CO_2 reduction reactions to various products in aqueous solution and energy consumption (from ElMekawy et al. 2016 [2], with permission of the publisher)

Product	Reaction	Energy expenditure			Value
		$\Delta G^{25°C}$ [J/mol]	Electricity [MWh/t]	Cost [EUR/t]	Market price [EUR/t]
Formic acid	$2\,CO_2 + 2\,H_2O \rightarrow 2\,CH_2O_2 + O_2$	271	1,64	164	515
Essigsäure	$2\,CO_2 + 2\,H_2O \rightarrow C_2H_4O_2 + 2\,O_2$	876	4,06	406	650
Oxalic acid	$4\,CO_2 + 2\,H_2O \rightarrow 2\,C_2H_2O_4 + O_2$	328	1,01	101	725
Methanol	$CO_2 + 2\,H_2O \rightarrow CH_4O + 2\,O_2$	673	5,85	585	410
Ethanol	$2\,CO_2 + 3\,H_2O \rightarrow C_2H_6O + 2\,O_2$	1328	8,02	802	445

(Table 29.2). Therefore, diversification should be applied to the targeted end products to avoid dependence on only one product and its market price [2].

Acknowledgements The authors would like to thank the INTERREG Flanders-Netherlands project EnOp (CO_2 as energy storage) for funding this work.

References

1. Apaydin DH, Schlager S, Portenkirchner E, Sariciftci NS (2017) Organic, organometallic and bioorganic catalysts for electrochemical reduction of CO_2. ChemPhysChem 18(22):3094–3116. https://doi.org/10.1002/cphc.201700148
2. ElMekawy A, Hegab HM, Mohanakrishna G, Elbaz AF, Bulut M, Pant D (2016) Technological advances in CO_2 conversion electro-biorefinery: a step toward commercialization. Bioresour Technol 215:357–370
3. Climate Launchpad (2015) CO_2ElectroRefinery – Electrochemistry driven conversion of carbon dioxide to chemicals. http://climatelaunchpad.org/finalists/co2electrorefinery/
4. Pant D, Van Bogaert G, Diels L, Vanbroekhoven K (2010) A review of the substrates used in microbial fuel cells (MFCs) for sustainable energy production. Bioresour Technol 101(6): 1533–1543
5. Pandey P, Shinde VN, Deopurkar RL, Kale SP, Patil SA, Pant D (2016) Recent advances in the use of different substrates in microbial fuel cells toward wastewater treatment and simultaneous energy recovery. Appl Energy 168:706–723
6. Rabaey K, Rozendal RA (2010) Microbial electrosynthesis - revisiting the electrical route for microbial production. Nat Rev Microbiol 8:706–716
7. Lovley DR, Nevin KP (2011) A shift in the current: new applications and concepts for microbe-electrode electron exchange. Curr Opin Biotechnol 22:441–448
8. Bajracharya S, Srikanth S, Mohanakrishna G, Zacharia R, Strik DP, Pant D (2017) Biotransformation of carbon dioxide in bioelectrochemical systems: state of the art and future prospects. J Power Sources 356:256–273
9. May HD, Evans PJ, LaBelle EV (2016) The bioelectrosynthesis of acetate. Curr Opin Biotechnol 42:225–233
10. Lu Q, Rosen J, Zhou Y, Hutchings GS, Kimmel YC, Chen JG, Jiao F (2014) A selective and efficient electrocatalyst for carbon dioxide reduction. Nat Commun 5:3242

11. Jhong HR, Ma S, Kenis PJ (2013) Electrochemical conversion of CO_2 to useful chemicals: current status, remaining challenges, and future opportunities. Curr Opin Chem Eng 2(2): 191–199

12. Mahmood MN, Masheder D, Harty CJ (1987) Use of gas-diffusion electrodes for high-rate electrochemical reduction of carbon dioxide. I. Reduction at lead, indium- and tin-impregnated electrodes. J Appl Electrochem 17:1159–1170

13. Kopljar D, Inan A, Vindayer P, Wagner N, Klemm E (2014) Electrochemical reduction of CO_2 to formate at high current density using gas diffusion electrodes. J Appl Electrochem 44:1107–1116

14. Mitacs (2022) Development and characterization of catalyst electrodes for electrochemical reduction of CO2. https://www.mitacs.ca/en/projects/development-and-characterization-catalyst-electrodes-electrochemical-reduction-co2. Accessed 13 Nov 2022

15. DNV - Det Norske Veritas (2007) Positionspapier. Carbon dioxide utilisation: electrochemical conversion of CO_2 – opportunities and challenges

Frank Kensy

Abstract

b.fab GmbH attaches particular importance to design of sustainable bioprocesses. With the formate bioeconomy, the basis for this has been created. From renewable energies, CO_2 and water, formic acid is used to create an almost infinitely available substrate for fermentation that can be converted by formatotrophic microorganisms into valuable products. b.fab uses synthetic biology to implement natural and synthetic metabolic pathways in established production microorganisms. The patented reductive glycine pathway represents a very simple and energy-efficient metabolic pathway, which has the advantage of being able to be operated both anaerobically and aerobically. The broad implementation of the reductive glycine pathway in potent host organisms for the establishment of ecological and at the same time economical bioproductions represents one of the goals of b.fab in the coming years.

Keywords

b.fab GmbH · Formic acid · Formatotrophic microorganisms · Formate bioeconomy

30.1 Introduction

Sugar or sugar-like substances have served for centuries as central substrates for fermentation in order to produce valuable products in bioprocesses. The larger the material flows for such applications become, the more the "tank versus plate" discussion comes into play. This refers to the competition between food and

F. Kensy (✉)
b.fab GmbH, Cologne, Germany
e-mail: kensy@bfab.bio

feedstock production for fuels and chemicals on conventional agricultural land. In a world with an increasing scarcity of natural resources and a foreseeable population growth from 7 to 10 billion people by 2050 [1], a rethink is required. The solution approach of b.fab GmbH is to break away from the dependence on agricultural land and to use an infinite feedstock for bioprocesses and the development of biotechnological value chains. The company relies on the use of renewable electricity, carbon dioxide (CO_2) and water (H_2O) as starting sources. The simplest chemical molecule that can be assembled from these starting sources is formic acid/formate. Formic acid can be produced highly efficiently electrochemically in an electrolyzer from CO_2 and H_2O. Faraday efficiencies of 80–90% and energetic efficiencies of 40–50% have been described [2]. This means that a substrate is available from unlimited sources and can be efficiently produced from renewable energy sources. The advantage of formic acid is also that it exists as a liquid at room temperature and thus represents a liquid hydrogen and at the same time a CO_2 carrier. Unlike gaseous products of electrochemical conversion, such as hydrogen or synthesis gas, formic acid can be stored and thus produced on a decentralized basis. Initially, point sources such as power, steel and cement plants are envisaged as CO_2 sources. With further efficiency improvement of air CO_2 adsorbers (direct air capture, DAC), even direct CO_2 capture from air is possible [3]. Renewable electricity can be generated using photovoltaics or wind power, with preference given to deserts, barren land, or built-up areas. It is even possible to combine photovoltaics and conventional agriculture (agro-photovoltaics) [4].

With this approach, increases in yields per unit area of 30–40 times compared to sugar production from sugar beets or sugar cane (e.g., in Brazil) are possible without actually using agricultural land in the case of formic acid (own calculations).

The use of formic acid as a fermentation substrate has been poorly described. In anaerobes, formatotrophic metabolic pathways are known, but little research has been done on them. The research group of Arren Bar-Even at the Max Planck Institute in Golm/Potsdam was the first to describe the formate bioeconomy and is focusing on the design and improvement of metabolic pathways to utilize formic acid (formate) as the sole source of carbon and energy [5].

30.2   Formate Bioeconomy

b.fab GmbH has set itself the task of establishing the formate bioeconomy commercially (Fig. 30.1). It focuses on the use of the reductive glycine pathway (RGP). This synthetic metabolic pathway uses formic acid as the sole source of carbon and energy. Through five enzymatic steps, formic acid is converted to pyruvate via glycine and serine. The pyruvate then serves as a substrate to form biomass or products. The reductive glycine pathway to pyruvate requires 2 ATP and 3 NAD (P)H. When considering various natural and synthetic metabolic pathways, the reductive glycine pathway appears to be one of the most efficient with only a few enzyme steps. Another advantage of this metabolic pathway is the oxygen tolerance

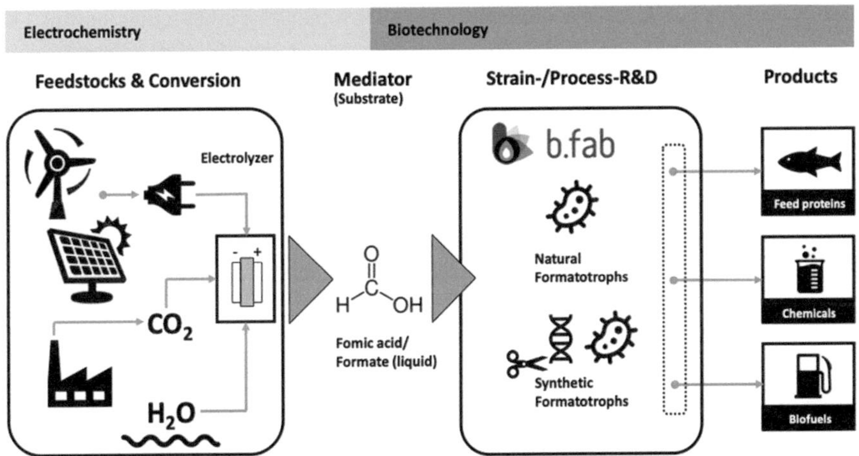

Fig. 30.1 Formate bioeconomy concept. Renewable electricity, CO_2 from industrial point sources (or DAC), and water form the infinite feedstock for the formate bioeconomy. These are converted to formic acid in an electrolyzer. Formic acid serves as a mediator between electrochemistry and biotechnology. Within b.fab, natural and synthetic formatotrophs are screened, designed and developed so that they produce feed, value-added chemicals or biofuels based on the substrate formic acid

of the enzymes used, so that the use of this pathway is possible in anaerobic as well as aerobic microorganisms [6].

b.fab therefore primarily uses the bacteria *Escherichia coli*, *Cupriavidus necator* and the yeast *Saccharomyces cerevisiae* as production organisms. These organisms can be easily manipulated by molecular biology, and the fermentation processes are already established on a large scale, so rapid development in strain generation and subsequent process transfer can be expected.

The use of formic acid as a substrate in fermentation offers the advantage over CO_2 and hydrogen (H_2) that it is indefinitely soluble in water and thus does not limit mass transfer in fermentation. Furthermore, it avoids explosion protection, which requires high costs for safety equipment when using H_2 and also restricts the operation window of the process. For large-scale application, formic acid also offers the advantage of being autosterile due to its strong acid activity. In addition, the production fermenter can be operated without sterilization because no formatotrophic organisms are expected in the natural production environment.

30.3 Potential of the Technology

The formate bioeconomy concept has the potential to revolutionize the manufacturing of new products via biotechnological processes. On the one hand, it relies on an almost infinite raw material of CO_2, water and renewable energies, and on the other hand, it offers a solution for reducing CO_2 emissions via carbon capture and utilization (CCU) at the same time. In order to achieve a net reduction in CO_2 emissions, it makes sense to replace chemicals and fuels from fossil raw materials or to incorporate the CO_2 into products with long-term/permanent benefits.

Initially, b.fab intends to use the technology for the production of animal feed. This is the simplest approach, as it is only necessary to grow biomass in the bioreactor. Bacteria and yeast biomass is very rich in amino acids, vitamins and peptides and is already used in many animal feeds. Other products will be biopolymers as they gain more consumer recognition and are produced in high tonnages. Polylactides and polymers based on succinic acid (succinate) are promising. At a later stage, biofuels are targeted. Here, it is advisable to use a life cycle analysis (LCA) as an accompanying measure at a very early stage of development and to constantly analyze the economic efficiency of the process in parallel, in order to keep the ecological footprint of the process as small as possible and to be able to offer the fuel at competitive prices in the end.

The long-term goal of b.fab is to establish a new biorefinery concept based on formate bioeconomy and to demonstrate that processes can achieve sun-to-product efficiencies of >5% using artificial photosynthesis. Nowadays, natural photosynthesis of plants reaches an efficiency of about 1% (sun-to-sugar) and loses further energy in the production of the final products (total efficiency < 0.2%) [7].

References

1. Abel GJ, Sander N (2014) Quantifying global international migration flows. Science 343(6178): 1520–1522
2. Agarwal AS, Zhai Y, Hill D, Sridhar N (2011) The electrochemical reduction of carbon dioxide to formate/formic acid: engineering and economic feasibility. ChemSusChem 4:1301–1310
3. Marshall C (2017) In Switzerland, a giant new machine is sucking carbon directly from the air. http://www.sciencemag.org/news/2017/06/switzerland-giant-new-machine-sucking-carbon-directly-air. https://doi.org/10.1126/science.aan6915
4. Goetzberger A, Zastrow A (1981) Kartoffeln unter dem Kollektor. Sonnenenergie 3:19–22
5. Yishai O, Lindner SN, Gonzalez de la Cruz J, Tenenboim H, Bar-Even A (2016) The formate bio-economy. Curr Opin Chem Biol 35:1–9
6. Bar-Even A (2016) Formate assimilation: the metabolic architecture of natural and synthetic pathways. Biochemistry 55:3851–3863
7. Hawkins AS, McTernan PM, Lian H, Kelly RM, Adams MWW (2013) Biological conversion of carbon dioxide and hydrogen into liquid fuels and industrial chemicals. Curr Opin Biotechnol 24: 1–9

R&D&I and Industry Examples: Industrial Gases as a Carbon Source for Terpene Production

31

Christian Janke

Abstract

Especially acetogenic microorganisms are biological catalysts with a high-potential biotechnological use for the conversion of exhaust gases into valuable chemicals like terpenes. No doubt, their catalytic potential should be part of a broader decarbonization program. The linear approaches published to date to produce industrially relevant quantities of the desired terpenes will most likely not lead to a successful industrial process. Metabolic and energetic limitations of currently known organisms will not allow this without more sophisticated process strategies. A linear approach in this case refers to the heterologous expression of genes of the mevalonate biosynthetic pathway and a terpene synthase that convert acetyl-CoA into the desired product. Regulatory and energetic integration of terpene biosynthetic pathways into prevailing complex cell metabolic interactions with adenosine triphosphate (ATP) regeneration systems represent the major challenge. The current results with production yields of only milligrams of product available so far indicate the need for more innovative approaches that use modern synthetic biology technologies. These include the development of novel enzymes and the design of complex artificial biosynthetic pathways with logic gates. Synthetic acetogenic cells would then be the final step for terpene production. Tools like CRISPR/Cas9 and available methods for DNA synthesis would allow complex genome wide modifications and state of the art DNA assembly methods have already demonstrated their ability to create new life from scratch. Above all, however, it will need projects who are willing to go beyond the current biological limits of acetogens and other CO_2-fixing microorganisms and compete with plants species, which are often the natural chassis for the production of terpenes today.

C. Janke (✉)
Stolberg, Germany

Keywords

ZeroCarbFP · CO_2 · H_2 · Acetate · Succinic acid · E. coli · Bioethanol

31.1 Technological Diversity as Solution to Better Control Climate Change

Solutions to cope with the anthropogenic climate change and carbon dioxide emissions have been an intensively discussed topic in politics, media, and science for decades, but still wait for fundamental breakthroughs. Real breakthroughs must be able to cope with the carbon cycles of the entire atmosphere itself. The molecule carbon dioxide (CO_2) has the property to absorb electromagnetic radiation in the infrared range, which explains its significant influence on the climate. Rapidly advancing industrialization over the past 100 years has increased the concentration of the greenhouse gas carbon dioxide in the atmosphere by about 100 ppm to its current level of over 410 ppm [1], whereas it has been consistently below 300 ppm for the past hundreds of thousands of years. There is little doubt that this rapid increase is caused by anthropogenic influences. Yet CO_2 is per se a harmless, odorless gas at these concentrations, which is why the consequences of the supposedly small increase in the atmosphere often seem difficult to comprehend. Since this numbers are dimensionless quantities, changes of about 100 ppm are gigantic amount of up to 5×10^{11} tons of CO_2 in the earth atmosphere, because the atmosphere has a total mass of over five quadrillion tons. A long-term reduction of the nutritional value of crops like rice by lowering the protein, vitamin and mineral content due to higher CO_2 concentrations is also discussed in the literature [2]. This would further increase the number of malnourished humans of 0.7–0.8 billion people in 2020 (see fao.org). From this example, one can see that climate change is a complex global problem. But industries and also humans cannot simply stop producing net CO_2 emissions overnight, and furthermore, carbon equilibria in nature have very slow response times of decades and centuries. Nevertheless, there is no question that—forced by the feared consequences of climate change and its self-amplifying effects—we must reach a fundamental new technological level. For years, therefore, research on various strategies for decarbonization has been deepening worldwide. CO_2 is understood as an almost unlimited raw material that can be used in various chemical and biological ways. Also in this regard diversity might be a fundamental value for a healthy and justice society of the future. Best general strategies are prevention e.g., through material substitution in cement industries or the direct capturing and conversion of CO_2 emissions, such as those from steel production, into valuable bulk and fine chemicals. Capturing CO_2 from the atmosphere after it has been emitted is much diluted and therefore energy intensive and probably impossible to accomplish by purely technical processes without the preservation and massive reforestation of land. But today both chemical and biological catalysts already enable the direct conversion of industrial waste gases. Nevertheless, the development of large-scale processes is still hardly advanced for many approaches and, above all, due to the availability of large quantities of inexpensive fossil raw materials (oil, gas, coal), they are not cost-competitive with the well-established petrochemical processes. Industrially relevant chemicals such as isoprene and farnesene are found in large numbers among terpenes and terpenoids. These molecules form a huge class of natural products with high industrial potential. In this subchapter, excerpts from current academic and industrial research will be used to discuss the extent to which the production of terpenes from exhaust gases in biological systems can be seen as realistic from an industrial point of view and what the current state of research is.

31.2 The Importance of Terpenes for our Modern Life

Terpenes and the terpenoids derived from them represent a very large and heterogeneous class of natural products with estimated up to 80,000 representatives [3]. Their industrial potential has so far only been partially tapped. From defense mechanism in plants against predators to the formation of condensation nuclei for the formation of clouds—their biological roles are divers and often not fully understood like the accumulation of isoprene in plants [4]. Industrial relevant representatives range from bulk chemicals, perfumes and flavorings to vitamins and pharmaceuticals. Molecules already in industrial use with a market value of millions to billions of US dollars per year include biofuel precursors such as farnesene, the flavoring agent and solvent limonene as byproduct of orange juice production, the polymerizable isoprene, vitamins such as β-carotene as food additives and various herbal medicines such as hexacyclic ginkgolides in ginkgo extracts as antidementives. Almost every perfume contains terpenes. With a price of several thousand euros per liter, rose oil stands out as a high-priced odorant for the perfume industry. In everyday life, we often encounter many industrially interesting terpenes and terpenoids quite inconspicuously: α-farnesene in apple peel, α-bisabolol in chamomile as an anti-inflammatory substance for the pharmaceutical and cosmetics industries, and carotenoids in vegetables and fruit. People are often unaware that terpenes surround us everywhere and also perform important functions in our bodies, such as the molecule cholesterol. Even the ability for visual perception is only made possible by the carotenoid retinal, in that it leads to the formation of a functional visual pigment through covalent bonding to the ε-amino group of lysine K256 in the opsin. There is another example in our everyday lives: natural rubber. Nearly everyone is familiar with the material. What is less well known is that natural rubber is not only an important raw material for car tires, but also a polyterpene. It consists of cis-1,4-linked isoprene units and is produced on a million-ton scale per year. The terpene artemisinin from the leaves and flowers of the annual mugwort (*Artemisia annua*) exhibits activity against the malaria pathogen *Plasmodium falciparum*. In 2015, Chinese scientist Youyou Tu was awarded the Nobel Prize in Physiology and Medicine for the first isolation of artemisinin. The cytostatic paclitaxel (Taxol) from the bark of the Pacific yew *Taxus brevifolia* play a major role as

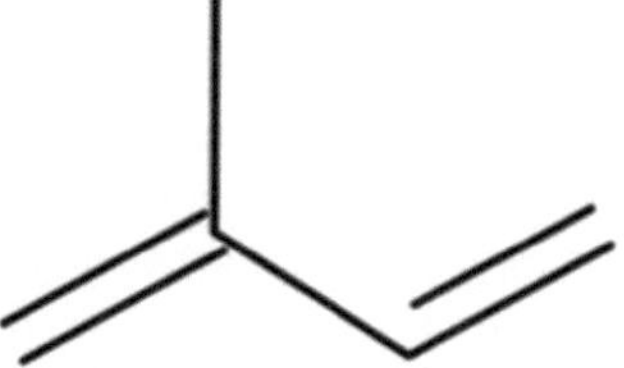

Fig. 31.1 Isoprene—a hemiterpene and at the same time the basic unit for all other terpenes and terpenoids. Their biosynthesis is realized from the phosphorylated C5 bodies dimethylallyl pyrophosphate and isopentenyl pyrophosphate

chemotherapeutic agents. Due to its high price of several hundred US$ per gram, taxol is therefore of economic interest, despite only low quantities required.

The basic structure of terpenes consists of isoprene units. The basic building block isoprene is a molecule of five carbon atoms (C5 body, Fig. 31.1). Therefore, a basic classification of terpenes is also based on the number of isoprene units: Hemiterpenes (C5), monoterpenes (C10), sesquiterpenes (C15), diterpenes (C20), ester terpenes (C25), triterpenes (C30), and tetraterpenes (C40). Terpenes are pure hydrocarbons, and thousands of terpenes exist. However, the synthesis of terpenes does not start from isoprene molecules themselves, but from the phosphorylated precursor molecule dimethylallyl pyrophosphate (DMAPP) and its isomer isopentenyl pyrophosphate (IPP). Both molecules are provided for terpene synthesis by the mevalonate pathway [5] or the methylerythritol phosphate pathway [6]. Terpenoids contain additional functional groups. Academically, these structurally diverse natural products found early appreciation. For their fundamental work in the field of terpenes, both Otto Wallach in 1910 and Leopold Ruzicka in 1939 were awarded the Nobel Prize in Chemistry.

The isoprene monomer itself has a strong ecological significance, as plants worldwide emit millions of tons of isoprene into the atmosphere every year [7]. This once led former US President Ronald Reagan to say "Trees cause more pollution than automobiles do" [8]. Even if this view is far too simplistic, it can be stated that terpenes play an important role in the carbon cycles of the atmosphere [9]. Nevertheless, till today isoprene is mainly produced as a byproduct of petroleum processing. However, not only plants, but even humans, without possessing a terpene synthase, emit trace amounts of isoprene in the parts-per-billion range [10]. The enzymatic synthesis of the hemiterpene isoprene is comparatively simple. Isoprene synthases catalyze an elimination reaction of pyrophosphate from the substrate DMAPP. Isoprene synthases have their pH optimum at slightly alkaline pH around 8 and a turnover number k_{cat} of $\approx$0.5–2 s^{-1}. Thus, in contrast to classical textbook examples such as carbonic anhydrase with a turnover number of up to 600,000 s^{-1}, these terpene synthases appear at the lower end of the enzyme scale. The catalytic efficiency of terpene synthases is generally considered low. In nature energy consumption has to be reduced to a maximum. Hence, enzymes are optimized for their biological role(s) and evolved in a way that allow adaption to environmental changes. From the biological point of view a highly active terpene synthase for wild type organisms might not be desirable, because of the energy-consuming metabolic pathways to DMAPP and IPP due to the high consumption of the cellular universal energy carrier adenosine triphosphate (ATP). Otherwise, the cell would be exposed to metabolic stress through high pools of the toxic isoprenoid precursors DMAPP and IPP [11] and an unbalanced consumption of ATP and reduction equivalents. Finally, this would then result in a negative selection pressure on terpene synthesis. For industrial processes mainly catalytic activity counts. Engineering of terpene synthases did not prove to be as successful as it has been for other industrial relevant enzymes [12] [13]. Therefore, future engineering has to create something new and better than natural evolution. Altering the substrate of known enzymes like e.g., the hydratase from *Elizabethkingia meningoseptica* by using isoprenol as precursor show that it is in principle possible to create alternative, energetically more favored routes to isoprene [14].

31.3 Production of Terpenes in Biological Systems

Naturally, terpenes are mainly produced by plants, which convert phototrophically CO_2 to glucose. Terpene containing extracts like resins from trees have been used by humans since the Stone Age as a glue and for religious rites. Extraction from plants is done for turpentine with its main volatile compounds α-pinene and β-pinene. One famous pharmaceutical relevant secondary metabolite is eucalyptol extracted from the tree *Eucalyptus globulus*. Due to the complexity of plants and their time-consuming (genetic) manipulation for the enhanced production of the terpene of choice, which is most often not established for many species, industry prefers to focus on common microbial model organisms such as *Escherichia coli* or *Saccharomyces cerevisiae*, which show heterotrophic growth only [15]. Genencor (Rochester, New York, USA) and Goodyear Tire & Rubber Company (Akron, Ohio, USA) have developed the product BioIsoprene. California-based Amyris (Emeryville, California, USA) developed processes for large-scale production of terpenes such as β-farnesene from cellulosic sugar, which can serve as a precursor molecule for various products. Full hydrogenation to farnesane produces a usable diesel and jet fuel substitute that was tested by Lufthansa Group (Frankfurt am Main, Germany) in 2011. In early 2018, Amyri's Brotas 1 production plant for β-farnesene in Brazil was purchased by Dutch company DSM (Heerlen, Netherlands) for US$96 million. The plant produces up to 24 kT per year. In addition, Amyris uses its know-how in terpene production in yeasts to manufacture high-quality fine chemicals. Another prominent example is the production of artemisinic acid, a precursor of the antimalarial drug artemisinin in yeast [16]. But besides all the fascination for the possibilities of genetic engineering and of bioorganic chemistry in general and the great scientific work that has been done, it must be soberly noted that up to now many processes with biological catalysts have often failed to meet economic expectations. Furthermore, efficient industrial waste gas utilization by *Escherichia coli* and *Saccharomyces cerevisiae* has not been realized so far, although the implementation of metabolic pathways for the efficient fixation of C1 molecules would theoretically be possible for these production strains.

Interestingly, classical production in recombinant yeasts is getting competition from a process using crude plant extracts [17]. The company ArtemiFlow is already using this new process. Next to the ArtemiFlow example just mentioned, another sustainable approach for terpene and terpenoid production has been developed by Fraunhofer IME, Westfälische Wilhelms-Universität Münster, the plant breeding company ESKUSA, Continental Reifen GmbH Germany and the Julius Kühn Institute for the isolation of natural rubber from the Russian dandelion *Taraxacum koksaghyz* [18]. In addition to the reduction of the CO_2 by large-scale cultivation of the rubber producing plants in Europe, the production yields of the rubber for the manufacturing of car tires seem to become economically competitive in the near future. Moreover, heterologous expression of known genes of the mevalonate pathway from *Arabidopsis thaliana* in dandelion leads to accumulation of the unsaturated triterpene squalene [19]. Since squalene is still also extracted from sharks [81], its large-scale production in plants would be an extremely environmentally friendly and promising process. Squalene and its hydrogenated form squalane also have great potential, including for use as transformer oil and lubricants. They are already used for medical and cosmetic applications [20]. The use of cyanobacteria is another alternative from the group of photosynthetic production organisms, and a feasibility has already been shown here [21]. Regarding the use of industrial waste gases as a carbon source, especially for plants and also for cyanobacteria, a direct conversion of industrial waste gases can be considered as a difficult way to reduce CO_2 emissions from industry, because the process control is extremely challenging due to the light and space requirements.

If one takes into account the consequences of climate change with water scarcity and an increase in the world's population to over 9 billion by 2050, then the presented biotechnological approaches seem to miss sustainable features for large scale terpene production. Either these processes are powered by carbon sources that compete in some way with human food chains, or they require not only an excessive amount of space and light, but also a lot of water, especially in the case of plants. Moreover, direct, desirable coupling of production processes with climate-damaging industries is often not readily possible. Existing chemical processes to convert these emissions have advantages here in terms of rapid implementation, such as Fischer-Tropsch synthesis, and their use in the future is very likely. But biological systems are more specific in synthesis and more adaptable to gas impurities compared to chemical catalysts. Furthermore, they need lower temperatures and therefore less energy for production. Certainly, to this day, nature is often still superior to our knowledge of chemistry when it comes to synthesizing terpenes. Moreover, methods like Fischer-Tropsch synthesis are not at all suitable for terpene production. For this reason, CO_2-fixing microorganisms have become a focus for biotechnological terpene production in the last 10–15 years. These often strictly anaerobic microorganisms cannot only convert CO_2 and CO to acetate and alcohols via i.a. the Wood-Ljungdahl pathway (WLP), but also enzymatically produce a desired, defined product around the clock without the need of light. Moreover, these organisms are not inhibited by high CO_2 concentrations in contrast to plants. Acetogens include many Clostridia such as *C. ljungdahlii*, *C. autoethanogenum*, and *C. carboxidivorans*. All known acetogens are in principle capable of producing terpenes. *C. carboxidivorans* is particularly suitable for the production of alcohols ranging from ethanol to octanol [22–26]. But also aerobic, carboxydotrophic strains such *Oligotropha carboxidovorans* [27] or *Hydrogenophaga pseudoflava*, which use the Calvin cycle for CO_2 fixation, have come into focus. With the WLP, one resorts to biological catalysts whose origin lies in a time of our planet when there was no oxygenated atmosphere [28]. It is believed that the common ancestor of cellular life possessed some form of WLP. Having made a basal contribution to the evolution of life on our planet, it is hoped that WLP can now be used to help shape the future of our planet. Extensive feasibility studies exploring its true potential using modern methods of molecular biology have been ongoing for several years and are the subject of discussion.

31.4 Feasibility Studies of Terpene Production from Exhaust Gases by Acetogens

Global players such as DuPont/Danisco [29] or Sekisui Chemicals [30] and companies such as LanzaTech [31], have so far attracted attention through patents and publications in this field. According to their publications it is possible to produce in the range of micro- to milligrams per liter of cell culture isoprene and farnesene from industrial waste gases if the genes for the mevalonate pathway and the corresponding synthases were introduced. So far, nothing is known about industrially relevant production quantities or even the planning of a pilot plant. Hence, the technological readiness level can be considered low. A publication in the Journal of Applied and Environmental Microbiology confirms the difficulty of terpene production from synthesis gas using isoprene [32]. Limitations in terpene production exist for acetogens due to the energetics of chemoautotrophic growth as well as through the genes and metabolic pathways to be introduced. As discussed by Bertsch and Müller 2015 [33], ATP supply is a fundamental limitation of acetogenic metabolism. The Wood-Ljungdahl pathway requires one molecule of ATP generated by acetate formation, so that net (almost) no ATP is generated by chemoautotrophic growth. Via an ion gradient (either Na^+ ions or protons) and ATP synthases, only ≈ 0.3 ATP/acetate is synthesized. Moreover, the mevalonate pathway directly competes with metabolism for acetyl-CoA, which the acetogenic microorganism laboriously obtains from CO/CO_2 [34]. Accordingly, one way to increase terpene production anyway would be to increase acetate production. In addition, a change in gas composition toward carbon monoxide would positively influence energetics. However, the charm of biological exhaust gas utilization would then be gone if the gas composition is so decisive and the carbon is mainly fixed in the acetate. Nitrate could also be added as an electron donor [35]. Evolutionary adaptation by classical strain improvement methods such as chemical mutagenesis, UV mutagenesis or genome shuffling to a desired gas composition for a production strain cannot be considered

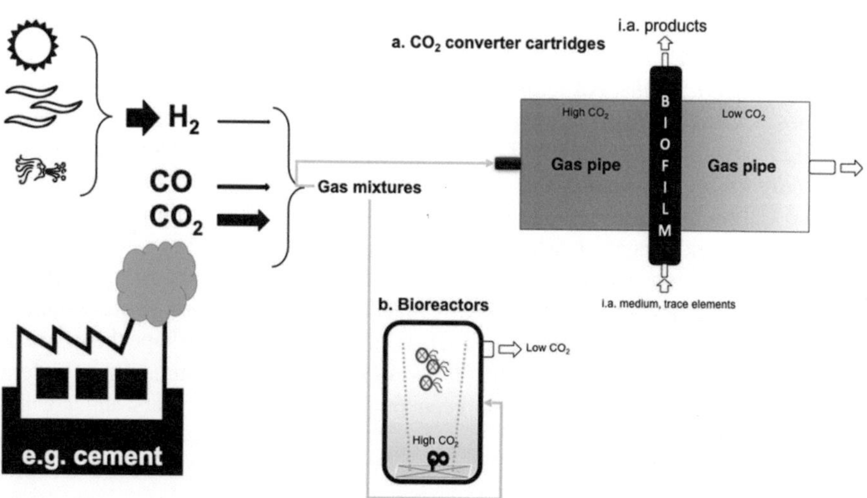

Fig. 31.2 Routes for terpene production by CO_2 metabolizing cell factories in bioreactors or biofilm-based CO_2 converter cartridges (simplified model)

purposeful, since experience has shown that there is a negative selection pressure on the heterologous expression of genes for terpene production in unicellular, non-compartmentalized model organisms. While multiple genetic engineering tools exist for yeasts and aerobic prokaryotes such as *Escherichia coli*, complex regulation with logic gates of a biosynthetic pathway in acetogens has so far only been possible in a rudimentary form. In addition, there is a problematic time consumption for process development. While in *Escherichia coli* an experimental cycle for genetic modification including product and sequence analysis takes five working days, the same process can take up to 1 month in acetogenic microorganisms if synthesis gas is used as a sole carbon source. To sum up, the production of terpenes in acetogens does not appear to make much sense at first glance. Nevertheless, acetogenic and other CO_2 metabolizing microorganisms offer a great potential for CO_2-neutral processes or even with a negative CO_2 balance. These organisms are very robust for gas fermentation and offer a useful physiological diversity.

An alternative to acetogens are aerobic organisms that use the Calvin cycle for CO_2 fixation. In *Hydrogenophaga pseudoflava*, the terpene (E)-α-bisabolene was successfully produced from syngas [36]. Moreover, the possibilities of metabolic engineering and process optimization are far from being fully exploited. These could start with the use of robust biofilms that capture CO_2 in conversion cartridges directly at the point of origin (Fig. 31.2) and might also produce terpenes and/ or other products like proteins or lipids. The biofilms could e.g., be placed between gas-permeable membranes. The design, arrangement and number of the stackable biofilms (e.g., lined up in a pipe system), matrix material and size would be variable, adaptable and scalable—depending on the application. Other adaptations of packed reactors would of course also be conceivable. Besides well-known representatives, sustainable plant polymers such as glucomannan could also be used as a matrix. Furthermore, the low solubility of hydrogen and therefore the local availability in the aqueous phases could be increased not only by increasing the pressure but through nanoparticles. In nature, substances are built up in microbial communities. Instead of trying to fix CO_2 and build terpenes in only one organism, it would be possible to use the potential of waste gas metabolization of e.g., acetogens to have the acetate used again as a carbon source by a production strain. Other intermediate stages are conceivable. Stepwise assembly of complex organic compounds by microbial catalysts in cocultivation or multi-step fermentation allows to circumvent the energetic limitation of the approach with recombinant acetogens.

In conclusion, classical linear biotechnological approaches, as presented in the feasibility studies of the companies, do not seem to be sufficient for the establishment of an industrial process for terpene production in acetogens and other CO_2 fixing microbes so far. Despite all the problems predicted by energetic considerations and already documented in publications and patents, the last word has not yet been spoken for terpene production from waste gases by microbial catalysts. Synthetic biology with its possibilities to create genomes, artificial enzymes and artificial metabolic pathways is only beginning to unfold its potential with CO_2-fixing microorganisms. Whether the lack of compartmentalization of bacteria can prevail over plant systems such as a hypothetical genetically engineered, vertically grown halophyte with numerous storage possibilities, only time will tell.

References

1. Keeling RF, Keeling CD (2017) Atmospheric monthly in situ CO2 data – Mauna Loa Observatory, Hawaii. In: Scripps CO2 program data. UC San Diego Library Digital Collections
2. Ebi KL, Ziska LH (2018) Increases in atmospheric carbon dioxide: anticipated negative effects on food quality. PLoS Med 15:e1002600
3. Christianson DW (2018) Correction to structural and chemical biology of terpenoid cyclases. Chem Rev 118(24):11795
4. Sharkey TD, Monson RK (2017) Isoprene research – 60 years later, the biology is still enigmatic. Plant Cell Environ 40:1671–1678
5. Miziorko HM (2011) Enzymes of the mevalonate pathway of isoprenoid biosynthesis. Arch Biochem Biophys 505:131–143
6. Zhao L, Chang WC, Xiao Y, Liu HW, Liu P (2013) Methylerythritol phosphate pathway of isoprenoid biosynthesis. Annu Rev Biochem 82:497–530
7. Koksal M, Zimmer I, Schnitzler JP, Christianson DW (2010) Structure of isoprene synthase illuminates the chemical mechanism of teragram atmospheric carbon emission. J Mol Biol 402:363–373
8. Radford T (2004) Do trees pollute the atmosphere? The Guardian. https://www.theguardian.com/science/2004/may/13/thisweekssciencequestions3. Zugegriffen: 2 Apr 2019
9. Kansal A (2009) Sources and reactivity of NMHCs and VOCs in the atmosphere: a review. J Hazard Mater 166:17–26
10. Cailleux A, Cogny M, Allain P (1992) Blood isoprene concentrations in humans and in some animal species. Biochem Med Metab Biol 47:157–160

11. Wang Q, Quan S, Xiao H (2019) Towards efficient terpenoid biosynthesis: manipulating IPP and DMAPP supply. Bioresour Bioprocess 6:6. https://doi.org/10.1186/s40643-019-0242-z
12. Leferink NGH, Scrutton NS (2022) Predictive engineering of class I terpene synthases using experimental and computational approaches. Chembiochem 23:e202100484
13. Zhang C, Hong K (2020) Production of terpenoids by synthetic biology approaches. Front Bioeng Biotechnol 8:347. https://doi.org/10.3389/fbioe.2020.00347
14. Li M, Nian R, Xian M et al (2018) Metabolic engineering for the production of isoprene and isopentenol by Escherichia coli. Appl Microbiol Biotechnol 102:7725–7738. https://doi.org/10.1007/s00253-018-9200-5
15. Wang C, Liwei M, Park JB, Jeong SH, Wei G, Wang Y, Kim SW (2018) Microbial platform for terpenoid production: Escherichia coli and yeast. Front Microbiol 9:2460
16. Paddon CJ, Westfall PJ, Pitera DJ, Benjamin K, Fisher K, McPhee D, Leavell MD, Tai A, Main A, Eng D, Polichuk DR, Teoh KH, Reed DW, Treynor T, Lenihan J, Fleck M, Bajad S, Dang G, Dengrove D, Diola D, Dorin G, Ellens KW, Fickes S, Galazzo J, Gaucher SP, Geistlinger T, Henry R, Hepp M, Horning T, Iqbal T, Jiang H, Kizer L, Lieu B, Melis D, Moss N, Regentin R, Secrest S, Tsuruta H, Vazquez R, Westblade LF, Xu L, Yu M, Zhang Y, Zhao L, Lievense J, Covello PS, Keasling JD, Reiling KK, Renninger NS, Newman JD (2013) High-level semi-synthetic production of the potent antimalarial artemisinin. Nature 496:528–532
17. Triemer S, Gilmore K, Vu GT, Seeberger PH, Seidel-Morgenstern A (2018) Literally green chemical synthesis of artemisinin from plant extracts. Angew Chem Int Ed Engl 57:5525–5528
18. Continental (2018) Continental eröffnet Forschungs- und Versuchslabor für Löwenzahnkautschuk "Taraxagum Lab Anklam". https://www.continental-reifen.de/autoreifen/media-services/archiv/archiv-2018/20181206-continental-eroeffnet-taraxagum-lab-anklam. Accessed 2 Apr 2019
19. Putter KM, van Deenen N, Unland K, Prufer D, Schulze Gronover C (2017) Isoprenoid biosynthesis in dandelion latex is enhanced by the overexpression of three key enzymes involved in the mevalonate pathway. BMC Plant Biol 17:88
20. Chaves JE, Melis A (2018) Biotechnology of cyanobacterial isoprene production. Appl Microbiol Biotechnol 102:6451–6458
21. Spanova M, Daum G (2011) Squalene - biochemistry, molecular biology, process biotechnology, and applications. Eur J Lipid Sci Technol 113:1299–1320
22. Richter H, Molitor B, Diender M, Sousa DZ, Angenent LT (2016) A narrow pH range supports butanol, hexanol, and octanol production from syngas in a continuous co-culture of Clostridium ljungdahlii and Clostridium kluyveri with in-line product extraction. Front Microbiol 7:1773
23. Fernandez-Naveira A, Veiga MC, Kennes C (2019) Selective anaerobic fermentation of syngas into either C2-C6 organic acids or ethanol and higher alcohols. Bioresour Technol 280:387–395
24. Doll K, Ruckel A, Kampf P, Wende M, Weuster-Botz D (2018) Two stirred-tank bioreactors in series enable continuous production of alcohols from carbon monoxide with Clostridium carboxidivorans. Bioprocess Biosyst Eng 41:1403–1416
25. Abubackar HN, Veiga MC, Kennes C (2018) Production of acids and alcohols from syngas in a two-stage continuous fermentation process. Bioresour Technol 253:227–234
26. Cotter JL, Chinn MS, Grunden AM (2009) Ethanol and acetate production by Clostridium ljungdahlii and Clostridium autoethanogenum using resting cells. Bioprocess Biosyst Eng 32:369–380
27. Paul D, Bridges S, Burgess SC, Dandass Y, Lawrence ML (2008) Genome sequence of the chemolithoautotrophic bacterium Oligotropha carboxidovorans OM5T. J Bacteriol 190:5531–5532
28. Weiss MC, Sousa FL, Mrnjavac N, Neukirchen S, Roettger M, Nelson-Sathi S, Martin WF (2016) The physiology and habitat of the last universal common ancestor. Nat Microbiol 1:16116

29. Beck ZQ, Cervin MA, Chotani GK, Peres Caroline M, Sanford KJ, Scotcher MC, Wells DH, Whited GM (2013) Compositions and Methods of producing isoprene and/or industrial bio-products using anaerobic microorganisms. WO/2013/181647
30. Masahiro F, Akihiro U, Koichiro I, Jennewein S, Fischer R (2013) Recombinant cell, and method for producing isoprene. US9783828B2
31. Chen W, Liew F, Köpke M (2013) Recombinant microorganisms and uses therefore. WO/2013/180584
32. Diner BA, Fan J, Scotcher MC, Wells DH, Whited GM (2018) Synthesis of heterologous mevalonic acid pathway enzymes in Clostridium ljungdahlii for the Conversion of fructose and of syngas to mevalonate and isoprene. Appl Environ Microbiol 84(1):e01723–e01717
33. Bertsch J, Müller V (2015) Bioenergetic constraints for conversion of syngas to biofuels in acetogenic bacteria. Biotechnol Biofuels 8:210
34. Emerson DF, Stephanopoulos G (2019) Limitations in converting waste gases to fuels and chemicals. Curr Opin Biotechnol 59:39–45. https://doi.org/10.1016/j.copbio.2019.02.004
35. Emerson DF, Woolston BM, Liu N, Donnelly M, Currie DH, Stephanopoulos G (2019) Enhancing hydrogen-dependent growth of and carbon dioxide fixation by clostridium ljungdahlii through nitrate supplementation. Biotechnol Bioeng 116(2):294–306. https://doi.org/10.1002/bit.26847
36. Grenz S, Baumann PT, Rückert C, Nebel BA, Siebert D, Schwentner A, Eikmanns BJ, Hauer B, Kalinowski J, Takors R, Blombach B (2019) Exploiting Hydrogenophaga pseudoflava for aerobic syngas-based production of chemicals. Metab Eng 55:220–230. https://doi.org/10.1016/j.ymben.2019.07.006

Introduction

Wolfgang Wach, Edda Höfer, Guido Meurer, Philip Weyrauch,
and Martin Langer

Abstract

ZeroCarbFP's goal is to establish bioprocesses for the conversion of carbon-rich waste streams into valuable products. Four CO$_2$ reduction pathways have been developed and intensively tested. The most promising pathway was selected to be further developed towards piloting. In the first step CO$_2$ and H$_2$ are converted into acetate by a specialized anaerobic microorganism. The second stage is flexibel to a certain extent as the used microorganism can be adapted in accordance to the desired chemical product. In the case of conversion of acetate into succinic acid we use an E. coli strain. The two-stage fermentation developed, converting CO$_2$ and hydrogen gas into acetate and succinic acid, can be coupled to an existing bioethanol plant providing CO$_2$.

Keywords

ZeroCarbFP · CO$_2$ · H$_2$ · Acetate · Succinic acid · E. coli · Bioethanol

Shrinking resources, increasing CO$_2$ emissions and rising waste production urgently call for a re-thinking and re-organization within the manufacturing industry. A sustainable, bio-based economy should establish resource-effective and waste-avoiding production processes to replace conventional production strategies and products. To support this transition, the German Federal Ministry of Education

W. Wach · E. Höfer (✉)
Südzucker AG, CRDS, Obrigheim (Pfalz), Germany
e-mail: wolfgang.wach@suedzucker.de; edda.hoefer@suedzucker.de

G. Meurer · P. Weyrauch · M. Langer
BRAIN Biotech AG, Zwingenberg, Germany
e-mail: gm@brain-biotech.de; philip.weyrauch@mailbox.org; ml@brain-biotech.de

© The Author(s), under exclusive license to Springer Nature Switzerland AG 2023
M. Kircher, T. Schwarz (eds.), *CO2 and CO as Feedstock*, Circular Economy and
Sustainability, https://doi.org/10.1007/978-3-031-27811-2_32

and Research (BMBF) already in 2012 started the "Industrial Biotechnology Innovation Initiative".

Under this framework several strategic alliances of academia and industry aim to tackle some of the most challenging topics of the twenty-first century in close cooperation. By developing innovative products and systems with high potential for added value the change from a fossil to a bio-based industry is promoted.

The use of alternative raw materials supports defossilization and the independence from fossil feedstock by having a positive effect on cost structures and competitiveness. Thereby, the alliances take up specific central innovation subjects for politics, economy and the society like protection of arable land for food production or sustainable use of waste streams by recycling carbon and thereby contributing to the implementation of a circular economy ("cradle to cradle"). Furthermore, a cross-alliance PhD program strengthens the future development of sustainable production routes, by implanting the idea of close cross-linking of academia and industry on the topic with the next generation of scientists.

32.1 Project Description

One of the selected innovative strategic alliances to boost this is "Zero-Carbon-FootPrint (ZeroCarbFP)". ZeroCarbFP's goal is to establish bioprocesses for the conversion of carbon-rich waste streams into valuable products. This BMBF-funded program started in July 2013 and is planned to be finished in March 2024. The time frame is split in three phases of 3 or 4 years, respectively, in order to perform research, development and piloting. The strategic alliance consisted initially of 12 partners from academia and rather diverse industries, who work together in five different sub-programs, each led by an initiative company (Fig. 32.1). Each sub-program addressed a specific carbon-rich side stream for the production of building blocks of higher value *e.g.* the conversion of used cooking oil for the provision of high tech lubricants. All approaches were evaluated within a sustainability assessment from a social, ecological and economical point of view, led by the nova-Institute, Cologne, Germany as the external expert (Fig. 32.2). In the third phase of the alliance two of initially five sub-programs are still active: "Bioplastics" (Südzucker, Mannheim) and "Additives 1" (Fuchs, Mannheim).

32.1.1 ZeroCarbFP "Bioplastics"

ZeroCarbFP sub-program "Bioplastics" aims with carbon dioxide as the sole carbon source a sustainable and biodegradable polymer chemistry by biotechnological production of monomers (mono- and dicarboxylic acids). In this set-up, Südzucker is coordinator, technology developer and raw material supplier. Why?—well, Südzucker has wide experience in bio-based feedstock and downstream processing. Next to established and biotechnologically driven processes like bioethanol and

Fig. 32.1 Strategic Alliance ZeroCarbFP

palatinose production, Südzucker has a long-term experience in working with hydrogen.

BRAIN Biotech, a specialist in industrial biotechnology and expert in the design and provision of (whole cell) biocatalysts has been supporting Südzucker right from the beginning of ZeroCarbFP. The company has created a bio-archive providing a huge number of microbial strains for a broad range of applications. Based on this shared know-how an innovative two-step microbial conversion process has been established.

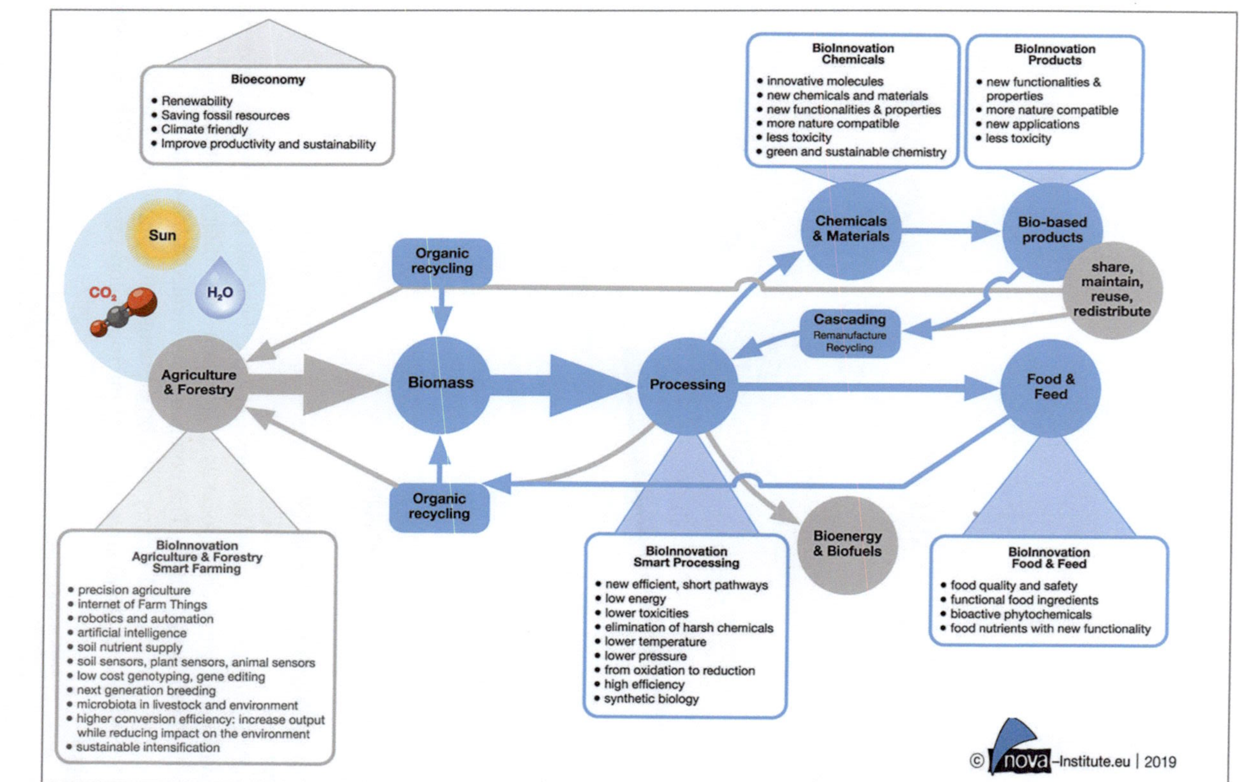

Fig. 32.2 Bioeconomy: More than circular economy

32.1.2 Feedstock

Südzucker's production sites started with the implementation of circular economy and the continuously extension of a biorefinery concept quite a long time ago. Hereby, the aim is to integrate co-products in the existing value chain next to the generation of new added value. At Zeitz, Saxony-Anhalt, a production site in the east of Germany, factories for the production of sugar and starch (Südzucker) and bioethanol (subsidiary CropEnergies) are co-located.

During bioethanol production carbon dioxide is released as a co-product. The CO_2 is of biogenic origin as it develops during fermentation of plant-based feedstocks like sugar beets and grain. In 2010 a joint venture with SOL Group was founded for the liquefaction of 100.000 t CO_2/year. Next to the provision of liquified CO_2 for the food industry in order to carbonize beverages, GHG emissions onsite can be simultaneously reduced. Besides Zeitz, there are further production sites in Europe providing biogenic CO_2 streams. Thus, measures for the material use of CO_2 are highly investigated. Within ZeroCarbFP the focus is on biotechnological process development. Apart from biocatalysis chemically catalyzed carbon capture and utilization technologies are considered as well. A purity of 99% makes the biogenic carbon dioxide unique and very attractive raw material for such further processing.

32.1.3 Product

Already in 2004 succinic acid was classified by DOE (US-Department of Energy) as one of the twelve most promising bio-based building blocks for the future [1]. Due to its linear and saturated carboxylic structure it can be converted to huge number of nowadays still fossil-based base chemicals. Companies have already started to produce succinic acid via fermentation from C6 sugars and there are various efforts applied to feed lignocellusic second generation sugars to succinic acid producing microorganisms as well. CO_2-based approaches are technically still under-represented. One special opportunity is seen in the provision of succinic acid as educt for polybutylensuccinate (PBS) and its co-polyesters (PBSA). PBS and PBSA are biodegradable and compostable bio-polymers, produced from succinic acid, 1,4-butanediol (1,4-BDO) and often a third monomer like an organic di-acid [2].

32.1.4 Technology

Within phase 1 (2014–2017) of ZeroCarbFP "Bioplastics" four CO_2 reduction pathways have been developed and intensively tested with the support of a number of academic partners like Karlsruhe Institute of Technology (KIT), Helmholtz Center for Environmental Research (UFZ) Leipzig, Technical University Darmstadt and Goethe University Frankfurt am Main. In phase 2 (2017–2020) the most promising pathway was selected to be further developed towards piloting in phase 3 (2020–2024). In the first step CO_2 and H_2 are converted into acetate by a specialized anaerobic microorganism. The second stage is flexibel to a certain extent as the used microorganism is genetically well understood and can be adapted in accordance to the desired chemical product. In the case of conversion of acetate into succinic acid we use an *E. coli* strain.

In order to optimize this *E. coli* strain for conversion of acetate into succinate, a protein system required for the uptake and activation of acetate (to intracellular acetyl-CoA) was up-regulated. Furthermore, an highly efficient and specific succinate export system was introduced to avoid intracellular accumulation of the product. Several genes contributing to metabolic diversions towards unwanted side-products like other organic acids were eliminated and the metabolic flux from acetyl-CoA to succinate via the glyoxylate bypass of the citric acid cycle was optimized. For this purpose, expression levels of the contributing genes were balanced in a combinatorial approach. The current production strain enables the bioconversion of acetate into sucinate as the sole product at a volumetric productivity of >2 g/(L·h) and an efficiency (▯) of up to 0.8.

In general, the two-stages process provides a modular platform where different CO_2 sources can be used as a feedstock. A key aspect of the current development is the coupling of the two process stages and the establishment of a continuous process, which can be efficiently operated on a routine basis. At the present laboratory process scale, recorded successes include the efficient use of authentic carbon dioxide from bioethanol production, first scale-up of the process from 1 to 20 L and optimization of biocatalysts' efficiencies to economically interesting and competitive numbers. For further proof of technology and for boosting TRL (technology readiness level) from 4 to 6, a further scale-up and operation in an authentic environment is essential as well as downstream process development. Especially the latter one is challenging due to lower concentrated product solutions in bio-based processes.

32.2 Results

Fermentative, large-scale industrial processes have so far primarily concentrated on the maximization of the main product yield for achieving sustainability criteria. However, the example of bioethanol production shows that not all carbon provided can be bound in one product. The liquefaction and utilisation of the fermentation-offgas in the beverage industry was an important step towards improving the profitability of such refinery concepts at Südzucker.

Within the sub-program "Bioplastics" of ZeroCarbFP microbial fixation of carbon dioxide is investigated. The two-stage fermentation developed, converting CO_2 and hydrogen gas into acetate and succinic acid, can be coupled to an existing bioethanol plant providing CO_2 (Fig. 32.3). This eliminates the need to compress and transport CO_2 and enables the cascading use of CO_2. Hydrogen can be produced onsite via electrolysis powered by renewable electricity (*i.e.* wind energy). Using waste and side streams as carbon source leads to an overall higher carbon utilization

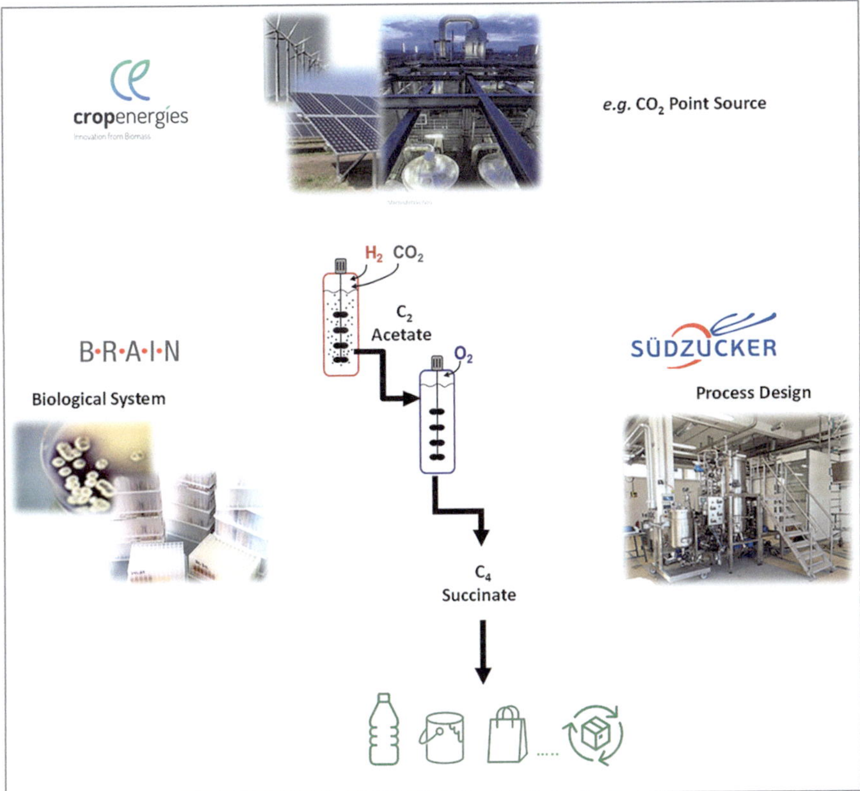

Fig. 32.3 Building block for bioplastic

efficiency, reduces the dependency on critical raw materials and increases the flexibility of production sites.

The replacement of fossil-based processes, which could potentially involve toxic substances and demand a lot of energy, leads to an improved safety for employees, less hazardous waste and easier waste disposal. Bio-based processes also replace fossil carbon, but require land for biomass cultivation. By using CO_2 as raw material the full biomass utilisation is maximized and a balanced demand for food, feed and materials can be achieved. It is anticipated that the acceptance of CO_2-based processes in the society generally can be rated higher since the demand for greener products increases. However, the feasibility of the herein described ZeroCarbFP CO_2-based process is still uncertain as its latest evaluation is still based on laboratory data.

Upscaling is necessary in order to obtain more reliable information towards efficient production, cost comparison as well as a deeper evaluation of the environmental impact. For an economic success four factors are thought to be most relevant:

- Price development of green hydrogen and CO_2
- Amount and necessity of process control chemicals
- Energy and water consumption and degree of recycling
- Efficacy of downstream processing and product purification

Concerning the utilization of exhaust gases for material use to produce high value biopolymers such as succinic acid, the following can be stated from a technical point of view:

- Biogenic CO_2 can be used without restrictions as carbon source
- Highly efficient utilisation of hydrogen
- Carbon derives to 100% from CO_2
- Succinate is the sole product
- No by-products
- High productivity: 80% of the target value
- High overall synthesis efficiency: $\eta = 0.8$
- High product purity: 99.7% purity

32.2.1 Fundamental Questions

- Does the development of an alternative production process for succinic acid (2-step fermentation of CO_2 and H_2/acetate) lead to a more sustainable production process than glucose-based fermentation?
- How can a CO_2-based process compete with long existing fossil-based processes?
- Can potential additional costs be justified by additional environmental and social benefits?

To come closer to an answer, VDI 4605, a new guideline published in November 2017 with the aim of introducing a comparative assessment of sustainability in processes and products, was applied in the second phase (2016–2019) of ZeroCarbFP. The VDI 4605 standard is based on a four shell model that depicts the three dimensions of sustainability and takes into account relevant VDI, ISO and DIN standards. Ecological, economic and social criteria have been determined in order to compare the CO_2-based with the fossil- and bio-(glucose)-based processes. Major output was, that with regard to ecological and social criteria the CO_2-based process outperforms the bio- and the fossil-based process. Concerning economic aspects the bio-based process is still better, but the fossil-based route has no advantage. To fully exploit the potential of CO_2-based processes and to underline the environmental, social and economic benefits by life cycle assessment (LCA), a further development and upscaling is obligatory. The outcome of the VDI assessment at that stage is shown in Fig. 32.4.

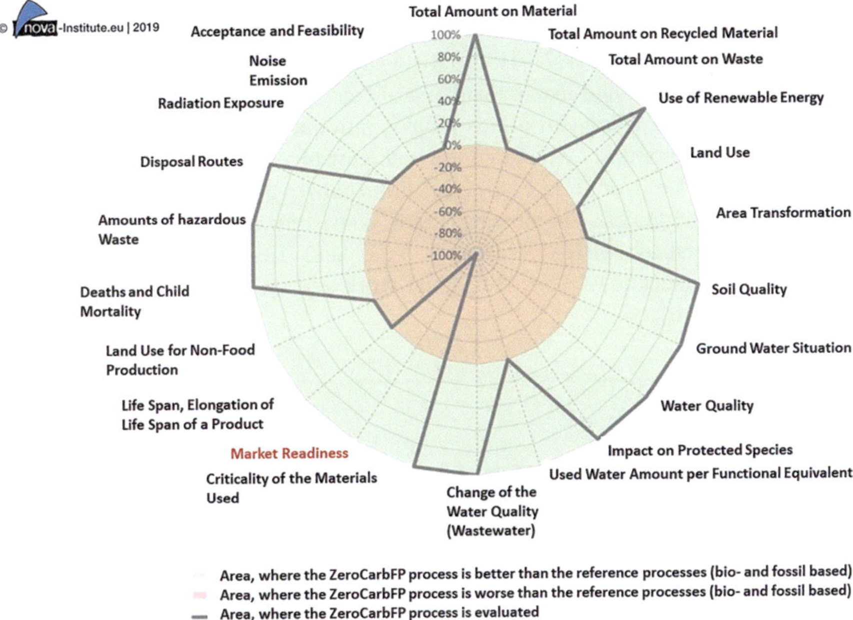

Fig. 32.4 VDI 4605 assessment

32.3 Conclusion

ZeroCarbFP's two-step microbial CO$_2$ conversion technology provides a process with a high overall efficiency already at lab scale. Its implementation can help to lower GHG emissions of existing production sites. Circular usage of carbon generates added value, yielding a 100% CO$_2$-based product. The technology could be licensable to further CO$_2$ providers to help reducing their emissions. The modular two-step concept ensures a high flexibility and provides further industrial partners a platform for the provision of various CO$_2$ based products. The feasibility of biotechnological conversion of acetate into value-added products has already been demonstrated for numerous examples [3].

To realize the provision of CO$_2$-based products the material use of CO$_2$ should be honored and vice versa fossil-based products should be taxed. A further instrument could be a label system to educate customers about the carbon footprint of the individual products. Next to this, a certain degree of sustainability in materials and products could be required. Last but not least to give new technologies a chance the CO$_2$ budget should be limited across all sectors and distributed via a CO$_2$ trading system.

Our special thanks go to the German Federal Ministry of Education and Research (BMBF) for initiating and for funding the strategic alliance ZeroCarbFP over almost 8 years (funding code: 031A217H; 031B0181E; 031B0898B).

Furthermore, we would like to thank our alliance partners from academia as well as from industry for being part of this project, introducing an entirely new route for the provision of products by material use and recycling of CO_2.

In addition we would like to take the chance to thank our ZeroCarbFP "Bioplastics" Bioprocess Engineering Team, consisting of Benjamin Lebküchner, Jens Peter Gersbach (BRAIN Biotech AG) and Heiko Schmittbetz, Marco Weinmann (Südzucker AG) for their great work and their support in establishing this innovative process.

References

1. Werpy T, Petersen G, U.S. Department of Energy, Energy Efficiency and Renewable Energy, Pacific Northwest National Laboratory (PNNL), National Renewable Energy Laboratory (NREL) (2004) Top value added chemicals from biomass, Volume I—Results of screening for potential candidates from sugars and synthesis gas
2. Chinthapalli R, Puente A, Skoczinski P, Raschka A, Carus M, nova Institut GmbH, Germany (2019) Succinic acid: From a promising building block to a slow seller – what will a realistic future market look like? https://renewable-carbon.eu/publications
3. Kiefer D, Merkel M, Lilge L, Henkel M, Hausmann R (2021) Review: from acetate to bio-based products: underexploited potential for industrial biotechnology. Trends Biotechnol 39(4): 397–411

R&D&I and Industry Examples: Ineratec's ICO2CHEM Project to Utilize CO$_2$

Thomas Bayer

Abstract

The industry sector remains as one of the largest emitters of greenhouse gases, especially CO$_2$. The aim of the ICO2CHEM project is to develop a new production concept for converting waste CO$_2$ to value added chemicals. By reutilization, the value of CO$_2$ is attempted to add in order to turn it to a commodity for a new generation of synthetic chemicals and fuels that will replace their fossil-based counterparts. ICO2CHEM specifically addresses a possible Carbon Capture and Utilization pathway by catalytically converting CO$_2$ into chemicals.

Keywords

ICO2CHEM · CO$_2$ · Chemicals · Fuels · Carbon Capture and Utilization (CCU) · Chemical CCU

33.1 ICO2CHEM Transforms CO$_2$ into Chemicals and Fuels

In the ICO2CHEM demonstration plant, CO$_2$ is transformed into precursors of consumer products and fuels. These are going towards what is called the circular economy with carbon neutral emissions. Within ICO2CHEM, a containerized chemical demonstration plant is installed and operated at the Infraserv Höchst Industrial Park in Frankfurt, Germany (Fig. 33.1). The demonstration plant converts CO$_2$ from a biogas upgrading plant and industrial H$_2$, a by-product of a chloro-alkali electrolyser plant, into highly valuable white oils and high molecular weight Fischer-Tropsch (FT) waxes.

T. Bayer (✉)
Provadis School of International Management and Technology AG, Frankfurt am Main, Germany
e-mail: thomas.bayer@provadis-hochschule.de

© The Author(s), under exclusive license to Springer Nature Switzerland AG 2023   381
M. Kircher, T. Schwarz (eds.), *CO2 and CO as Feedstock*, Circular Economy and
Sustainability, https://doi.org/10.1007/978-3-031-27811-2_33

Fig. 33.1 Containerized pilot plant Industriepark Höchst, Frankfurt am Main, Germany

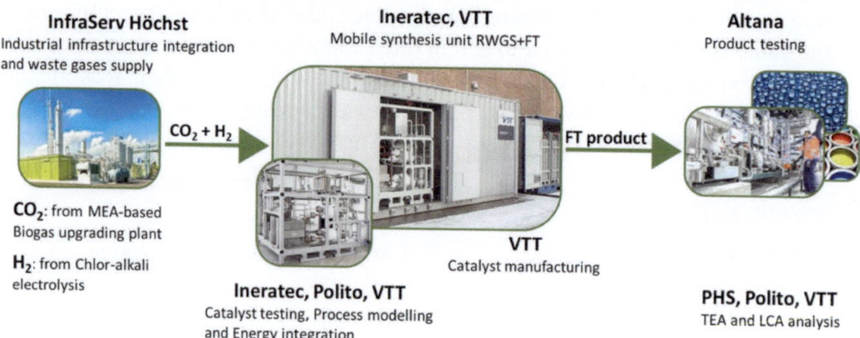

Fig. 33.2 ICO2CHEM concept

The technological core of the project consists of a Reverse Water Gas Shift (RWGS) reactor followed by an innovative modular Fischer-Tropsch reactor. In the RWGS step CO_2 and H_2 is converted into a synthesis gas mixture. This synthesis gas is converted into the chemical products in the following FT reaction step. The FT reactor is loaded with a novel Co-based catalyst with enhanced selectivity and lifetime developed by VTT. Altana analyses the utilization of the products in existing and new consumer products. For instance, the white oils and wax emulsions is utilized as raw material for chemical products, such as coatings and sealant materials, and the properties of the raw material are benchmarked against the fossil-based material currently in use. ICO2CHEM not only aims to develop a novel feasible CO_2 economic route but also to gain further knowledge about the entire process as well as the FT catalyst (Fig. 33.2).

During the first period, the work focused on the preparation of the demonstration run, which takes place at Industrial Park Höchst in Frankfurt. In the demonstration, the whole production chain from industrial raw materials (CO_2 and H_2) via syngas to FT products is demonstrated in a mobile synthesis unit (MOBSU). To achieve this target the process design to define the battery limits, feed gases and utilities are in essential role. The safety analysis for MOBSU was finished by INERATEC and Infraserv. During the second period an official environmental permit was required and obtained to operate the demonstration plant at Höchst. Plant installations were carried out, but the plant acceptance by an approved inspection agency is needed before the demonstration run can be started.

For Fischer-Tropsch reaction to take place, CO_2 needs to be converted to CO. During the first period INERATEC designed and built the first version of an RWGS reactor. Development work was continued during the second period and an advanced RWGS reactor was integrated to the MOBSU.

Catalyst development was necessary to achieve the desired products and utilize enhanced heat transfer properties of the INERATEC microstructure FT reactor. The development work at VTT was based on a conventional cobalt catalyst recipe modified with several different methods. The target in catalyst development was to produce catalyst with high intrinsic activity and selectivity towards heavy hydrocarbons (i.e. α-value). Those qualities were pursued by adding promoters by traditional incipient wetness method and atomic layer deposition. The suitability of the catalyst to the microstructure reactor was verified by lab-scale in a microstructure reactor. A 1 kg catalyst batch was manufactured at VTT and packed to MOBSU FT reactor by INERATEC for the demonstration run at Höchst.

Linked to the catalyst development activities, kinetic experiments and modelling were performed by Polito. The kinetic model was integrated into a process model to study different process concepts and their efficiencies and yields of desired product fractions. As the RWGS step requires heat, whereas the FT reaction is strongly exothermic heat integration is essential to achieve high efficiencies. The evaluation of heat integration possibilities and cost effectiveness were started in the first period and continued throughout the second reporting period. This task will continue until the end of the project, as Höchst experimental campaign will provide pilot scale concept data for more detailed process modelling.

The mass and energy balances obtained from the process modelling were used in the evaluation of the techno-economics of the process. Estimates for both capital and operational costs of an industrial scale plant that could be located at Höchst were prepared. Evaluated end-products included FT waxes and solvents. The findings from the first LCA studies indicate that if renewable electricity is available, it is possible to exceed the 20% GHG emission reduction target with the production of waxes and chemicals via the studied ICO2CHEM concept. However, the results are sensitive to applied system boundaries and assumed energy production profiles. During the final period, evaluated concepts will be updated and the economic and environmental impacts of alternative locations for the production will be studied.

In the MOBSU, the FT product attained is divided to a liquid fraction and a wax fraction. Further separation is required to achieve product fractions used in different applications. The first product separations in lab-scale by short path distillation were made at VTT and Altana. However, during this reporting period, no suitable samples were produced to application tests. VTT and Altana started discussion with outsourced partner for MOBSU wax distillation. This distillation will be conducted after wax is available from Höchst experimental campaign.

ICO2CHEM aims to demonstrate a concept to produce white oils and high molecular weight waxes from industrial CO_2 sources. Currently both of these hydrocarbons are produced from fossil raw materials coal, natural gas or crude oil. More specifically the hard wax is produced from fossil FT-wax and white oil from crude oil distillation. By using CO_2 as raw material, at least 20% reduction in CO_2 emissions is targeted compared to fossil-based products based on life cycle assessment.

To use the point sources of CO_2 and H_2, intensified FT synthesis technology by INERATEC plays an essential role in achieving the required capital costs of the synthesis for distributed plant concepts matching with industrial CO_2 sources. The excellent heat transfer properties of the reactor allow the use of highly active catalyst developed during the project. In traditional reactor technologies, the catalyst activity needs to be limited to control the heat produced during the reaction. The catalyst development investigates advanced catalysts profiting from ALD coating technology.

The ICO2CHEM project involves industry, SME, research institute and universities. This combination will allow the widespread exploitation of the project results and increase the social impact of the project by increasing the competitiveness of the new European technologies, providing job opportunities in research and production as well as creating trainee and thesis opportunities for young people.

33.2 Industrial Power-to-Liquid Pioneer Plant 2022 in Germany

INERATEC plans an industrial pioneer plant for the production of sustainable synthetic fuels starting in 2022. Up to 3500 tons or 4.6 million liters of INERATEC e-Fuels can be produced annually from up to 10,000 tons of biogenic CO_2 and renewable electricity. The pioneer plant will be the largest one yet and will serve as an enabler for further worldwide power-to-liquid projects of INERATEC.

An essential step towards the broad availability of sustainable, synthetic fuels in Germany is the implementation of Power-to-Liquid (PtL) technology on an industrial scale. With Industriepark Höchst in Frankfurt am Main, INERATEC has identified a site that provides access to low-cost, renewable hydrogen (H_2) and a biogas plant as a CO_2 source. In 2022, the world's first PtL plant with a production capacity of up to 3500 tons, i.e. 4.6 million liters of fuel per year, is to be commissioned there. Every renewable liter of INERATEC e-Fuel produced replaces one liter of fuel from fossil sources. The pioneer plant recycles up to 10,000 tons of CO_2 annually, which is currently released unused into the atmosphere (Fig. 33.3).

The pioneer plant produces a range of sustainable fuels. The clean diesel has drop-in quality and can be used directly in cars and trucks. The naphtha and wax

Fig. 33.3 INERATEC plant at Industriepark Höchst

fractions can be refined into CO_2-neutral gasoline and e-kerosene as well as synthetic chemicals. Especially the sustainable e-kerosene is considered a promising solution for climate protection in the aviation industry. An expansion of additional generation capacities for PtL kerosene is therefore being sought. For this purpose, separate talks are being held on the application for federal subsidies.

The sustainable products are marketed through a tendering process. Interested parties can apply for the offtake (minimum purchase quantity 500,000 l/a) directly via INERATEC.

Since 2016 INERATEC supplies modular chemical plants for Power-to-X and Gas-to-Liquid applications as well as sustainable fuels and products. With the pioneer plant in Frankfurt Höchst, INERATEC introduces its standard product for industrial applications. In further projects, larger capacities are achieved by multiplying the modules. The renewable application of the processes is enabled by the innovative chemical reactors. With the first-time application of the pioneer plant in Frankfurt Höchst, a demonstration of the technology is implemented in Germany, which serves as a pioneer for numerous other projects worldwide.

Surely the price for renewably produced hydrocarbons is very much higher today (>4 €/Litre fuel without taxes) than for fossil-based ones.

With increasing taxes for the emission of CO_2 and decreasing prices for renewable energies, the gap will become smaller in the future. Studies assume achievable prices in the future of around €1 per litre. Based on electricity prices of 1–2 ct/kWh for wind or solar power, which is expected to be achievable at least in some areas of the world [1].

Reference

1. The future cost of electricity-based synthetic fuels (2018). Agora Verkehrswende, Berlin. www.agora-verkehrswende.de/en/publications/the-future-cost-of-electricity-based-synthetic-fuels/

Piloting, Scale-Up, and Demonstration

34

Koen Quataert, Ellen Verhoeven, Karel De Winter,
and Hendrik Waegeman

Abstract

Although CO_2 is often considered a problematic waste gas, a lot of promising research indicates that it can be adopted as a carbon source for fermentation processes, resulting in a wide array of valuable biochemicals. However, to make products from biological carbon capture and utilization (CCU) technologies cost-competitive on a commercial scale, many challenges must still be overcome. Slow gaseous mass transfer rates result in low productivities and the use of combustible gasses, such as hydrogen, methane, and carbon monoxide, complicate operations and increase capital investment requirements. Furthermore, industrial C1-gasses typically convey components that potentially inhibit the microorganism's performance. Here we describe potential solutions to overcome the major hurdles in gas fermentation and highlight the role of validation of gas fermentation processes on real industrial gas, acting as a stepping-stone to bring the technology to scale.

Keywords

Capture and utilization (CCU) · Biological CCU · Gaseous mass transfer rate · Productivity · H_2 · Methane · CO · CO_2 · Inhibitors · Gas fermentation

The utilization of carbon dioxide (CO_2) as a resource to produce our everyday materials, is suggested to be the cornerstone of a truly circular economy. And still, despite being the most elementary carbon-containing molecule, its industrial, large-scale conversion appears to be far from trivial. In the chemical realm, several CO_2 conversion processes have been industrialized, producing chemicals such as urea,

K. Quataert (✉) · E. Verhoeven · K. De Winter · H. Waegeman
Bio Base Europe Pilot Plant, Gent, Belgium
e-mail: koen.quataert@bbeu.org; hendrik.waegeman@bbeu.org

© The Author(s), under exclusive license to Springer Nature Switzerland AG 2023
M. Kircher, T. Schwarz (eds.), *CO2 and CO as Feedstock*, Circular Economy and Sustainability, https://doi.org/10.1007/978-3-031-27811-2_34

methanol, formaldehyde, formic acid, etc. on a multi-ton scale but this is less true for current biological conversion routes [1]. Because of the many advantages of bio-CCU routes, such as operating at ambient temperatures and pressures, high product selectivity, flexibility, and robustness relating to feed gas composition and contaminants, an increasing amount of endeavours are set in motion to harness their potential. Since the publication of the first C1-converting gas fermentation studies in the early 1900s [2, 3], more exhaustive research in this field has been conducted in the last three decades. Many improvements have been made, but as long as the (petro)chemical industry can provide cheap, fossil-based products, the competition for bio-manufacturing is immense. This is especially true when commodities or drop-in chemicals are targeted. Since those commodities, such as ethanol, acetic acid, 2,3-butanediol, 1,3-butanediol, isobutylene, acetone, isopropanol, and lactic acid have a huge potential for abating CO_2 because of their large market volumes and as they can be produced efficiently by several C1-utilizing bacteria, they are often the target for gas fermentation technologies [4, 5]. On the other hand, the biotechnological revolution did successfully penetrate markets that were untouched by chemistry before, such as microbially produced antibiotics, industrial enzymes, vaccines, etc. Building on those foundations, an increasing amount of gas fermentation technologies are under development for the production of such highly-functionalized target products including single-cell proteins (SCP) for food and feed, bioplastics, food additives, amino acids, etc. Before gas fermentation technologies can become cost-competitive with their fossil-based counterparts, many challenges such as low space-time yields or product titers must be tackled.

As gas fermentation is an emerging domain within the world of industrial biotechnology, this section will first describe important factors related to the scale-up of traditional aerobic fermentation processes and existing analogies between gas fermentation and traditional fermentation. The role of piloting will be discussed, emphasizing the essence of demonstrating gas fermentation technologies in an industrial environment. Finally, the developments in the field of gas fermentation are described, aiming to provide an overview of the crucial aspects of a successful scale-up.

34.1   Scale-Up of Industrial Microbial Processes

Industrial biotechnology has been projected as a potential manner of reducing our dependence on fossil-based feedstocks and optimization of fermentation processes on an industrial scale has been a priority ever since. However, the scale-up of biotechnological processes has often proved to be a challenging undertaking and it is generally known that processes that perform well on a laboratory scale, do not necessarily work well on a production scale. Because of a different mixing behavior, hydrostatic pressure and gas transfer at a larger scale compared to bench-scale which lead to nutrient concentration and thermal gradients, and because of the low genetic stability of engineered strains, issues related to contamination, and other challenges, industrial biotechnology has not developed as fast as was expected [6]. Economically feasible fermentative production processes are characterized by high space-time yields, or productivity, combined with achieving high yields, and product titers [7]. Those metrics can be achieved by both strain improvements and process improvements. The latter combines optimization of fermentation parameters and conditions (pH, temperature, nutrient composition, dissolved oxygen or DO content), feeding regimes (constant, linearly or exponentially ramped, pulsed, DO-based, etc), fermentation modes (batch, fed-batch, continuous), as well as bioreactor technology and set-up. Scaling up a fermentation process demands a careful and conscientious translation of process parameters across scales a couple of orders of magnitude apart, aiming to maintain or even increase the productivity and titer. Many analogies exist between gas fermentation and aerobic fermentation processes, of which the most important aspects are described here.

As suggested by the terminology, aerobic fermentations must be sparged continuously with air, at a rate that can sustain the metabolic requirements or oxygen uptake rate (OUR) of the microbial culture. The rate at which oxygen can be dispersed in a bioreactor, the so-called, oxygen transfer rate (OTR) is described as follows:

$$OTR = \frac{dC}{dt} = k_L a * (C^* - C)$$

where k_L is the mass transfer coefficient, a is the specific exchange area of the gas bubbles, $k_L a$ is the volumetric oxygen transfer coefficient, C^* is the dissolved oxygen saturation concentration and C is the actual dissolved oxygen concentration. The k_L factor relates to the resistance for the transfer of gas to liquid and is typically combined with the specific exchange area to measure the aeration capacity of a fermentation vessel. The $k_L a$ is the most vital factor for achieving good oxygen transfer and depends on the agitation system and power input, agitation speed, aeration speed, vessel geometry, and composition of the fermentation broth in terms of viscosity and density, etc. An inadequate supply of oxygen, that fails to sustain the increasing metabolic activity of a well-producing microbial culture, is often the limit of a fermentation's productivity. Adjusting fermentation parameters to match the OTR with the increasing OUR is often the main topic of process control and is typically achieved by increasing the stirrer speed, aeration rate, or working under slightly elevated pressures. In well-aerated stirred tank reactors (STR), a $k_L a$ value of 1000/h can be achieved, resulting in a maximally achievable OTR of about 250 mmol/L/h. This can be further improved by increasing the partial pressure of oxygen in the gas inlet feed. According to Henry's law, in case pure oxygen is used, the OTR is enhanced fivefold. While this strategy could boost the fermentation's performance, the potential benefits often do not exceed the additional costs. Moreover, running processes using pure oxygen is inherently more dangerous due to its highly oxidizing character.

Generally speaking, mixing and aeration are the most important parameters when it comes to the scale-up of aerobic fermentation processes, as both parameters have a tremendous influence on the fermentation conditions. To realize similar aeration and agitation conditions when increasing the scale of a fermentation process, a general rule of thumb is to maintain a constant k_La and shear. The former scale-up parameter is achieved by balancing the volumetric power consumption and volumetric airflow rate, whereas the latter may be achieved by maintaining a constant tip speed (m/s). Adjusting aeration and agitation between minimal and maximal values gives rise to the so-called 'scale-up window'. Outside of this window, the scaled-up fermentation process could suffer from low oxygenation levels, inferior bulk mixing, CO_2 accumulation, excessive shear stress, foaming, high operation costs, or a combination thereof.

Though scale-up based on several critical scale-up parameters is often an applicable strategy, a holistic approach considering key metabolic stress factors, physical parameters that influence productivities, and important process control aspects, is essential for a successful scale-up. Next to this approach, requiring extensive experience and expertise in the field of bioprocess engineering, an increasing amount of research focuses on the use of data-driven frameworks linking Computational Fluid Dynamics (CFD) and metabolic flux analysis [8]. Those strategies aim to predict the effect of process conditions on fluid dynamics and cell physiology as a means of understanding and predicting how scale-up affects fermentation's performance. A particularly interesting development in the field of CFD is Freesense's 3D sensor platform, consisting of free-floating sensors that relate actual DO and pH to a position in the bioreactor, enabling experimental validation of the models [9]. As computation power is increasing fast, it is expected that the role of modeling in fermentation scale-up will play an important role in the future [10]. Combined, raw bioprocess engineering and data-driven modeling, could potentially secure the economic success of an industrial fermentation. Additionally, a key message for new fermentation processes is to develop the strategy while keeping industrial and economic feasibility in mind: parameter optimization in the lab on 10 L, must be translatable to bioreactors with volumes of 100,000 L or more. Such a scale-down approach reduces the risk of unexpected complications during scale-up and improves the chances for success [11].

In a laboratory environment, with high experimental throughput for minimal resources, the challenges related to product titers and productivities can often be tackled. However, to prove that the technology is ready for large-scale deployment, it must first be demonstrated at the pilot level. Piloting can bring essential insight into the process running at a larger scale, deliver an adequate amount of product samples for application testing, and provide crucial processing data that can serve as input for techno-economic assessment (TEA) and life cycle assessment (LCA). In this piloting phase, both the technological risks and financial implications are at a high, therefore the willingness from investors is at a low, which is why this phase is often referred to as the 'Valley of Death'. For technology developers to build their demonstration facilities, CAPEX easily surpasses ten million EUR, with OPEX also running into millions. Besides, the design and construction phase at least costs 2 or more years. To save money (and time), businesses can collaborate with open-access piloting facilities that offer equipment and—equally important—expertise, to demonstrate their proprietary technologies on the pilot scale. Several shared pilot facilities exist in

Europe, including Bio Base Europe Pilot Plant (BE), Centre for Process Innovation (UK), VTT Technical Research Centre (FI), Bioprocess Pilot Facility (NL), Innovhub SSI (IT), ARD (FR), Acies Bio (SI) and Fraunhofer CBP (DE). Demonstrating the feasibility on a pilot level, lowers the barriers for future investors and could act as a steppingstone for large-scale deployment and commercialization.

34.1.1 Scale-Up of Gas Fermentation

Specifically for gas fermentation, some challenges are far more of an issue to bring the processes to full scale compared to traditional aerobic fermentation. This is primarily because (1) the technology must be validated with real industrial off-gasses, because of potential inhibiting compounds in the gas stream and potential fluctuations in the gas composition, and (2) gas fermentation pilot and demo units often require unique designs with an increased gas transfer, leading to both higher CAPEX and OPEX when compared to the infrastructure needed for traditional fermentation processes. On top, only a few open-access pilot and demonstration facilities offer gas fermentation equipment, slowing down technology developers wishing to scale up and validate their processes.

As discussed in the previous section, a general scale-up framework for fermentation does not exist which is mainly due to inherent differences between processes. The umbrella term gas fermentation, likewise, covers a variety of distinct processes that are highly diversified in terms of gaseous feedstock, microbial hosts, metabolisms, sources of energy, targeted products, etc. Similar to the scale-up of aerobic fermentation, there is no one solution that fits all. The most extensively studied group of gas fermentation microorganisms are the anaerobic acetogens, using the Wood-Ljungdahl pathway to grow on solely CO, CO and H_2, CO_2, and H_2, or a combination thereof. The hydrogen-oxidizing aerobic Knallgas bacteria utilize the Calvin cycle to grow autotrophically on CO_2 with H_2 as the energy source. Finally, methanotrophic bacteria are able to grow on CH_4 as their sole source of carbon and energy, in an aerobic or anaerobic manner. Other C1-utilizing organisms, such as the photoautotrophic microorganisms or the aerobic carboxydotrophs, are not included here.

Similar to aerobic fermentation, productivities in gas fermentation are generally limited by gas to liquid mass transfer rate [12, 13]. Gas fermentation even requires more moles of gas to be dispersed into the culture broth per amount of carbon equivalent and the low solubilities of the pure gases H_2 and CO, compared to O_2 (Table 34.1), further complicate the transfer of gas in gas fermentation.

The severity of this limitation is dictated by the metabolic requirements of the organisms (the Calvin cycle f.e. requires at least 6 moles of H_2 to convert 1 mole of CO_2) as well as the partial pressure of the supplied gasses. Following the law of Henry, high partial pressures result in higher solubilities, which act as the driving force for the gas transfer rate. In a laboratory environment, these partial pressures can be set whereas in an industrial environment, volumetric ratios depend on the point source, hugely influencing the potential productivities, and more importantly, the C1

Table 34.1 Properties of different gasses including estimation of kLa for a given bioreactor. Henry's constants, solubilities, and diffusivity of CH4 are given at 25 °C, diffusivities of O_2, H_2, CO_2, and CO at 37 °C. Whereas diffusivity of the gasses change with temperature, the ratio of diffusivities is approximately temperature-independent

	$H_i \times 0^{-6}$	$D_i \times 10^{-9}$	S_i		$k_La_i{}^a$
	[mol.m^{-3}.Pa^{-1}]	[m^2.s^{-1}]	[mol.m^{-3}]	[mg.L^{-1}]	[h^{-1}]
O_2	12.0 [b]	3.05 [c]	1.22	38.91	1000
H_2	7.8 [d]	6.48 [c]	0.79	1.59	1458
CO	9.7 [b]	3.26 [c]	0.98	27.53	1034
CO_2	330.0 [e]	2.74 [c]	33.44	1471.57	948
CH_4	14.0 [b]	1.49 [b]	1.24	22.75	699

[a] Calculated using formula $k_La_i = k_La_j \sqrt{\frac{D_i}{D_j}}$

[b] [14]

[c] [15]

[d] [16]

[e] [17]

carbon conversion rate. The importance of mass transfer rate or MTR on the carbon conversion rate for different gas fermentation processes can be exemplified through theoretical calculations. Similarly to the equation for OTR, the MTR for each gas 'i' can be described by

$$MTRi_i = \frac{dCi}{dt} = k_Lai * (Ci* - Ci)$$

where the factor k_La_i depends on the diffusion constant of the gas 'i' and Ci* on the constant of Henry for gas 'i'. The k_La_i of a pure gas can be derived by transforming the k_La for O_2 of a given bioreactor, by multiplying it with the square root of the ratio of the diffusion coefficients [18, 19]. When fed with air (containing 21% of O_2), an aerobic fermentation process on glucose in a bioreactor with a k_La_{O2} of 1000/h could theoretically support an OTR of 250 mmol/L/h, which could be roughly correlated to a glucose consumption up to 7.6 g/L/h or 3.1 g/L/h of carbon equivalents. For an acetogenic fermentation on CO_2 and H_2, the latter will be the limiting substrate. In an identical bioreactor, the k_La_{H2} would become 1458/h (attributed to its higher diffusivity). Taking into account the metabolic requirements of acetogens fed with CO_2 and H_2, the partial pressure of H_2 must amount to 45%, to obtain a similar carbon equivalent conversion rate of 3.1 g/L/h. Increasing the partial pressure of H_2 thus serves as a powerful tool to cope with mass transfer limitations and boost productivity up to levels comparable with aerobic heterotrophic fermentation processes. For aerobic Knallgas fermentation adequate supply of O_2 is typically the limiting factor: for safety reasons, its partial pressure must be kept below the minimal oxygen concentration (MOC), which is defined as the O_2 concentration below which combustion of an explosive gas cannot occur [20]. With H_2 as the combustible gas, depending on conditions such as the characteristics of inert gasses present, temperature, and pressure, the MOC for Knallgas fermentation is 4.3% of O_2

[21]. With this O_2 concentration and a k_La_{O2} of 1000/h, a carbon conversion rate of only 0.31 g/L/h can be supported in the same bioreactor. To improve these metrics, one strategy would be to operate the process in explosion-proof ATEX equipment, allowing O_2 concentrations above the MOC. In such a case, the potential gains in productivity should of course outweigh the increased CAPEX of an industrial installation.

The transformation of k_La's in the above example only holds under the condition that other parameters influencing the k_La remain unchanged, which is typically not the case for the total gas flow sparged into the bioreactor. In an aerobic fermentation process, the excess of unconverted gasses is allowed to be exhausted, whereas gas fermentation processes often aim to achieve high gas conversion, limiting total gas flow rates, and concurrently gaseous mass transfer. Strategies such as recycling the exhaust gasses [22], could allow the use of high gas flow rates, while still achieving acceptable gas conversion rates.

Finally, the use of elevated vessel pressures has been an increasingly investigated approach to improve gaseous mass transfer [23]. Again, according to Henry's Law, gaseous solubility increases with partial pressure, hence the solubility of all present gasses increases proportionally with reactor pressure. Unlike often believed, a doubling in reactor pressure does not necessarily mean a doubling of MTR, as reactor pressure decreases the superficial gas velocity, therefore k_La [24]. Taking this into account, an increase in reactor pressure mostly results in an increased MTR rate, be it not completely linear. The principle of slightly elevated pressure (up to 1.5 barg) in industrial aerobic fermentation is well-practiced, and even higher pressures (up to 10 barg) have shown positive influences on OTR and productivities in f.e. *E. coli* and *Y. lypolytica* [25, 26]. For syngas fermentation, pressures up to 12 barg have been shown to increase the biomass production of *C. ljungdahlii* [27] and positive effects of elevated pressures on aerobic methanotrophy have been demonstrated by Soni et al. [28] using slight overpressures and Wendlandt et al. [29] using *Methylocystsis* up to 3 barg. More recently, elevated pressures were applied for the cultivation of *C. necator* on Knallgas by Yu and Munasinghe [30] resulting in increased growth rates up to 3 barg.

Although microorganisms typically tolerate pressures remarkably higher than atmospheric pressure, in some cases elevated pressures were shown to negatively impact fermentation performance. The indirect effect of pressure, causing increased dissolved gas concentrations, could have a profound impact on the cell's physiology. Even in aerobic fermentation, increased dissolved CO_2 concentration, as the main by-product of metabolic activity, was shown to decrease the intracellular pH of the cells which consequently resulted in increased maintenance energy requirements [31]. In methanotrophic fermentation, it was shown that increased CH_4 and O_2 concentrations inhibited some of the screened organisms [28]. In acetogenic fermentation, elevated dissolved CO concentrations are known to inhibit hydrogenases [32], whereas in Knallgas fermentation, elevated dissolved O_2 concentration potentially forces the central autotrophic enzyme RuBisCO to exhibit increased oxygenase activity at the expense of CO_2 fixation activity [33, 34]. Coping with these toxic effects requires further research elucidating the impact of each gas on the metabolic activity, as well as extremely diligent process control to balance the dissolved gas concentration above the threshold of the organism's metabolic requirement, without entering the toxic region. These findings illustrate the importance of in-line, and fast-response sensors for dissolved gas concentration measurements [35]. Whereas such probes are commonplace in aerobic fermentation for dissolved O_2 follow-up and dissolved CO_2 probes are becoming commercially available, the existence of dissolved H_2 and CO for industrial biotech is lacking to date, hindering R&D efforts as well as industrial-scale commercialization.

Next to difficulties in MTR, another hurdle specifically for large-scale knallgas or methanotrophic fermentation is the high amounts of metabolic heat generated. Similar to aerobic fermentation processes, O_2 acts as the final electron acceptor in the metabolism, improving the amount of ATP generated, but simultaneously resulting in a huge release of metabolic energy. Yu and Lu [36] compared biomass formation of *C. necator* utilizing glucose or CO_2 as the carbon source and found that for the same amount of C-mol cell mass formed, the autotrophic growth mode released double the amount of heat. Therefore once C1 fermentation performance approximates aerobic fermentation in terms of space-time yields, engineering efforts should be able to address heat removal requirements.

34.1.2 Bioreactor Technology

Stirred tank reactors (STR) are traditionally considered the preferred reactor design for smaller-scale experimental work, as they ensure homogenous mixing conditions and are easy to use in terms of process control. Also in commercial applications at relatively lower scales, f.e. when high-value molecules are the targeted product, they remain the bioreactor design of choice. When large-scale manufacturing of commodity products in large volumes is envisioned, however, the power input required to ensure favourable mixing conditions in STR bioreactors becomes uneconomical. That is why on a commercial scale, other bioreactor designs such as bubble columns and airlift reactors gain interest. The main advantage of such reactor designs is that they allow adequate mixing at considerably lower power input. Moreover, their mechanical simplicity positively impacts investment and maintenance costs. For aerobic fermentation processes, starting from 50–100 m^3, bubble columns and airlift reactor designs are typically selected [37]. In the field of gas fermentation, because of additional difficulties related to gaseous mass transfer and cooling capacities, bubble columns or airlift reactors might even become more attractive at volumes <50 m^3. Different reactor designs in terms of vessel geometry, internal, or external draught tubes, additional spargers, or even hybrids between STRs and the airlift reactors are investigated to optimize gaseous mass transfer [23]. Another interesting bioreactor design, so-called tubular loop fermenters, have been used to ensure high oxygen transfer in yeast fermentation [38, 39] and were further adapted for gas fermentation by pumping gas and liquid through a vertical U-shaped pipe, fitted with static mixers [40]. Such U-loop reactors have been commercialized for gas fermentation by Calysta and Unibio for the production of microbial protein from CH_4.

Hollow-fibre membrane bioreactors provide a high gas-liquid surface area and hence enable high k_La's with low power input [41]. Their advantages have been demonstrated on the lab scale but commercial-scale data does not exist. Similarly, trickling-bed reactors (TBR) have shown increased gaseous mass transfer rate in the lab [42], but have thus far not been used on an industrial scale for gas fermentation.

Besides the choice of bioreactor design, some gas fermentation-specific aspects that hugely impact the economic feasibility of a scaled-up gas fermentation process should be taken into consideration. First, the flammable character of H_2, CO, and CH_4, imposes compliance with the EU legislation concerning equipment to be used for potentially explosive atmospheres (ATEX). Particularly in the case of Knallgas or aerobic methanotrophy, where O_2 is co-fed, diligent follow-up of safety measures is key and the demand for explosion-proof equipment hugely impacts the equipment costs. When elevated pressure fermentation is envisioned, additional costs may be expected due to the requirement of pressure-resistant bioreactors, as well as peripheral equipment such as pneumatic valves and pumps. One aspect that might reduce CAPEX stems from the fact that gas fermentation processes are generally less prone to contamination, reducing the requirement of capital-intensive high hygienic graded equipment and steam generators. Contamination-proof processes also benefit from low failed batch rates and omit laborious sterilization procedures, also positively impacting operational expenditures. Furthermore, it enables the possibility to run processes in a continuous mode, which is typically not the case due to increased contamination risks when running aerobic fermentation processes over extended periods. This increases the economically feasible manufacturing of primary metabolites such as ethanol, acetic acid, or even microbial protein via gas fermentation.

34.2 Demonstration of C1 Technologies in an Industrial Environment

Another important aspect in pilot or industrial processes is the grade of the raw materials. Industrial grade raw materials must be validated in piloting trials ahead of further scale-up, as they may have a huge influence on the fermentation performance. As the most important raw material in gas fermentation are gasses which vary in composition depending on their source, validation of the process on industrial gasses is key. Gas fermentation processes that perform well on synthetic gas mixes, could completely fail when real industrial gasses are adopted. Especially waste gasses originating from industries such as steel manufacturing, cement production, power plants, biomass and waste gasification, etc. contain a vast amount of impurities such as tars, dust, aromatic compounds, hydrogen cyanide, ammonia, ethylene, acetylene, nitrogen oxides, oxygen, carbonyl sulfide, etc. Many of those compounds are known to be inhibiting to different gas fermenting organisms with Ineos Bio as an infamous example, having to seize operations of their waste-to-ethanol plant due to high hydrogen cyanide concentrations [43]. Such unexpected performance losses can be anticipated by executing cultivation experiments with

Fig. 34.1 Bio base Mobile Pilot Plant (BBMPP) installed at the site of Arcelor Mittal in Ghent (BE)

known contaminants. Some strategies include supplementing the contaminants to the cultivation medium or preparing a synthetic gas mixture mimicking the real gas. While such experimental trials could be a valuable indication of the toxicity of individual or multiple compounds early on in the development, they cannot guarantee successful cultivation on the industrial C1 waste gasses at hand. Compressing and collecting gasses at the site of the emitter and transferring them to the lab is yet another validation approach. The equipment required for this compression must be corrosion resistant, to cope with the condensation of corrosive compounds during compression, making it quite an expensive approach. Moreover, the condensation of certain compounds again puts a limit to the validity of such cultivation trials.

In gas fermentation, a particularly interesting manner to effectively validate fermentation performance on C1 waste gas is to perform cultivation trials on the site of the industrial emitter. This strategy was pioneered by LanzaTech and is now also adopted by other technology developers such as Deepbranch Biotech at the site of UK's largest power plant Drax [44] in the framework of the publicly funded project REACT-FIRST, and Algiecel, targeting conversion of CO_2 from biogas production, power plants, and fermentation plants. Bio Base Europe Pilot Plant designed and constructed an open-access containerized gas fermentation pilot plant, enabling stand-alone operation at any industrial site, given a supply of power and a source of industrial gas (Figs. 34.1 and 34.2).

It was designed to host a variety of gas fermentation microbes each requiring different gaseous substrates. In cases where the caloric value of the industrial gas is

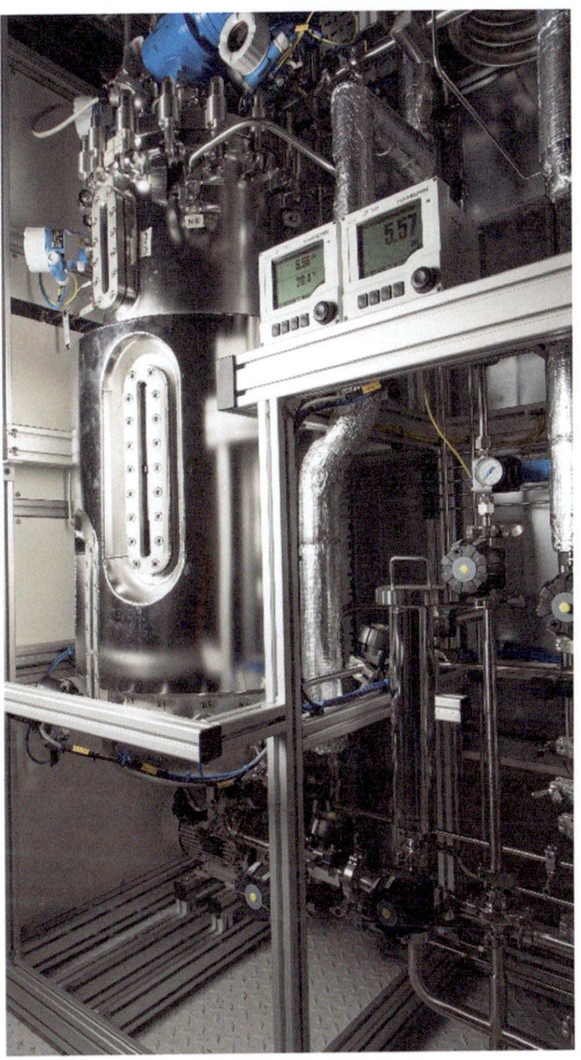

Fig. 34.2 BBMPP's 150 L pressurisable gas fermentation bioreactor

insufficient, a built-in hydrogen generator can act as an external energy source, further increasing the flexibility of the installation. Piloting gas fermentation technologies on industrial gasses has two main beneficiaries as (1) the emitter gets valuable insight into the potential valorization of their C1 by-products, and (2) gas fermentation technology developers get access to highly specialized equipment enabling the validation of their processes in a real industrial environment.

On top of the many strain developments, metabolic insights, process optimizations, and advances in reactor technology, on-site piloting could be the next step to take gas fermentation out of the lab and demonstrate the technology in the real world. Nowadays, an increasing number of pioneers are on their way to bringing their technology to scale (Table 34.2).

The company Lanzatech, founded in 2005, is the market leader in terms of bio-CCU, owning the technology for the world's first waste gas to ethanol plant on a commercial scale since 2018. The plant produces 46,000 tons of ethanol per year and is situated at the facilities of Shougang Jingtang United Iron & Steel Co., Ltd. They continue to partner with C1 gas emitting industrial players, such as Arcelor Mittal, Suncor Energy, and Shell, and have several commercial-scale plants currently under construction. While the first focus of their technology development was the production of ethanol, the big industrial player has also shifted focus to other commodity chemicals, often in collaboration with other companies such as Global Bioenergies [73]. Evonik initially joined forces with Lanzatech to produce 2-hydroxybutyric acid and has now also constructed its testing facility in Germany with a 2000 L bioreactor to develop and scale up gas fermentation processes [59]. The diversification of gas fermentation technologies has also been picked up by smaller, newer companies. A recent trend is the production of single-cell proteins (SCPs) investigated by companies such as Deep Branch, Avecom, NovoNutrients, and Air Protein, the latter being a spin-off from Kiverdi. Deep Branch is constructing a pilot-scale facility at Brightlands Chemelot Campus in The Netherlands and envisions going to commercial scale in 2023. NovoNutrients intends to shift to an industrial demonstration scale (20,000 L) in the near future. The most advanced technology regarding SCPs has been developed by UniBio, having opened the first commercial plant with a U-loop reactor in Denmark in 2016, producing up to 80 tonnes of SCP per year. In 2021, they initiated their first commercial agreement with a large aquaculture company. Another trend in gas fermentation is the production of bioplastics and precursors such as lactic acid and polyhydroxyalkanoates (PHAs). Mango Materials has validated its technology in an cooperation with a waste water treatment plant in California, and NewLight Technologies has advanced their PHA production technologies to a commercial level having opened a new production-scale facility in 2020, targeting the production of >500 kta. For the production of polylactic acid (PLA), the production of the monomer, lactic acid, by gas fermentation has been targeted mostly by Calysta in a collaborative R&D project with the biopolymer producer NatureWorks LLC. The technology has been validated in the lab but needs further development to reach an industrial or commercial scale.

The pioneering work in gas fermentation is of tremendous value for the (bio)-CCU world, elucidating the framework of current and future CO_2-based value chains. When these pioneering projects prove to be successful and are able to compete with fossil-based alternatives, they can bring about a disruptive deployment of similar technologies. Furthermore, successful demonstration projects serve as an example, boosting the confidence in the technology while promoting the cooperation between researchers, technology developers, entrepreneurs, CO_2-emitting industries, and capital investors. With so many developments underway, it is not a matter of if, but when biological CO_2 transformation could become a fundamental part of our manufacturing value chains.

Table 34.2 Without claiming to be exhaustive, this table gives an overview of companies including main contributors to the commercialization of gas fermentation processes. An estimation of the TRL level of each technology was made based on information acquired from scientific literature, websites, press releases, etc.

Feedstock	Company	Product	TRL	Remarks	Reference
CH_4	Atel	SCP	7–9	Trademark owner of Gaprin®	[45]
	Calysta	SCP	7–9	First plant (Calysseo) foreseen to produce 20,000 t/a of FeedKind® in collaboration with Adisseo	[46]
	Mango Materials	Polyhydroxyalkanoates	4–6	Pilot facility producing PHA on biogas from wastewater treatment plant	[47]
	NatureWorks (Calysta)	Lactic acid	4–6	Laboratory to develop commercial-scale process; looking into possibilities to construct $50 million demonstration plant	[48]
	Newlight Technologies	Polyhydroxyalkanoates	7–9	Commercial facility (Eagle 3) for the production of AirCarbon ™ operational since 2019	[49–51]
	MBP Titan (Precigen, former: Intrexon)	Isobutanol	4–6	500 L pilot plant operational	[52]
	UniBio	SCP	7–9	Commercial Uniprotein® plant of 6000 t/a; possibility to scale to 20,000 t/a	[53, 54]
CO_2	Air Protein (Kiverdi)	SCP	4–6	Demonstrated at 50 L scale; targeting scale up and marketing of technology and products	[55]
	Avecom	SCP	4–6	Pilot facility with 400 L bioreactor	[56]
	Deep Branch	SCP	7–9	Building scale-up facility for production of Proton™, aim for commercial production in 2023	[57]
	Econutri	SCP	4–6	Construction of 300 L reactor ongoing	[58]
	Evonik	Butanol/hexanol	4–6	Pilot facility with 2000 L bioreactor in place	[59, 60]
	Kiverdi	Fatty acids	4–6	Technology demonstrated at pilot scale; targeting scale-up of technology	[61]
	NovoNutrients (Oakbio)	SCP	7–9	Demonstration bioreactor of 1000 L; plans to construct 20,000 L plant	[62]
		Bioplastics	4–6	Technology validated in laboratory and field	
	Solar Foods	SCP	7–9	Target commercial production of Solein® by 2023	[63]

(continued)

Table 34.2 (continued)

Feedstock	Company	Product	TRL	Remarks	Reference
Syngas/CO	Jupeng Bio	Ethanol	7–9	Demonstration plant (20,000 t/a) completed; plans for commercial plant (200,000 t/a)	[64]
	Sekisui chemical	Ethanol	7–9	Collaboration with LanzaTech with demonstration plant producing >5000 t/a	[65]
	LanzaTech	Ethanol	7–9	Industrial waste gas to ethanol plants operational at various scales up to 300 MT/a	[66–68]
		2,3-butanediol	7–9	In scale-up phase	[69]
		Isobutylene	4–6	Development in collaboration with global Bioenergies	[70]
		Acetone	4–6	Performed in 80 L bioreactor	[71]
		Isopropanol	4–6	Performed in 120 L bioreactor	[5]
		SCP	7–9	10,000 ton/a facility	[72]

References

1. Alper E, Yuksel O (2017) CO2 utilization: developments in conversion processes. Petroleum 3(1):109–126. https://doi.org/10.1016/j.petlm.2016.11.003
2. Starkey RL (1925) Concerning the carbon and nitrogen nutrition of thiobacillus thiooxidans, an autotrophic bacterium oxidizing sulfur under acid conditions. J Bacteriol 10(2):165–195. https://doi.org/10.1128/jb.10.2.165-195.1925
3. Waksman SA, Starkey RL (1922) Carbon assimilation and respiration of autotrophic bacteria. Proc Soc Exp Biol Med 20(1):9–14. https://doi.org/10.3181/00379727-20-3
4. Köpke M, Mihalcea C, Liew FM, Tizard JH, Ali MS, Conolly JJ, Al-Sinawi B, Simpson S (2011) 2,3-Butanediol production by acetogenic bacteria, an alternative route to chemical synthesis, using industrial waste gas. Appl Environ Microbiol 77(15):5467–5475. https://doi.org/10.1128/AEM.00355-11
5. Liew FE, Nogle R, Abdalla T, Rasor BJ, Canter C, Jensen RO, Wang L, Strutz J, Chirania P, De Tissera S, Mueller AP, Ruan Z, Gao A, Tran L, Engle NL, Bromley JC, Daniell J, Conrado R, Tschaplinski TJ, Giannone RJ, Hettich RL, Karim AS, Simpson SD, Brown SD, Leang C, Jewett MC, Köpke M (2022) Carbon-negative production of acetone and isopropanol by gas fermentation at industrial pilot scale. Nat Biotechnol 40(3):335–344. https://doi.org/10.1038/s41587-021-01195-w
6. Chen GQ (2012) New challenges and opportunities for industrial biotechnology. Microb Cell Factories 11:111. https://doi.org/10.1186/1475-2859-11-111
7. Formenti LR, Nørregaard A, Bolic A, Hernandez DQ, Hagemann T, Heins AL, Larsson H, Mears L, Mauricio-Iglesias M, Krühne U, Gernaey KV (2014) Challenges in industrial fermentation technology research. Biotechnol J 9(6):727–738. https://doi.org/10.1002/biot.201300236
8. Delvigne F, Takors R, Mudde R, van Gulik W, Noorman H (2017) Bioprocess scale-up/down as integrative enabling technology: from fluid mechanics to systems biology and beyond. Microb Biotechnol 10(5):1267–1274. https://doi.org/10.1111/1751-7915.12803
9. Wireless 3D Bioreactor Sensor Platform (2022) Do you really understand the conditions inside your bioreactor? https://www.freesense.dk/. Accessed 27 Jun 2022
10. Noorman HJ, Heijnen JJ (2017) Biochemical engineering's grand adventure. Chem Eng Sci 170:677–693. https://doi.org/10.1016/j.ces.2016.12.065
11. Yang X (2014) Scale-up of microbial fermentation process. In: Manual of Industrial Microbiology and Biotechnology, pp 669–675. https://doi.org/10.1128/9781555816827.ch47
12. Drzyzga O, Revelles O, Durante-Rodríguez G, Díaz E, García JL, Prieto A (2015) New challenges for syngas fermentation: towards production of biopolymers. J Chem Technol Biotechnol 90(10):1735–1751. https://doi.org/10.1002/jctb.4721
13. Worden RM, Bredwell MD, Grethlein AJ (1997) Engineering issues in synthesis-gas fermentations. ACS Symp Ser 666(5):320–336. https://doi.org/10.1021/bk-1997-0666.ch018
14. Warneck P, Williams J (2012) The atmospheric chemist's companion. https://doi.org/10.1007/978-94-007-2275-0. Accessed 27 June 2022
15. Lide DR, Data SR, Board EA, Baysinger G, Chemistry S, Library CE, Berger LI, Goldberg RN, Division B, Kehiaian HV, Kuchitsu K, Rosenblatt G, Roth DL, Zwillinger D (2020) Section 4 - Properties of the elements and inorganic compounds. In: CRC handbook of chemistry and physics, pp 722–875. https://doi.org/10.1201/b17118-9
16. Fernández-Prini R, Alvarez JL, Harvey AH (2003) Henry's constants and vapor-liquid distribution constants for gaseous solutes in H2O and D2O at high temperatures. J Phys Chem Ref Data 32(2):903–916. https://doi.org/10.1063/1.1564818
17. Sander SP, Abbatt J, Friedl RR, Barker JR, Burkholder JB, Golden DM, Huie RE, Kolb CE, Kurylo MJ, Moortgat GK, Orkin VL, Wine PH (2011) Chemical kinetics and photochemical data for use in atmospheric studies evaluation Nb. 17. JPL Publication 10-6, Jet Propulsion Laboratory, Pasadeba, 17. http://jpldataeval.jpl.nasa.gov
18. Asher WE, Pankow JF (1991) Prediction of gas/water mass transport coefficients by a surface renewal model. Environ Sci Technol 25(7):1294–1300. https://doi.org/10.1021/es00019a011

19. Poling BE, Prausnitz JM (2001) The properties of gases and liquids. https://doi.org/10.1036/0070116822. Accessed 27 June 2022
20. Perry RH, Green DW, Maloney JO (1997) REF 19 from REF 5 from REF 10 Chemical Engineers ' Handbook Seventh. In: Society, vol 27). http://www.ketab.ir/DataBase/BookPdf/88320019.pdf
21. Molnarne M, Schroeder V (2017) Flammability of gases in focus of European and US standards. J Loss Prev Process Ind 48:297–304. https://doi.org/10.1016/J.JLP.2017.05.012
22. Bredwell MD, Srivastava P, Worden RM (1999) Reactor design issues for synthesis-gas fermentations. Biotechnol Prog 15(5):834–844. https://doi.org/10.1021/bp990108m
23. Van Hecke W, Bockrath R, De Wever H (2019) Effects of moderately elevated pressure on gas fermentation processes. Bioresour Technol 293(June):122129. https://doi.org/10.1016/j.biortech.2019.122129
24. van't Riet K, Tramper J (1991) Basic Bioreactor Design. https://doi.org/10.1201/9781482293333. Accessed 27 June 2022
25. Ingo K, Regestein L, Schauf J, Steinbusch S, Büchs J (2011) Linear correlation between online capacitance and offline biomass measurement up to high cell densities in Escherichia coli fermentations in a pilot-scale pressurized bioreactor. J Microbiol Biotechnol 21(2):204–211. https://doi.org/10.4014/jmb.1004.04032
26. Lopes M, Gomes N, Mota M, Belo I (2009) Yarrowia lipolytica growth under increased air pressure: influence on enzyme production. Appl Biochem Biotechnol 159(1):46–53. https://doi.org/10.1007/s12010-008-8359-0
27. Gaddy JL (1997) 5593886 Clostridium strain which produces acetic acid from waste gases. J Clean Prod. https://doi.org/10.1016/s0959-6526(97)82527-9
28. Soni BK, Conrad J, Kelley RL, Srivastava VJ (1998) Effect of temperature and pressure on growth and methane utilization by several methanotrophic cultures. Appl Biochem Biotechnol 70–72:729–738. https://doi.org/10.1007/BF02920184
29. Wendlandt KD, Jechorek M, Brühl E (1993) The influence of pressure on the growth of methanotrophic bacteria. Acta Biotechnol 13(2):111–115. https://doi.org/10.1002/abio.370130205
30. Yu J, Munasinghe P (2018) Gas fermentation enhancement for chemolithotrophic growth of Cupriavidus necator on carbon dioxide. Fermentation 4(3):63. https://doi.org/10.3390/fermentation4030063
31. Eigenstetter G, Takors R (2017) Dynamic modeling reveals a three-step response of Saccharomyces cerevisiae to high CO2 levels accompanied by increasing ATP demands. FEMS Yeast Res 17(1):1–11. https://doi.org/10.1093/femsyr/fox008
32. Wilkins MR, Atiyeh HK (2011) Microbial production of ethanol from carbon monoxide. Curr Opin Biotechnol 22(3):326–330. https://doi.org/10.1016/j.copbio.2011.03.005
33. Bowien B, Schlegel HG (1981) Physiology and biochemistry of aerobic hydrogen-oxidizing bacteria. Annu Rev Microbiol 35(1):405–452. https://doi.org/10.1146/annurev.mi.35.100181.002201
34. King WR, Andersen K (1980) Efficiency of CO2 fixation in a glycollate oxidoreductase mutant of Alcaligenes eutrophus which exports fixed carbon as glycollate. Arch Microbiol 128(1):84–90. https://doi.org/10.1007/BF00422310
35. Dang J, Wang N, Atiyeh HK (2021) Review of dissolved CO and H2 measurement methods for syngas fermentation. Sensors 21(6):1–32. https://doi.org/10.3390/s21062165
36. Yu J, Lu Y (2019) Carbon dioxide fixation by a hydrogen-oxidizing bacterium: biomass yield, reversal respiratory quotient, stoichiometric equations and bioenergetics. Biochem Eng J 152 (June):107369. https://doi.org/10.1016/j.bej.2019.107369
37. Lübbert A (2009) Bubble columns and airlift bioreactors. In: Encyclopedia of life support systems, pp 1–13
38. Russell TWF, Dunn IJ, Blanch HW (1974) The tubular loop batch fermentor: basic concepts. Biotechnol Bioeng 16(9):1261–1272. https://doi.org/10.1002/bit.260160909

39. Ziegler H, Meister D, Dunn IJ, Blanch HW, Russell TWF (1977) The tubular loop fermentor: oxygen transfer, growth kinetics and design. Biotechnol Bioeng 19(4):507–525. https://doi.org/10.1002/bit.260190406

40. Petersen LAH, Villadsen J, Jørgensen SB, Gernaey KV (2017) Mixing and mass transfer in a pilot scale U-loop bioreactor. Biotechnol Bioeng 114(2):344–354. https://doi.org/10.1002/bit.26084

41. Elisiário MP, De Wever H, Van Hecke W, Noorman H, Straathof AJJ (2021) Membrane bioreactors for syngas permeation and fermentation. Crit Rev Biotechnol 42(6):856–872. https://doi.org/10.1080/07388551.2021.1965952

42. Bengelsdorf FR, Beck MH, Erz C, Hoffmeister S, Karl MM, Riegler P, Wirth S, Poehlein A, Weuster-Botz D, Dürre P (2018) Bacterial anaerobic synthesis gas (syngas) and CO2+ H2 fermentation. Adv Appl Microbiol 103:143–221

43. Treasure Coast (n.d.). https://eu.tcpalm.com/story/money/business/2016/09/02/ineos-bio-selling-indian-river-county-biofuel-plant/89823970/. Accessed 26 May 2022

44. Deep Branch (2021a) Technology – Deep Branch. https://deepbranch.com/technology/. Accessed 26 May 2022

45. ATEL (2022) Biotechnology. https://Atelgroup.Org/En/Biotechnology. https://atelgroup.org/en/biotechnology Accessed 24 June 2022

46. Marcellin E, Angenent LT, Nielsen LK, Molitor B (2022) Recycling carbon for sustainable protein production using gas fermentation. Curr Opin Biotechnol 76:102723. https://doi.org/10.1016/j.copbio.2022.102723

47. Mango Materials (2022) Scalable solutions. https://www.mangomaterials.com/innovation/ Accessed 24 June 2022

48. NatureWorks (2022) About natureworks. https://www.natureworksllc.com/About-NatureWorks. Accessed 24 June 2022

49. Bioplastics Magazine (2020, September 25) Newlight Technologies opens new commercial-scale AirCarbon production facility. https://www.bioplasticsmagazine.com/en/news/meldungen/20200925-Newlight-Technologies-opens-new-commercial-scale-AirCarbon-production-facility.php#:~:text=Sep 2020-,Newlight Technologies opens new commercial-scale AirCarbon production facility,Southern Cal. Accessed 24 June 2022

50. Renstrom R (2014, June 24) Newlight ramping up commercial greenhouse-gas-to-plastic production. https://www.plasticsnews.com/article/20140624/NEWS/140629963/newlight-ramping-up-commercial-greenhouse-gas-to-plastic-production Accessed 24 June 2022

51. Koller M, Mukherjee A (2022) A new wave of industrialization of PHA biopolyesters. Bioengineering 9(2):74. https.//doi.org/10.3390/bioengineering9020074

52. F+L Daily (2016, April 14) Intrexon's isobutanol pilot plant now up and running https://www.fuelsandlubes.com/intrexons-isobutanol-pilot-plant-now-up-and-running/. Accessed 24 June 2022

53. De Guzman D, Green CB (2018, September 19) Methane-to-protein plant starts in Russia. https://greenchemicalsblog.com/2018/09/19/video-methane-to-protein-plant-starts-in-russia/. Accessed 24 June 2022

54. UniBio (2021, October 6) Press Release - UniBio Makes its First Commercial Shipment of UNIPROTEIN® in Europe to Danish Agro. https://www.unibio.dk/press-release-unibio-makes-its-first-ever-commercial-shipment-of-uniprotein-in-europe-to-danish-agro/. Accessed 24 June 2022

55. Megan Poinski (FOODDIVE) (2022, March 2) How Air Protein is making the journey from concept to product. https://www.fooddive.com/news/air-protein-journey-from-concept-to-product/619374/. Accessed 24 June 2022

56. Avecom (2022) Power-to-protein. https://avecom.be/research/power-to-protein/#:~:text=The single-cell protein is,than that of plant protein. Accessed 24 June 2022

57. Deep Branch (2021, March 16) Deep Branch Announces €8M Series A Round led by Novo Holdings and DSM Venturing. https://deepbranch.com/2021/03/16/series-a/. Accessed 24 Jun. 2022

58. Jasmin Speer (T&N) (2022, April 4) Econutri: Grazer Spin-off will das CO2 der Industrie essbar machen. https://www.trendingtopics.eu/econutri-grazer-spin-off-will-das-co2-der-industrie-essbar-machen/#:~:text=Econutri%3A Grazer Spin-off will das CO2 der Industrie essbar machen. Accessed 24 June 2022

59. Evonik, Siemens (2020, September 21) For a climate-friendly industry: Using carbon dioxide and hydrogen as raw materials for sustainable chemicals [Press release]. https://assets.new.siemens.com/siemens/assets/api/uuid:719c80ea-dc74-4a51-aeb2-4e5ffc58ccc1/HQGPPR202009185995EN.pdf. Accessed 24 Jun. 2022

60. Siemens, Evonik (2019, October 10) CO2 for a Clean Performance: Rheticus Research Project Enters Phase 2. https://press.siemens.com/global/en/pressrelease/research-project-rheticus. Accessed 24 Jun. 2022

61. Reed J, Geller J, McDaniel R (eds) (2017) CO2 conversion by knallgas microorganisms - evaluation of products and processes. California Energy Commission, Energy Research and Development Division, Sacramento

62. NovoNutrients (2021, July 29) NovoNutrients Scaling Up with Project Funding Plus $9M in Equity. https://www.prnewswire.com/news-releases/novonutrients-scaling-up-with-project-funding-plus-9m-in-equity-301344278.html. Accessed 24 June 2022

63. Solar Foods (2021, October 26) Solar Foods takes concrete steps to enter the market: construction of Factory 01 begins, set to produce the world's most sustainable protein in 2023. https://solarfoods.com/solar-foods-takes-concrete-steps-to-enter-the-market-construction-of-factory-01-begins-set-to-produce-the-worlds-most-sustainable-protein-in-2023/. Accessed 24 June 2022

64. Jupeng Bio (2018) Jupeng Bio, Lu'An Group and Shanghai Advanced Research Institute sign Research and Development collaboration agreement. http://www.jupengbio.com/blog/jupeng-bio-lu-an-group-and-shanghai-advanced-research-institute-sign. Accessed 24 June 2022

65. Sekisui Chemical CO. L (2022, April 11) 1/10th scale "waste to ethanol" demonstration plant completed in Kuji city. https://www.sekisuichemical.com/news/2022/1373480_38754.html. Accessed 24 June 2022

66. Fackler N, Heijstra BD, Rasor BJ, Brown H, Martin J, Ni Z, Shebek KM, Rosin RR, Simpson SD, Tyo KE, Giannone RJ, Hettich RL, Tschaplinski TJ, Leang C, Brown SD, Jewett MC, Kodiepke M (2021) Stepping on the gas to a circular economy: accelerating development of carbon-negative chemical production from gas fermentation. Annu Rev Chem Biomol Eng 12: 439–470. https://doi.org/10.1146/annurev-chembioeng-120120-021122

67. Köpke M, Simpson SD (2020) Pollution to products: recycling of 'above ground' carbon by gas fermentation. Curr Opin Biotechnol 65:180–189. https://doi.org/10.1016/j.copbio.2020.02.017

68. Takors R, Kopf M, Mampel J, Bluemke W, Blombach B, Eikmanns B, Bengelsdorf FR, Weuster-Botz D, Dürre P (2018) Using gas mixtures of CO, CO2 and H2 as microbial substrates: the do's and don'ts of successful technology transfer from laboratory to production scale. Microb Biotechnol 11(4):606. https://doi.org/10.1111/1751-7915.13270

69. Simpson S (2018) CCU-now: fuels and chemicals from waste. https://ec.europa.eu/energy/sites/ener/files/documents/25_sean_simpson-lanzatech.pdf. Accessed 24 Jun. 2022

70. GlobeNewswire (2016) Globar Bioenergies and Lanzatech strengthen cooperation. https://www.google.com/search?q=GLOBAL+BIOENERGIES+and+LANZATECH+strengthen+cooperation&rlz=1C1GCEA_enBE964BE964&oq=GLOBAL+BIOENERGIES+and+LANZATECH+strengthen+cooperation&aqs=chrome..69i57j0i546l2j69i60l3.159j0j4&sourceid=chrome&ie=UTF-8. Accessed 22 June 2022

71. Simpson, S. D., Abdalla, T., Brown, S., Canter, C., Conrado, R., Daniell, J., Asela, D., Gao, A., Jensen, R. O., Köpke, M., Leang, C., Leiw, F. E., Nagaraju, S., Nogle, R., Tappel, R. C., Tran, L., Charania, P., Engle, N., Giannone, R., . . . Yang, Z. (2019). Development of a sustainable green chemistry platform for production of acetone and donwstream drop-in fuel and commodity products directly form biomass synsgas via novel energy conserving route in engineered acetogenic bacteria. http://osti.gov/. Technical Report, 1–31. https://www.osti.gov/search/semantic:10.2172/1599328. Accessed 11 Nov 2022

72. Science and Technology Daily (2021, November 11) Carbon monoxide becoming a new source of protein. http://www.stdaily.com/English/ChinaNews/2021-11/11/content_1231169.shtml. Accessed 22 June 2022
73. Teixeira LV, Moutinho LF, Romão-Dumaresq AS (2018) Gas fermentation of C1 feedstocks: commercialization status and future prospects. Biofuels Bioprod Biorefin 12(6):1103–1117. https://doi.org/10.1002/bbb.1912

Final Evaluation and Summary

35

Manfred Kircher

Abstract

A technical carbon cycle based on C1 gases is possible and exemplarily established, and numerous processes for energy sources, basic chemistry and high-value chemical products are under development. Most processes are highly energy intensive, so regions with a structural energy surplus offer advantages. The energy carriers produced in this way can assume a function as energy storage, thus coupling the energy and chemical sectors. However, high raw material and energy costs hinder competitiveness. In addition, the current framework conditions do not support technical carbon cycles and should therefore be adjusted accordingly.

Keywords

C1 gases · CO_2 · CO · Methane · Energy carriers · Basic chemistry · Chemical products · Energy storage · Competitiveness · Framework conditions

35.1   Introduction

All authors see the concept of a technical carbon cycle presented in Chap. 1 as an option worth pursuing to contribute to circular value creation with the material recycling of C1 gases. Which gas sources are identified by the authors, which technical conversion technologies are available, which markets present themselves, which framework conditions support competitiveness, and which other impacts can be expected are summarized below.

M. Kircher (✉)
KADIB, Frankfurt am Main, Hessen, Germany
e-mail: kircher@kadib.de

M. Kircher, T. Schwarz (eds.), *CO2 and CO as Feedstock*, Circular Economy and Sustainability, https://doi.org/10.1007/978-3-031-27811-2_35

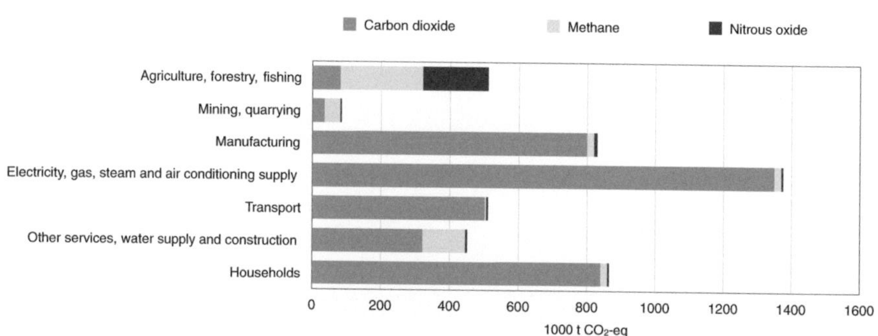

Fig. 35.1 Greenhouse gases by sector and gas (EU-28, 2015). © European Union, 1995–2013

One of the C1 gases that lends itself is CO_2, which is present in our planet's atmosphere at a volume of about 3000 Gt [1]. However, its concentration of 0.04% is so low that technical utilization is not considered by any of the authors. Gas streams of higher CO_2 concentration are considered to be more suitable for industrial utilization.

Figure 35.1 [2] shows CO_2 emissions by sector. Significant is the emission in the manufacturing industry and in power plants. There, large-volume sources of CO_2, so-called point sources, can be expected. Transportation, households and services are also significant CO_2 emitters, but the individual emission sources are comparatively small-volume and distributed over the area, and therefore less attractive for economic C1 conversion.

In Europe, especially in Germany, the Benelux countries, Great Britain, and Poland, about 2000 large industrial plants emit 1800 MT of CO_2 annually (Chaps. 2 and 16). High-volume, high-concentration emission streams with up to 100% CO_2 are generated in the production of hydrogen, ethylene oxide, and ammonia. Such very attractive point sources add up to a total volume of 45.6 MT of CO_2 annually in Europe today, containing 12.3 MT of carbon. This is equivalent to the carbon content of around 14.5 MT of crude oil. In this context, the comparison with the petroleum demand of the German chemical industry of 15.5 MT (2017, [3]) makes it clear that these are definitely industrially relevant quantities. In addition, the advantageous circumstance is noted that these gas streams occur at chemical sites, so their further processing could take place at the same site (Chaps. 2 and 6). Emission gases with 20–30% CO_2 are generated in the steel and cement industries (Chaps. 13 and 14). Further large emission streams with up to 12% CO_2 are generated in power plants (Chap. 15). Natural gas production, in Europe mainly in Norway, Great Britain, and the Netherlands, is also a CO_2 source of high concentration because CO_2 is captured from the raw natural gas. However, even though natural gas is part of the EU's long-term planned energy mix, a potential decline should be factored in for European production sites. The Netherlands has decided to phase out gas production by 2030 and has banned gas heating in new buildings since July 2018. Overall, it can be assumed that with the energy turnaround, the change in raw materials in the chemical industry and the loss of Russia as a supplier of natural gas since 2022, the European chemical industry's CO_2 emissions from fossil sources will decline by 2050.

In contrast, the CO_2 supply from biotechnological plants will increase (Chap. 17). Raw biogas, for example, contains an average of 55% CO_2. More than 90% of the world's 12,000 plants are located in Europe, including about 8000 plants in Germany. The technical potential for biogas in Germany alone is estimated to be up to 16.6 billion m3 in 2030 [4], which would contain 14.9 MT of CO_2 with about 4 MT of carbon. Other biotechnological processes also emit CO_2. Such plants tend to be small in volume compared to the fossil-fueled sites mentioned above, and therefore require either infrastructure to combine the gas in central processing facilities or conversion processes that allow for corresponding small-volume operations.

Today, a very small portion of the CO_2 gas streams mentioned above is used for applications in the food industry in fire extinguishers and refrigerants or as a protective gas. CO_2 transport from the point of origin to the consumer is technically established in pipelines; worldwide there is a CO_2 pipeline network of 6500 km [5].

35.2   Synthesis Gas

While CO_2 is an unavoidable by-product of the above-mentioned processes, carbon monoxide (CO) is a C1 molecule that is produced on an industrial scale as the main component of synthesis gas (30–60% CO, 25–30% hydrogen, CO_2, methane) specifically for a wide variety of conversions. According to the state of the art, synthesis gas is nowadays preferably produced by gasification of crude oil and natural gas. Since basically any organic material is suitable for gasification, i.e., also any biomass, biobased production side streams, sewage sludge, and organic industrial and municipal waste, gasification represents a method for standardizing complex materials (Chaps. 5, 18 and 20). Thus, even though fossil feedstocks have been pushed back, syngas has a potentially even growing future. Distribution of syngas in pipelines is state of the art (Chap. 23); thus, syngas is a carbon source that fits into the existing infrastructure.

35.3   Methane

Methane is the main component of biogas (average 55% methane, 45% CO_2) produced by anaerobic fermentation of biodegradable materials. Suitable materials include residual and waste materials from agriculture, including animal husbandry, the food industry, waste management, and biotechnological processes (Chaps. 17 and 20). This fermentation can thus also play a key role in the standardization of complex substances. In addition, methane is the main component of sewage gas, which is produced by spontaneous microbial degradation in wastewater treatment plants. It can be separated and contributes 1.5 TWh (2017) to electricity generation in Germany.

Biogas is also considered to have a central function in a technical carbon cycle because it can use the existing storage and transport infrastructure of natural gas, whose energy source is methane of fossil origin.

35.4 Conversion of C1 Gases

High-carbon gases can be converted chemically-catalytically or biotechnologically into fuels, basic chemicals, and higher-value chemical products (Chaps. 6–9). The chemocatalytic utilization of fossil methane (natural gas) and, with the consumption of hydrogen, of CO (synthesis gas based on fossil raw materials) has long been established on a large scale. In the chemical industry, syngas is a significant carbon source for the production of ammonia (Haber-Bosch process), hydrocarbons (Fischer-Tropsch synthesis), aldehydes (oxosynthesis), and methanol. Methanol may gain considerable importance because this C1 alcohol is said to have the potential to replace ethylene as today's most important basic organic chemical in the course of raw material change (Chaps. 6 and 18; for further literature, see Bertau et al. 2014; [6]). Both gases, methane and syngas, are also accessible from renewable carbon sources and can be further processed into the known products using established chemocatalytic methods. All these chemocatalytic methods have in common that they are economically more suitable for large-scale plants and thus for the production of bulk products of basic chemistry.

New chemocatalytic processes for the utilization of CO_2 are under development and have already reached demonstration scale in some cases (Chap. 6). Biological systems are also capable of utilizing CO and CO_2. The fact that plants photosynthetically bind and convert atmospheric CO_2 is common knowledge and should be mentioned here only for the sake of completeness. Microalgae also bind CO_2 via photosynthesis (Chap. 11). In technical plants, they can also utilize CO_2 fluxes whose concentration is higher than in the atmosphere and produce biomass, fatty acids, vitamins, dyes, and other high-value metabolites.

CO_2 and CO are accessible to certain microorganisms via further metabolic pathways (Chap. 7). They also form biomass from these carbon sources and are able to form both simple (e.g. methane, ethanol) and complex products (e.g. polymers) (Chaps. 8–10). Processing from the aqueous culture media is laborious (Chap. 12). Depending on the process, feeding of hydrogen is necessary. In this way, for example, the CO_2 emission of a biogas plant can be converted to methane, thus reducing CO_2 emissions from an average of 55% to a few percent. Biotechnological plants tend to be small-scale compared to chemocatalytic processes and are therefore also suitable for coupling with other relatively small-scale point sources, such as biogas plants (biogenic CO_2) or the decentralized gasification of municipal waste (biogenic and fossil CO_2) to synthesis gas. In total, considerable production capacities could be built up in this way.

The conversion of CO_2 and CO basically requires reducing agents. Bio- and electrochemical methods are currently being developed (Chaps. 3 and 4), but the state of the art is the use of hydrogen. Chapter 3 therefore presents the production of hydrogen by steam reforming, electrolysis, and biotechnological hydrogen production, which is still at an early stage of research. Reforming today starts from natural gas, petroleum, or coal and is associated with high CO_2 emissions. At some chemical sites, hydrogen surplus, e.g. from chlor-alkali electrolysis, can be used. In principle, biomass or the methane contained in biogas are also suitable. Hydrogen can be produced emission-free by means of water electrolysis if the required electricity is also generated emission-free. Water electrolysis is therefore one of the interfaces between C1 utilization and the energy sector.

35.5 Energy Storage

Because of their high energy content, hydrogen and CO and CO_2 derivatives produced with the aid of hydrogen are to be regarded as substances that can be used to store energy. However, it must be taken into account that considerable energy losses of up to 80% have to be accepted in the conversion of the gases (Chap. 6). Conversion therefore only makes sense if electricity consumption "plays no role," so to speak. This is precisely where the energy turnaround and the increasing share of renewable energies, especially solar and wind energy, come into play. Both energy sources are volatile, i.e., they temporarily generate more electricity than is needed. These so-called power peaks represent situations in which consumption "doesn't matter." On the contrary, the use for electrolysis of water and, if necessary, subsequent conversion of CO and CO_2 is a viable option for storing energy.

Thus, for energy carriers resulting from C1 conversion to form a relevant building block in the future energy system, it is necessary to couple the energy sector with the chemical sector (Chap. 6). Indeed, a wind turbine from which hydrogen, rather than electricity, is discharged will be commissioned in the Netherlands in 2019, and a hydrogen interconnection network is under preparation at the port of Amsterdam, which will be used, among other things, for the reduction of steel mill gases. Vattenfall has also announced its intention to become more involved in the production of hydrogen. Thus, industry is preparing to use hydrogen as an energy storage and reduction agent, and many authors of this book also recommend this route.

These authors also unanimously recommend the use of renewable energies, i.e., base-load capable hydropower, geothermal energy, and volatile wind and solar energy. However, hydropower is largely exhausted in Germany (at least for industrial purposes). For volatile energies, expectations vary widely and range from "far from sufficient" (Chap. 23) to "almost unlimited" (Chap. 22). Therefore, a sober look at the current status of renewable energies will be taken. Figure 35.2 shows the share of renewable energies in electricity generation in Germany [7]. In 2017, it amounted to 33%. In particular, the areas of heat generation and transport are still essentially fossil-based, but are to be powered by electricity in the future, among

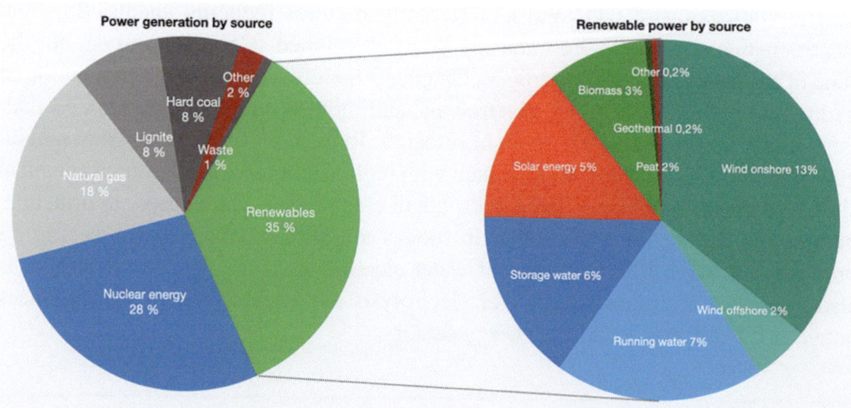

Fig. 35.2 Gross electricity generation by source (EU, 2021)

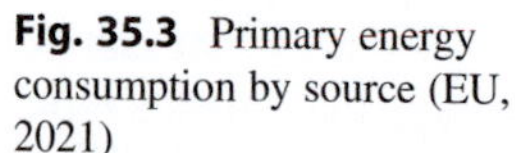

Fig. 35.3 Primary energy consumption by source (EU, 2021)

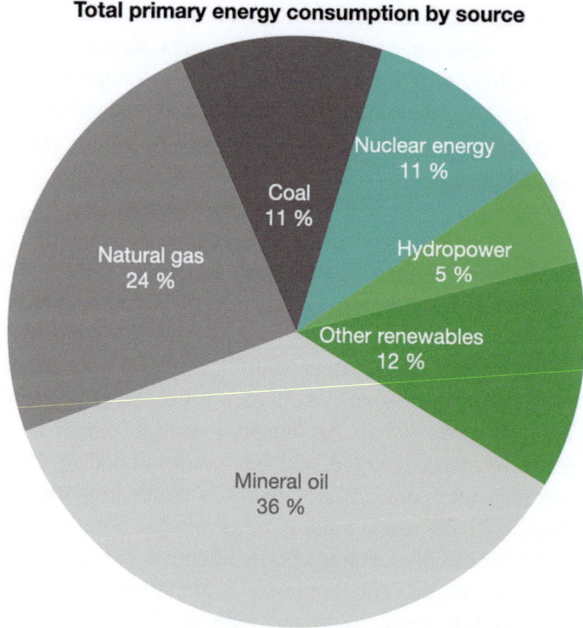

other things. Thus, for the share of electricity in the energy mix, a strong increase in demand can be assumed simply by taking over energy applications that are supplied differently today, and the extensive conversion of C1 gases would let the electricity demand grow even further. It is therefore advisable to look at the overall energy mix. Here, the share of renewables was 13.2% in 2017 (Fig. 35.3; [8]); volatile solar thermal, photovoltaic, and wind power together contributed 4.1%. These figures highlight the still very limited supply of renewable energies in Germany today. This is even more true for consumers that, like the technical carbon cycle, additionally enter the electricity market.

This leads to the consideration of whether the energy-intensive technologies discussed here should be integrated into the larger market of European emission-free energies. The EU, which reported a 22.1% share of renewables in the total energy mix in 2021 [9], expects electricity consumption to at least double by 2050. The demand for electricity, fuel, and heat should be met by a mix of fossil, renewable, and nuclear energy [10]. At the same time, electricity storage capacity is to be expanded by a factor of 6 by 2050. This also includes the explicitly mentioned energy storage in hydrogen and derived products. Examples of this are discussed in Chap. 3.

35.6 Climate Protection

Chapters 1, 22, and 23 cite sustainable production methods in general and climate protection in particular, which according to the Paris Climate Agreement requires a combination of greenhouse gas reduction and sink creation, as drivers for C1 recycling. Quite undoubtedly, recycling CO_2 directly from an industrial emission stream is a contribution to the reduction of climate gases. The priority for climate protection therefore remains the phase-out of fossil fuels in the energy industry and the raw material switch from fossil to renewable carbon sources in the chemical industry.

35.7 Bioeconomy

Renewable carbon sources (agricultural, forestry, marine biomass and their derivatives) are produced and processed by the bioeconomy, which is therefore cited by the EU as an indispensable building block of a climate-neutral economy, along with the circular economy [10]. The bioeconomy (for further reading, see Thrän, Moesenfechtel 2022; [11]) can provide energy and chemical products just like the fossil-based economy. However, the feedstock potential is limited due to ecological limits, which is why the conversion of C1 gases is also of great importance in the bioeconomy. The importance of methane for the integration of different carbon cycles in agriculture has been shown Chaps. 20 and Chap. 5 has addressed, among other things, the standardization of complex biomasses by gasification to synthesis gas.

The utilization of C1 gases thus not only directly reduces CO_2 emissions, but also contributes to the creation of sinks (CO, methane) and to relieving the production of biomass for industrial purposes (Chap. 1). This applies equally to the use of C1 gases of fossil and biogenic origin. For example, the consistent biotechnological utilization of fossil-based steel mill gases to produce ethanol would be equivalent to today's global production of bioethanol, a volume that ties up 40% of the corn harvest in the USA, for example (Chap. 27).

35.8 Competitiveness

The costs of conversion processes are discussed in Chap. 6 as inhibiting a comprehensive industrial realization of C1 utilization. Relevant cost blocks are the carbon feedstock and the conversion energy. The costs of CO_2 capture alone are estimated at more than 30 EUR per ton (Chaps. 14–16). Another important cost factor is the cost of emission-free energies. A decreasing trend is assumed for the electricity production costs, with photovoltaics expected to be the cost leader with 2–4 EURcent/kWh by 2034, even in comparison with fossil energies. Nevertheless, the very high energy input of many C1 conversion technologies poses an economic challenge, especially since electricity costs in Germany are significantly higher than the EU average. For this purpose, Chap. 6 suggests importing renewable energies or material energy sources from regions with a structural energy surplus. However, the regionally different potential for renewable energies could also lead to the migration of energy-intensive industries to these regions. The energy turnaround, the raw material change, and the conversion of C1 gases therefore also require traditional industrial centers to develop their potentials (Chap. 23). Economic policy should shape both options: the further development of a European network of emission-free energies and the regional structural transformation of energy-intensive industries, both with the aim of advancing the establishment of the technical carbon cycle.

Framework Conditions

The framework conditions (Chap. 24) for the emission and utilization of greenhouse gases are set by the European Emission Trading Scheme Directive ([ETS] 2003/87/EC), the Renewable Energy Directive [RED]2009/28/EC, and the Energy Efficiency Directive [EED] 2012/27/EU. Among these regulations, ETS is the competitive control instrument that captures and prices 45% of Europe's greenhouse gas emissions. In fact, since the end of 2017, the price has increased from EUR 5 to around EUR 50 (Fig. 35.4; [12]), and with the decreasing supply of allowances in the market, it is expected to increase further. The ETS price is thus approaching the cost of capturing CO_2.

As it stands, however, ETS not only does snot promote the technical carbon cycle, it inhibits it. The directive prices any fossil-based greenhouse gas emission, and does so whether or not it is reused. Only storage (CCS) is recognized as reducing emissions. This places an additional emissions trading cost on the recycling of CO_2 from captured sources. After all, the ETS-accompanying innovation fund, NER300, also promotes the use of carbon capture and storage (CCU).

Biogenic carbon emission is not even considered by ETS. Therefore, mitigating CO_2 emissions from biotech plants currently offers no economic advantage, although their share will increase as we shift from fossil to renewable resources. Renewable carbon sources are considered climate neutral because the carbon they contain has been removed from the atmosphere through photosynthesis, but it would still make sense to keep the resulting CO_2 in the technical cycle whenever possible to conserve biomass production capacity.

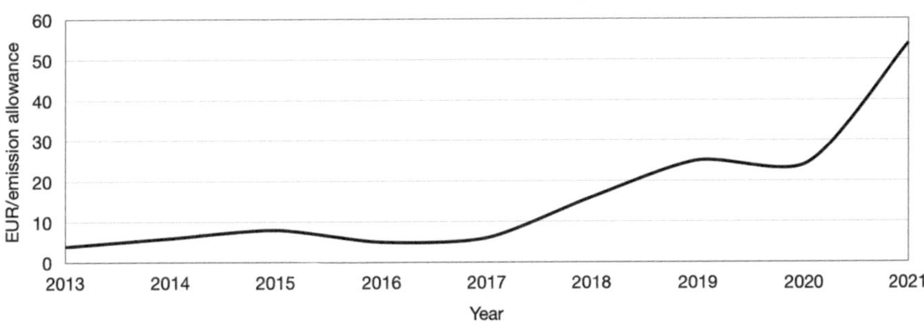

Fig. 35.4 Price development of EU emission allowances since 2015

In order to promote the use of C1 emission competitively, Chap. 6 therefore propose a revision of the ETS system. This also corresponds to the request of an expert group of the EU Commission from May 2018 [13].

The coming decades will be a time of transition from fossil to zero-emission energies and from fossil-based to renewable carbon sources for the chemical and fuel sectors. "Old" and "new" raw materials will continue to be used in parallel for a long time. For the establishment of the circular carbon value creation called for in Chap. 1, both feedstocks are therefore of equal value today, even if renewable primary sources are to be preferred. The framework conditions should allow this openness, because the technical carbon cycle is fundamentally desirable quite independently of the origin of the carbon.

References

1. Wikidot. Mass of atmospheric carbon dioxide. http://igss.wikidot.com/co2mass. Accessed 6 Nov 2022
2. European Commission/eurostat (2018) Greenhouse gas emissions by economic activity and by pollutant, EU-28, 2015. https://ec.europa.eu/eurostat/statistics-explained/index.php?title=File: Greenhouse_gas_emissions_by_economic_activity_and_by_pollutant,_EU-28,_2015_(thousand_tonnes_of_CO2_equivalents)_YB17.png. Accessed 6 Nov 2022
3. VCI (2019) Daten und Fakten – Rohstoffbasis der chemischen Industrie. https://www.vci.de/vci/downloads-vci/top-thema/daten-fakten-rohstoffbasis-der-chemischen-industrie.pdf. Accessed 6 Nov 2022
4. Fraunhofer-Institut für Umwelt-, Sicherheits- und Energietechnik UMSICHT (2006) Analyse und Bewertung der Nutzungsmöglichkeiten von Biomasse. http://publica.fraunhofer.de/dokumente/N-54117.html. Accessed 6 Nov 2022
5. IEAGHG (2014) CO2 pipeline infrastructure. https://ieaghg.org/docs/General_Docs/Reports/2013-18.pdf. Accessed 6 Nov 2022
6. Bertau M, Offermanns H, Plass L, Schmidt F, Wernicke H-J (2014) Methanol: the basic chemical and energy feedstock of the future. Springer, Heidelberg/New York/Dordrecht/London
7. Statista (2022) Anteil der Energieträger an der Nettostromerzeugung in der EU im Jahr 2021. https://de.statista.com/statistik/daten/studie/182159/umfrage/struktur-der-bruttostromerzeugung-in-der-eu-27/. Accessed 6 Nov 2022

8. World Energy Council (2022). Energie in der Europäischen Union. https://www.weltenergierat. de/publikationen/energie-fuer-deutschland/energie-fuer-deutschland-2021/energie-in-der-europaeischen-union-zahlen-und-fakten/?cn-reloaded=1. Accessed 6 Nov 2022
9. European Commission (2022). Renewable energy statistics. https://ec.europa.eu/eurostat/ statistics-explained/index.php?title=Renewable_energy_statistics&action=statexp-seat& lang=de. Accessed 6 Nov 2022
10. European Commission (2018) Communication (COM(2018)773)'A Clean Planet for all: A European strategic long- term vision for a prosperous, modern, competitive and climate neutral economy
11. Thrän D, Moesenfechtel U (2022) The bioeconomy system. Springer, Berlin, Heidelberg. https://doi.org/10.1007/978-3-662-64415-711
12. Trading Economics (2022) Carbon permits. https://tradingeconomics.com/commodity/carbon. Accessed 6 Nov 2022
13. European Commission (2018) Novel carbon capture and utilisation technologies. https://ec. europa.eu/research/sam/pdf/sam_ccu_report.pdf. Accessed 6 Nov 2022

Index